VLSI SPECIFICATION, VERIFICATION AND SYNTHESIS

THE KLUWER INTERNATIONAL SERIES IN ENGINEERING AND COMPUTER SCIENCE

VLSI, COMPUTER ARCHITECTURE AND DIGITAL SIGNAL PROCESSING

Consulting Editor

Jonathan Allen

Other books in the series:

Introduction to VLSI Silicon Devices: Physics, Technology and Characterization,
B. El-Kareh and R. J. Bombard.
ISBN 0-89838-210-6.
Latchup in CMOS Technology: The Problem and Its Cure,
R. R. Troutman.
ISBN 0-89838-215-7.
Digital CMOS Circuit Design,
M. Annaratone.
ISBN 0-89838-224-6.
The Bounding Aproach to VLSI Circuit Simulation,
C. A. Zukowski.
ISBN 0-89838-176-2.
Multi-Level Simulation for VLSI Design,
D. D. Hill and D. R. Coelho.
ISBN 0-89838-184-3.
Relaxation Techniques for the Simulation of VLSI Circuits,
J. White and A. Sangiovanni-Vincentelli.
ISBN 0-89838-186-X.
VLSI CAD Tools and Applications,
Wolfgang Fichtner and Martin Morf, Editors.
ISBN 0-89838-193-2.
A VLSI Architecture for Concurrent Data Structures,
W. J. Dally.
ISBN 0-89838-235-1.
Yield Simulation for Integrated Circuits,
D. M. H. Walker.
ISBN 0-89838-244-0.

VLSI SPECIFICATION, VERIFICATION AND SYNTHESIS

edited by

Graham Birtwistle
University of Calgary

and

P.A. Subrahmanyam
AT&T Bell Laboratories

KLUWER ACADEMIC PUBLISHERS
Boston/Dordrecht/Lancaster

Distributors for North America:
Kluwer Academic Publishers
101 Philip Drive
Assinippi Park
Norwell, Massachusetts 02061, USA

Distributors for the UK and Ireland:
Kluwer Academic Publishers
MTP Press Limited
Falcon House, Queen Square
Lancaster LA1 1RN, UNITED KINGDOM

Distributors for all other countries:
Kluwer Academic Publishers Group
Distribution Centre
Post Office Box 322
3300 AH Dordrecht, THE NETHERLANDS

Library of Congress Cataloging-in-Publication Data

VLSI specification, verification, and synthesis / by Graham Birtwistle and P.A. Subrahmanyam [editors].
p. cm.—(The Kluwer international series in engineering and computer science ; SECS 35)
A collection of papers presented at a workshop held in Calgary, Canada, Jan. 12–16, 1987.
ISBN 0-89838-246-7
1. Integrated circuits—Very large scale integration—Design and construction. I. Birtwistle, G. M. (Graham M.) II. Subrahmanyam, P. A. III. Series.
TK7874.V564 1988
621.395—dc19

Printed in the United States

Contents

Contents

Preface

VLSI Specification, Verification and Synthesis

Proceedings of a workshop
held in Calgary from 12-16 January 1987.

The collection of papers in this book represents some of the discussions and presentations at a workshop on hardware verification held in Calgary, January 12-16 1987. The thrust of the workshop was to give the floor to a few leading researchers involved in the use of formal approaches to VLSI design, and provide them ample time to develop not only their latest ideas but also the evolution of these ideas.

In contrast to simulation, where the objective is to assist in detecting errors in system behavior in the case of some selected inputs, the intent of hardware verification is to formally prove that a chip design meets a specification of its intended behavior (for all acceptable inputs). There are several important applications where formal verification of designs may be argued to be cost-effective. Examples include hardware components used in "safety critical" applications such as flight control, industrial plants, and medical life-support systems (such as pacemakers). The problems are of such magnitude in certain defense applications that the UK Ministry of Defense feels it cannot rely on commercial chips and has embarked on a program of producing formally verified chips to its own specification. Hospital, civil aviation, and transport boards in the UK will also use these chips. A second application domain for verification is afforded by industry where specific chips may be used in high volume or be remotely placed. The cost of a flawed telecommunications chip design might mean a prohibitive replacement program. Simply because of their locations, sensor chips installed on pipelines and surveillance chips in polar regions would be tremendously expensive to replace should their designs be faulty. A third application area is product redesign. Technology advances mean that chips have to be reworked into a new technology every two or three years, or even less. If a design has a verified design development tree, redesign to account for changes in technology will typically involve a replacement of the lower subtrees. There need be no change to the overall chip specification if the behaviors of the replacement designs conform to the specification of their plug-in slot in the new tree. In this case, the new versions of the same part can be inserted into existing systems with a guarantee that they will not introduce new bugs.

The major themes of the workshop are presented here in four sections. First, the state of the art in hardware verification and specification is covered with two papers on the VIPER chip and its proof. VIPER is the first production microprocessor chip to be formally verified and implemented in silicon. The second section presents Mike Gordon's

Higher Order Logic System (HOL) which has been used for several substantial verification efforts. The papers explain how to construct proofs in HOL and give several examples of HOL in use as a hardware description language describing circuits ranging is size from leaf cells through sub-systems all the way through to a complete (but small) microprocessor. The third section discusses some ideas for automated CAD systems using formal specifications. The final section covers transistor level modelling, a proof system for a development of Mossim, the verification of a dynamic CMOS circuit design style, and the use of abstraction mechanisms in verification.

The VIPER chip and its proof. The first paper, by John Cullyer, describes the design and development of the VIPER microprocessor. In order to satisfy certification authorities of the correctness of the processor's implementation, semi-formal mathematical methods were used, both to specify the overall behavior of the processor and to prove that two different gate level realizations conform to this top level specification. The second paper on VIPER by Avra Cohn (not read at the workshop) complements Cullyer's overall description and design insights. It details formal mechanization of the first two levels of the semi-formal proof developed by RSRE (The Royal Systems and Radar Establishment in UK). The mechanical verification was a formidable task, which revealed several small flaws with the original hand proof. Three of the flaws revealed were of a more serious nature: a phase error in the fetch cycle, and some incomplete checks for illegal instructions due to unforeseen conditions. Cohn's work would seem to vindicate completely formal methods.

HOL. VIPER demonstrates that existing design and verification techniques can be applied to complex synchronous circuits in a practical manner. Future automated verification tools need to be interfaced to industrial VLSI systems and transform specifications to layout without human intervention. The second section concentrates upon Mike Gordon's HOL (a mechanization of Higher Order Logic). Many of the largest proofs completed to date have been constructed using either Mike's LCF-LSM system or its successor HOL. HOL has already shown (by construction) that it can handle circuits, sub-systems and complete architectures. Thus one notation can span the whole VLSI spectrum. What is lacking at the moment is the ability to incorporate electrical properties into HOL definitions; constraints fit are readily incorporated into the HOL pattern. Gordon's paper gives an introduction to the properties of the HOL language, its proof support apparatus, and small examples of an interactive proof in HOL. Jeff Joyce discusses his reworking in HOL of Gordon's original LCF-LSM proof of a toy microprocessor.

VLSI CAD environments. The Synapse environment is intended to support the development of provably correct designs, proceeding from high level behavioral specifications and optional performance criteria. Its features include various flavors of expert system support, as well as support for formal manipulations; ways of incorporating order of magnitude constraints along dimensions such as area and response time; and a leaf cell synthesizer that has generated novel CMOS circuits. An important issue in the context of such high level design tools is that of *system timing,* which deals with the relative sequencing of subcomputations, the detailed behavior of signal voltages over time, and overall system performance. The design of a VLSI system typically involves many decisions relating to several levels of timing detail. Examples include choices relating to timing disciplines, architectural and circuit design paradigms, the number of clock phases, the number of amplification stages, the sizes of specific transistors, and the temporal behavior demanded of specific input signals (such as stability criteria they must obey). Subrahmanyam's paper introduces a framework for dealing with system timing issues in the context of very high level silicon compilers. The paradigm considered allows the details of system timing to be *gradually* introduced as a design progresses. During the early stages of a design, no timing details need be explicit, and the only temporal constraints arise due to functionality, causality and stability criteria. As a design evolves, and decisions relating to resource allocation, computation schedules, timing and clocking disciplines are made, the temporal constraints on signals evolve correspondingly. If the external environment in which a system resides imposes any *a priori* temporal constraints, such constraints can be explicitly included in the initial specification, and accounted for in the synthesis process. The framework supports synchronous and asynchronous (self-timed) disciplines, multiphase-clocks, as well as techniques that bear on optimizations of performance metrics such as power and area.

The paper by Musser *et al* outlines an approach to verification based on the formalism of abstract data types, and as implemented in the Affirm system. The paper by Narendran *et al* describes work on the Interactive VHDL workstation (IVW) project that uses the Rewrite Rule Laboratory (RRL) theorem prover. The main goals of this project are to furnish the user with textual and graphical capabilities to enter and modify his behavioral and structural specifications, and tools to relate these specifications to formulas for input to the Rewrite Rule Laboratory theorem prover. The final paper in this section, by George Milne, sketches a technique where specifications, designs and contextual information are all described in a common language. Contextual knowledge is captured as just another expression in the language and may be used to simplify the representation of a design in a rigorous manner.

Modelling and abstraction. Tom Melham addresses abstraction issues in specification. Abstract specifications of design components are used to derive more compact descriptions of behavior. By doing this at each level of a hierarchically structured design, the size and complexity of a large proof can be controlled. Thus, abstraction mechanisms complement structural hierarchy in handling proofs of large systems. Glynn Winskel discusses several models of MOS circuits and points out some of their shortcomings. Bryant's model, used in Mossim II, tackles such issues in reasonable generality and has been used as the basis of several MOS switch-level simulators; however, this model is not compositional. Winskel presents a compositional model for the behavior of MOS circuits when the input is steady, and shows how this leads to a formal logic. He also indicates the difficulties in providing a full and accurate treatment for circuits with changing inputs. Inder Dhingra formalizes some rules of thumb for popular CMOS design styles, and uses them to analyze the rules of the NORA design style. In this design style, there are two complementary clock lines, and p- and n-type gates. The design rules govern how the gates may be connected, what clock lines they are driven by, and if the output is to be guaranteed. In order to make the design style synchronous, a CMOS latch is used as a dynamic register. This further complicates the rules but gives rise to a design style that generates smaller and faster circuits as compared to standard CMOS. Dhingra goes on to prove that the NORA design style can fail for large circuits, and develops CLIC, a refinement of NORA using a two phase non-overlapping clocking scheme that is guaranteed. Dhingra also gives a formal motivation for the design style using a simple transistor model with charge storage capability. Steve Johnson discusses work-in-progress on the automation of digital-design synthesis from functional specifications via transformations. He describes work concerned with systematically obtaining correct physical implementations from iterative specifications, and with building aids to facilitate the process. The motivation is to develop a practical methodology with an algebraic flavor. Christian Berthet presents a methodology for comparing switch-level circuits with their high-level specifications wherein circuits, user behavior and input constraints are modelled as asynchronous machines. The model is based on characteristic functions. A Boolean algebra of asynchronous machines is defined. Machine composition and internal variable abstraction are shown to correspond respectively to the product and sum operations of the algebra. Internal variables can be abstracted under the presence of a domain constraint, provided that the I/O behavior is preserved. The verification of a speed-independent fair arbiter is presented to exemplify the techniques.

Acknowledgements. The workshop was funded by the Natural Sciences and Engineering Research Council of Canada through an Operating Grant, by the University of Calgary, and by the Department of Computer Science at the University of Calgary. The Alberta Microelectronics Center kindly furnished the location and kept us going with coffee throughout the week. The workshop organizers would like to thank those who made the job easy: the speakers who provided excellent written material beforehand, and the participants for many lively discussions which never got (completely) out of hand.

Graham Birtwistle, *University of Calgary.*
P.A.Subrahmanyam, *AT&T Bell Laboratories.*

1

Implementing Safety-Critical Systems: The VIPER Microprocessor [1]

W. J. Cullyer

Computing Division
Royal Signals and Radar Establishment
St. Andrews Road, Malvern
Worcs WR14 3PS, England

Introduction

During the last five years industry has progressively spread the use of commercial microprocessors into many areas of real-time control. This has been particularly marked in military projects and has happened in a number of civil applications, such as control of nuclear reactors and railway signalling. Some of these applications are 'safety-critical' in the sense that design errors could result in accidents and loss of life. Good examples of such critical applications are found in the avionics fitted in civil and military aircraft [5].

Based on work at RSRE it has been concluded that commercial microprocessors do not from a sound basis for equipments to be used in the most hazardous environments. The methods used to design mass produced general purpose devices are relatively informal and more often than not the resulting implementation contains design errors. Handbooks describing operation codes can be ambiguous and there is no guarantee that second and third sourced devices will have the same functionality as the parent. For these and other related reasons there is a requirement for high integrity processors and support chips produced by the most formal methods available. Such devices may find only a limited market but their use in vital 'black boxes' could be a major contribution to safety.

Overall, the RSRE research programme has three main limbs. Chronologically the first area tackled was the verification of computer programs using automated tools based on representation of the program text as directed graphs [1]. This has resulted in two separate sets of analysers for validation of programs, called MALPAS and SPADE, which are available commercially from UK companies. This topic is dealt with briefly in this paper, insofar as it affects the checking of programs written for the new VIPER microprocessor.

The second limb of the programme is the research on hardware verification, described in this paper. Currently the techniques in use employ Michael Gordon's LCF-LSM, invented at Cambridge [9] and the ELLA logic description language, invented at RSRE [10]. The first chip produced by this means is the VIPER 32-bit microprocessor [6]. This has been conceived using the most formal methods available in order to satisfy certification authorities and equipment designers of the correctness of the practical devices. Two different realisations are being produced, based on specifications issued by RSRE.

This paper describes the development of VIPER and points out some of the practical problems encountered over the two years of the project. Informal proofs of correctness using LCF-LSM at the higher levels and ELLA nearer to the circuit level have been carried out at RSRE [7]. These methods are outlined in Section 2. Fully formal proofs, using Higher Order Logic (HOL) are being carried out by Avra Cohn and Michael Gordon at Cambridge, under contract to RSRE. It is intended to place these proofs in the public domain so that other scientists can check on the validity of the work.

The final and most recent limb of the RSRE programme is the study of safety-critical systems, up to the level of a complete 'black-box', which may control some vital actuator. These studies embrace the whole area of n-channel systems, (n = 2, 3...) and are intended to demonstrate how high integrity processors can be combined with comparators and error detection on memories to produce certifiable systems. Clearly the support (or 'glue') chips need to be formally specified and designed and this is itself an area of research. RSRE has placed a contract with Cambridge Consultants Ltd, Cambridge, UK, for work on these system topics.

In the end all of this activity is intended to produce safer equipment. So far there have been few deaths or injuries due to design faults in the hardware or software of computer-based systems. However, if nothing is done to improve the standards of design and certainty in some application areas it is probable that the casualty rate due to flawed microelectronics will rise over the next decade.

It will be demonstrated in this paper that current methods of hardware verification are mature enough to be used for practical problems, on the scale of a simple 32-bit microprocessor. Based on this realisation there is every reason to suggest to certification authorities and to industry that formal methods should now be applied to critical devices. By paying equally close attention to the correctness of software it is believed that in the 1990's it should be possible to produce ultra-safe equipment for those regimes in which large numbers of human lives are at stake. However, the ability to do this will depend in part on scarce human resources, since knowledge of the requisite techniques has not spread far from the originating laboratories.

1 The Specification and Design of VIPER

The key feature about the new VIPER processor, Figure 1, is its simplicity. A more complex architecture would have taken the proofs beyond the current

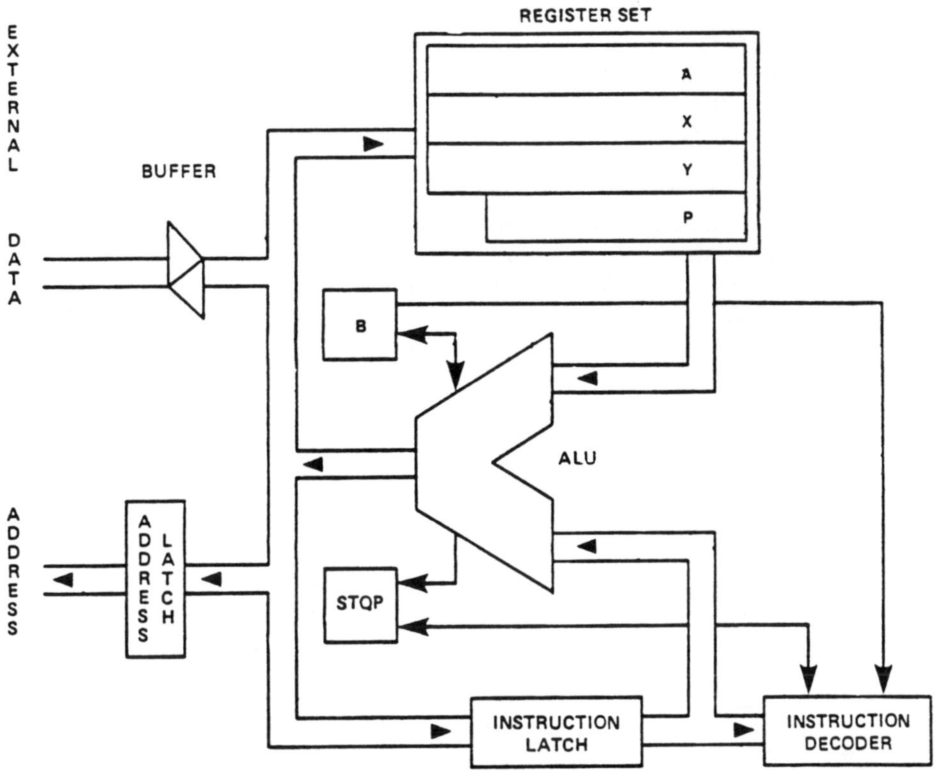

FIG 1 ***VIPER*** MICRO-PROCESSOR ARCHITECTURE

state of the art. As will be seen from Figure 1, the device has an accumulator, A, two 'index registers', X and Y and a program counter, P. There is a single-bit register, B, which holds the results of comparisons and can be concatenated with registers in shift operations. Unique to VIPER is the 'stop' signal from the ALU. Any illegal operation, arithmetic overflow or computation of an illegal address causes the device to stop and raise an exception on an external pin so that corrective action can be taken by the rest of the system.

All instructions have an identical format, shown in Figure 2. The logic of instruction decoding is outlined in Figure 3. Note that comparisons subsume all other operations. Table 1 gives a list of the comparison and ALU functions available in the prototype chips. The mf field of the instruction is used to select shift instructions when $cf = 0$ and $ff = 12$. This minor degree of overloading is acceptable since shifts are applied to registers and there is no 'memory read' in this context.

Table 1: Comparison and Alu Operations

(r = 32 bits from register, m = 32 bits from memory)

ff	COMPARISON, $cf = 1$	NOT COMPARISON, $cf = 0$
0	$r < m$	NEGATE m
1	$r \geq m$	CALL
2	$r = m$	READ from PERIPHERAL
3	$r \neq m$	READ from MEMORY
4	$r \leq m$	$r + m$, $b := carry$
5	$r > m$	$r + m$, stop on overflow
6	unsigned $r < m$	$r - m$, $b := borrow$
7	unsigned $r \geq m$	$r - m$, stop on overflow
8	$(r < m)$ OR b	r XOR m
9	$(r \geq m)$ OR b	r AND m
10	$(r = m)$ OR b	r NOR m
11	$(r \neq m)$ OR b	r AND NOT m
12	$(r \leq m)$ OR b	$mf = 0$: SHIFT RIGHT, copy sign
		$mf = 1$: SHIFT RIGHT, through b
		$mf = 2$: SHIFT LEFT, stop on oflo
		$mf = 3$: SHIFT LEFT, through b
13	$(r > m)$ OR b	illegal
14	(unsigned $r < m$) OR b	illegal
15	(unsigned $r \geq m$) OR b	illegal

Some of the rationale for this architecture is as follows. One of the primary requirements was for the instruction decoder to be as simple as possible.

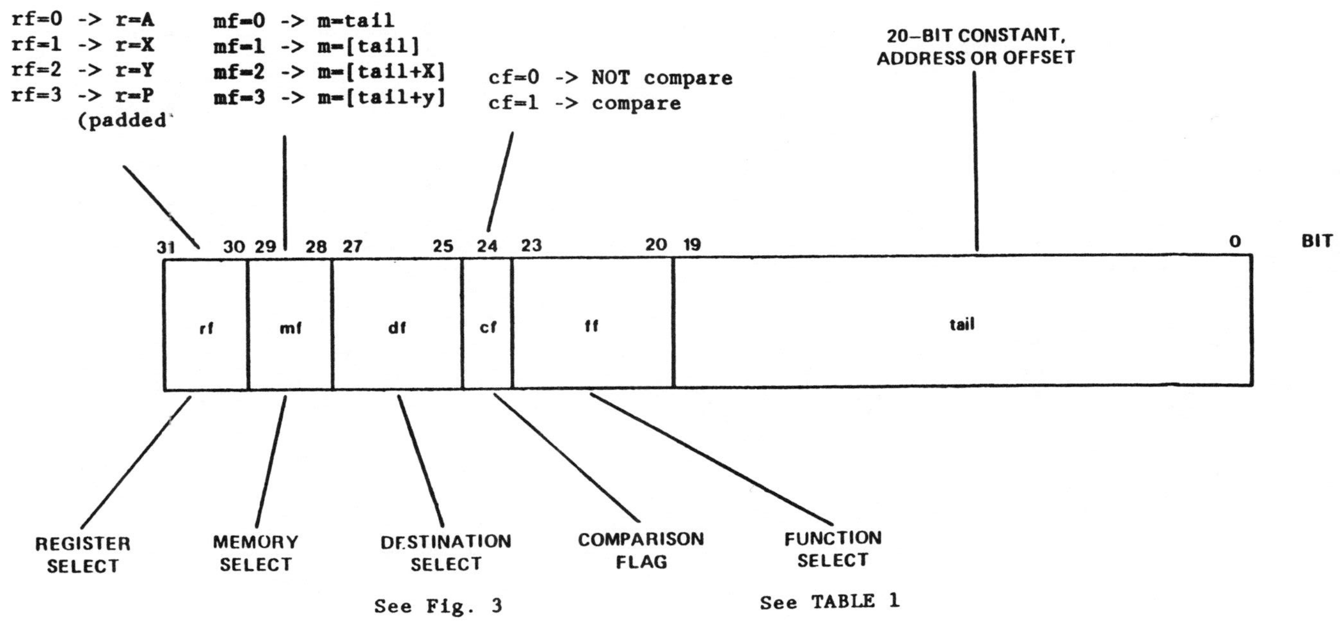

FIG 2 INSTRUCTION FORMAT

```
df=7 -> Write to memory
df=6 -> Write to I/O
df=5 -> IF B THEN no-op ELSE destination = P
df=4 -> IF B THEN destination = P ELSE no-op
df=3 -> destination = P (unconditional)
df=2 -> destination = Y
df=1 -> destination = X
df=0 -> destination = A
```

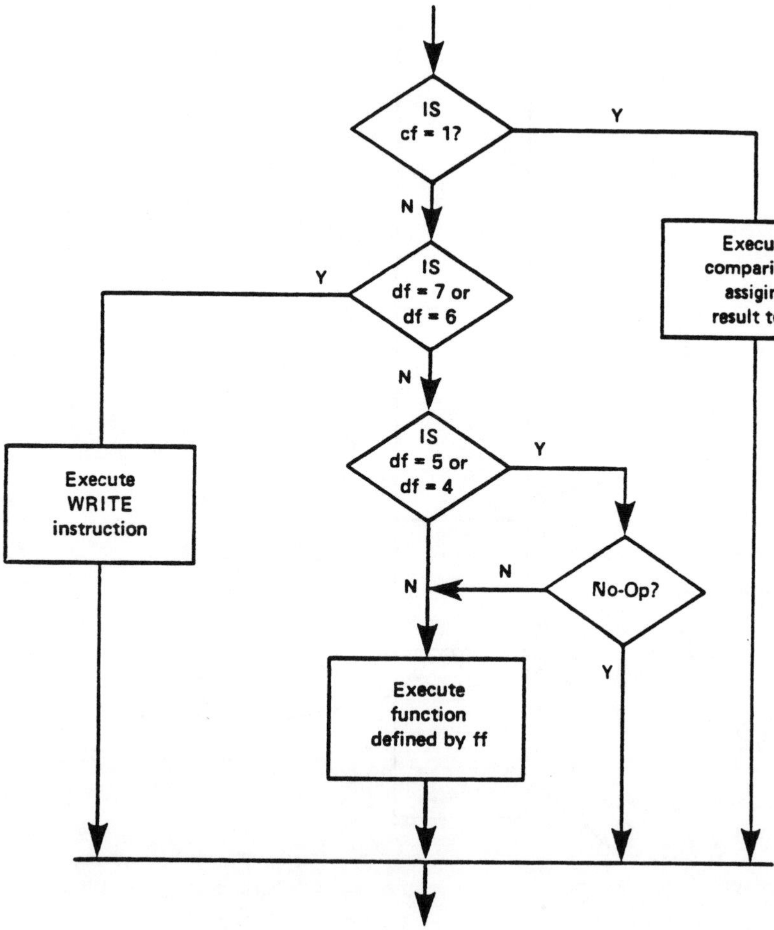

FIG 3 LOGIC OF INSTRUCTION DECODING

Secondly, a wide data word was needed for high resolution in digital filter applications. Thirdly, a rich set of comparison functions was desirable, based on the setting or clearing of a single flag, B. No support has been provided for interrupts or for the manipulation of a stack. Both of these facilities are potentially hazardous. Interrupts mean that program validation is not feasible to the same depth as in a non-interruptable program. Stacks have the difficulties of overflow or trying to POP from an empty stack and are best avoided in safety-critical work.

As regards address range, 20 bits may seem excessive, since safety-critical programs should be short. However, the I/O space may need as much as one million words and to preserve uniformity the main memory space is made the same size, selection between the two being by a one-bit signal (MEM/IO) which essentially becomes a 21st address bit. In practice the ROM and RAM used with VIPER processors will probably be widely separated in the main memory address space, e.g. trusted ROMs at the lower end of memory and RAM at the top. Only static RAMs should be used with VIPER and the ommission of any means of providing refresh to dynamic memories is deliberate.

The top level specification for VIPER was written in LCF-LSM and is reproduced as Annex A. This specification is based on the transformations of a vector:

$$state = (ram, P, A, X, Y, B, stop)$$

where *ram* implies the current contents of both address spaces, P, A, X, Y and B are the current contents of registers in VIPER and *stop* is a Boolean which indicates if the machine has stopped. The function NEXT() defined in Annex A gives all possible transitions for VIPER in the form:

$$\mathsf{NEXT}(state) \rightarrow state$$

The *state* which forms the argument is that just before the next instruction fetch. Notice that as illustrated in Annex A it proved possible to provide a formal specification of this processor in just six pages of A4, namely five pages of auxiliary functions and a single page definition of NEXT.

Hierarchically, the next layer of documentation defines the 'major state machine' illustrated in Figure 4. This combination of elementary functions must be shown to conform precisely with the top level specification. The node 'dummy' has been included solely to act as the 'primary node' for the purposes of proof. It has to be established that all possible journeys round this diagram, from dummy back to dummy, produce the transformations required in the top level specification.

To do this requires the augmented vector:

$$major = ((ram, P, A, X, Y, B, stop), (t, inst))$$

where t represents the contents of an internal register not accessible to the programmer and *inst* the current 12 bits of the instruction fields. Each of the

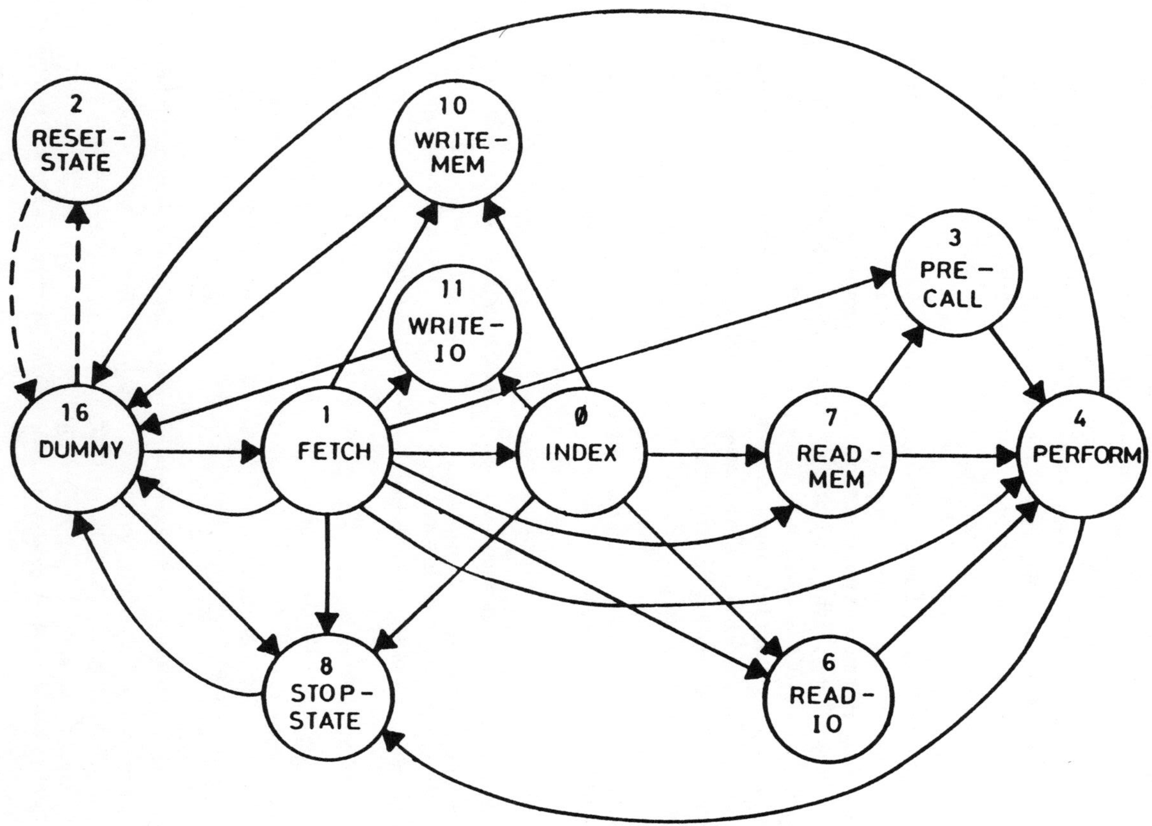

FIG.4 STATE TRANSITIONS IN MAJOR STATE MACHINE

nodes in Figure 4 produces the mapping:

$$\mathsf{MACHINE}(major) \rightarrow (major, next\_node)$$

where *next_node* is the number of one of the other nodes in the diagram.

The third layer of documents specifies the so-called 'electronic block model' shown in Figure 5. This is the lowest level of documentation issued by RSRE. Those wishing to fabricate VIPER in practice have to utilise their own in-house VLSI technology to produce each of the blocks shown. The specification for each of these blocks was defined initially in LCF-LSM and subsequently in the logic description language ELLA. There were a number of reasons for changing language. As the design moves closer towards circuit level, the definition of a Boolean variable has to be extended to allow for values such as x for 'irrelevant' and i for 'illegal'. ELLA allows this. Secondly, ELLA has been interfaced to various VLSI CAD systems using a package called 'ELLANET' which enables 'net lists' for chips to be transported to alien CAD environments.

In practice the translation of LCF-LSM into ELLA texts was straightforward. A library of LCF-LSM primitive functions was written in ELLA. This enabled LCF-LSM functions to be compiled in ELLA with only minor syntactic changes [11]. As a result there is considerable confidence that the two sets of texts are functionally identical. As regards complexity, the building blocks shown in Figure 5 can be specified completely in about thirty pages of A4. In addition, the industrial teams producing the prototype chips had a gate level design for each block, produced at RSRE, and this aided their task in understanding the functions of each module.

The bottom level of documentation is the 'circuit'. This level of documentation was produced by Ferranti, using their own design language (FDL) and Marconi using HILO. In order to achieve comparison with the RSRE block model these texts were translated into ELLA, using software written at RSRE. This was a vital step in the whole process of verification of the gate level design and is considered further in Section 2.3.

The above description gives the impression of 'top down' design. This was not the case in practice. There were several iterations of the top level specification before the final text of Annex A was agreed. Since no proofs were available in the early stages of the project, confidence was built up in the various texts largely by simulation, as described in Section 2.4. Based on this experience it is fair to conclude that at least three levels of documentation (top level specification, major state machine and electronic block model) have to be iterated in harmony to achieve a practical and utilitarian design. Detailed design at silicon level should be postponed until these three layers of documents have been frozen.

It is interesting to note that the manufacturers of the prototype VIPER chips made comparatively little use of the top level specification or the major state machine descriptions. Their designs were completed largely on the basis of the electronic block model. This suggests that the higher levels of specification can be left in the hands of small teams of specialists, with detailed design and

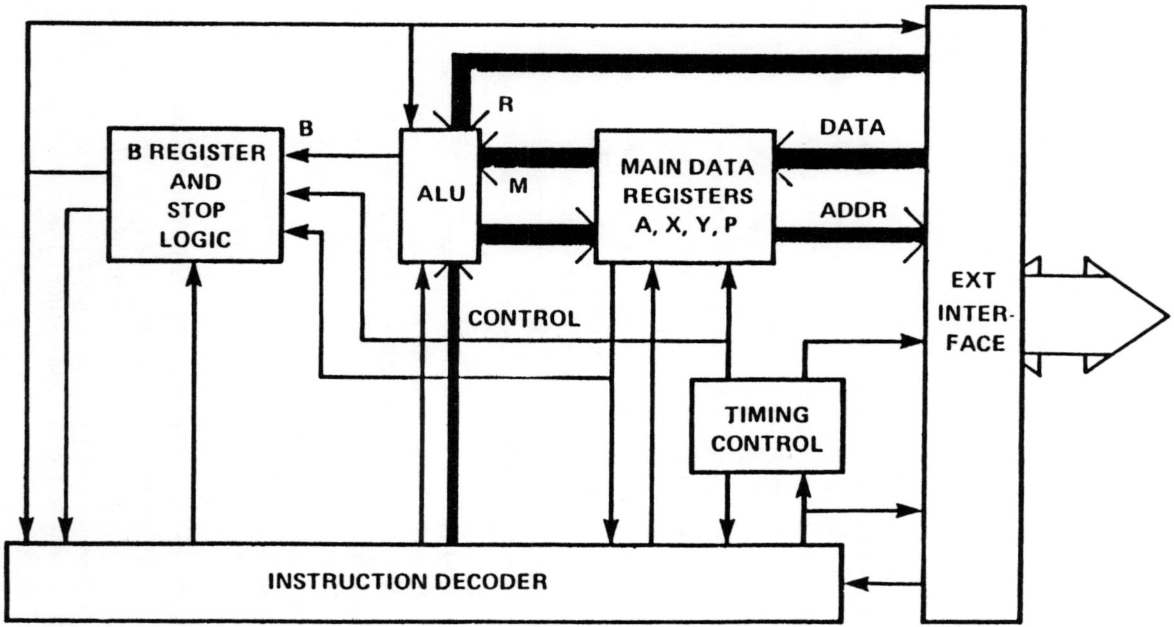

FIG 5 ***VIPER*** ELECTRONIC BLOCK MODEL

fabrication of chips being devolved to VLSI companies who do not necessarily need to employ experts in formal methods.

2 Verification of the VIPER Design

2.1 Major State Machine to Top Level Specification

The informal proof of correspondence was done by the author at RSRE and took about three man weeks, by hand analysis, i.e. without the use of the Cambridge LCF-LSM tools. By deriving the spanning tree of all possible transitions of the major state machine, Figure 6, it proved possible to conduct a 24-limb proof in first-order logic. This established with a reasonable degree of certainty that the major state machine conformed with the top level specification.

Subsequently, Avra Cohn and Michael Gordon, working with HOL versions of the texts at Cambridge and using their automated theorem prover have shown that this informal proof was flawed in some minor respects but overall was sound. This exercise has shown the importance of formal analysis using well founded axioms, rules and tactics to reveal the unexpected logical features of a specification. As this formal verification work in Cambridge has been completed it is safe to conclude that the major state machine is an accurate model which forms a rigorous basis for the lower levels of design [4].

2.2 Electronic Block Model to Major State Machine

This layer of proof is too extensive to carry out by hand and definitely requires automated theorem proving. At the time of writing (December 1986), this work is about to be started in Cambridge. Avra Cohn and Michael Gordon will be using their HOL tools on a SUN workstation to establish the correspondence of the electronic building blocks of Figure 5 with the major state machine. Inevitably this set of proofs will be voluminous if published in full but it is intended to produce a detailed summary for review, probably as a Monograph.

The technique to be used relies on carrying out case analysis for the various node numbers in Figure 4. Details of how this was done for the simpler example of a 6-bit counter are given in Reference [3], in terms of the ML code defining the rules and tactics.

2.3 Circuit to Electronic Block Model

This proof uses the ELLA simulator and a technique devised by Clive Pygott of RSRE. The essence of the method, which is explained in full in Reference [11], is to carry out the equivalent of exhaustive simulation on the ELLA description of one block, such as the instruction decoder, in parallel with applying the same test vectors to an ELLA description of the implementation. To avoid the combinatorial explosion and totally unrealistic run times, extensive use is made

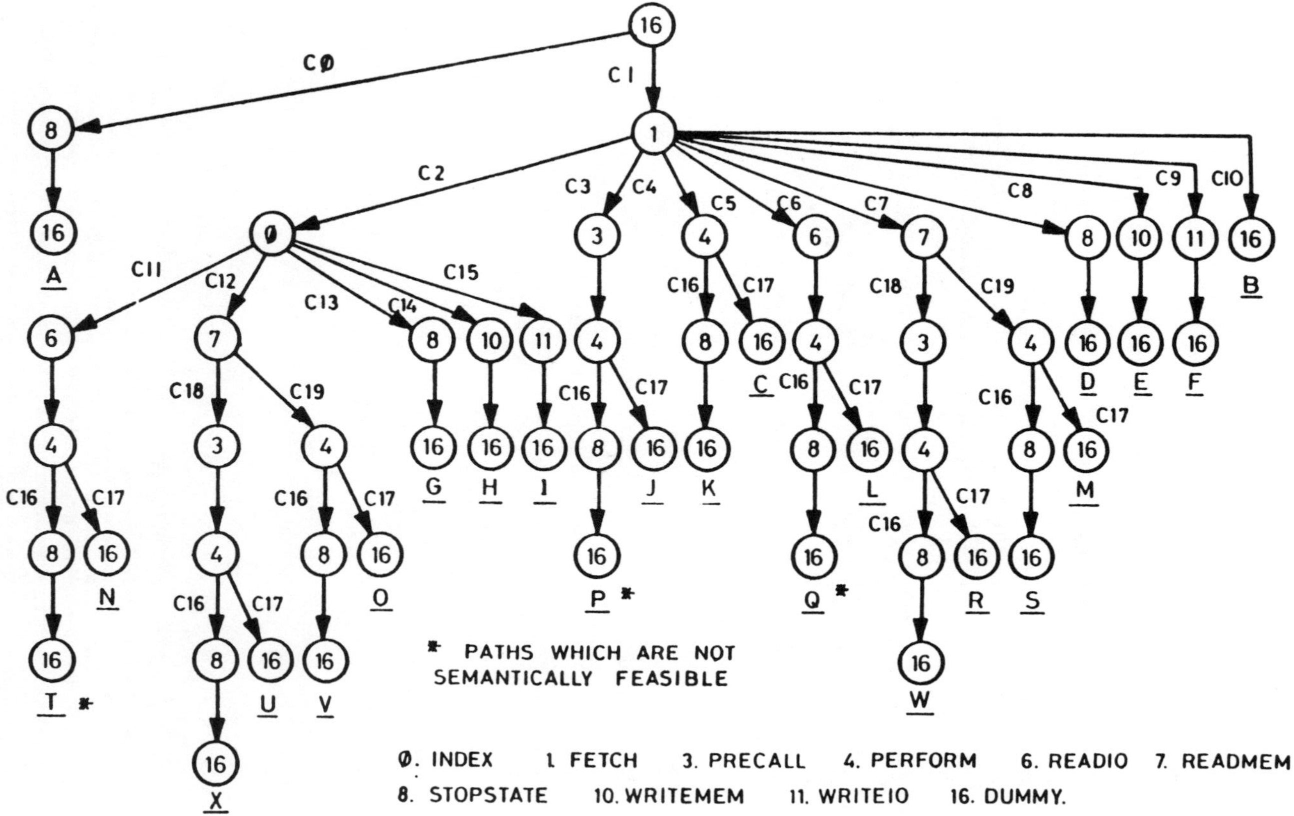

FIG. 6. SPANNING TREE OF POSSIBLE STATE TRANSITIONS

of the x (i.e. irrelevant) value of Boolean for all components of a test vector which do not affect the outcome of a particular test. By this means the quarter of a million tests which would have been required for the VIPER instruction decoder were thinned to just over 900 runs of the ELLA simulator. Indeed, the whole of the implementation of VIPER can be checked with only 5,500 or so such vectors. It is worth noting that all of the tests for one chip can be completed, on heavily loaded VAX 8600's in about three hours of elapsed time.

In practice, both of the contractors fabricating the prototype chips made mistakes early on at circuit level when implementing particular blocks. The method outlined above proved to be powerful enough to detect the discrepancies. Both designs were corrected before net-lists were finalised and masks generated, with substantial savings in overall project timescales. It is believed that this method of 'intelligent exhaustive simulation' can be refined and turned into a practical tool for the VLSI industries. Work is in hand at RSRE towards this end.

At Cambridge this level of proof has been done in HOL. As an interesting comparison, Reference [3] shows how HOL can be used to prove the circuit of a 6-bit counter correct whereas Reference [11] shows the same 'proof' done using the ELLA simulator. It is a relevant question to ask if a logic description language, such as ELLA, is needed as a stepping stone from the abstraction of HOL to current VLSI design languages. Experience on VIPER suggests that the introduction of ELLA was essential to gain industrial acceptance of the methods in use.

2.4 The Use of Simulation

In the absence of formal proofs of correspondence, extensive use was made of simulation to build up confidence in the various levels of documentation. Specifically, the following animations or simulations exist, running on VAX/VMS :

- An animation of the LCF-LSM top level specification, written in ALGOL68, based on a library of LCF-LSM primitives (WORDn, BITSn, etc.);
- An animation of the major state machine, written in ALGOL68;
- A simulation of the electronic block model in ELLA, based on a library of LCF-LSM primitives written in ELLA.

Software has been written to compare the outputs from these programs automatically when the three programs are driven with the same VIPER object code, as illustrated in Figure 7. Note also the links to the industrial VLSI CAD systems, which proved to be vital when interacting with circuit designers.

It is important to comment on the relative speeds of these simulations. On VAX 8600 it is about 50 VIPER instructions per second at the top level, about 12 instructions per second for the major state machine and only 1 or 2 instructions per second for the ELLA simulation of the electronic block model.

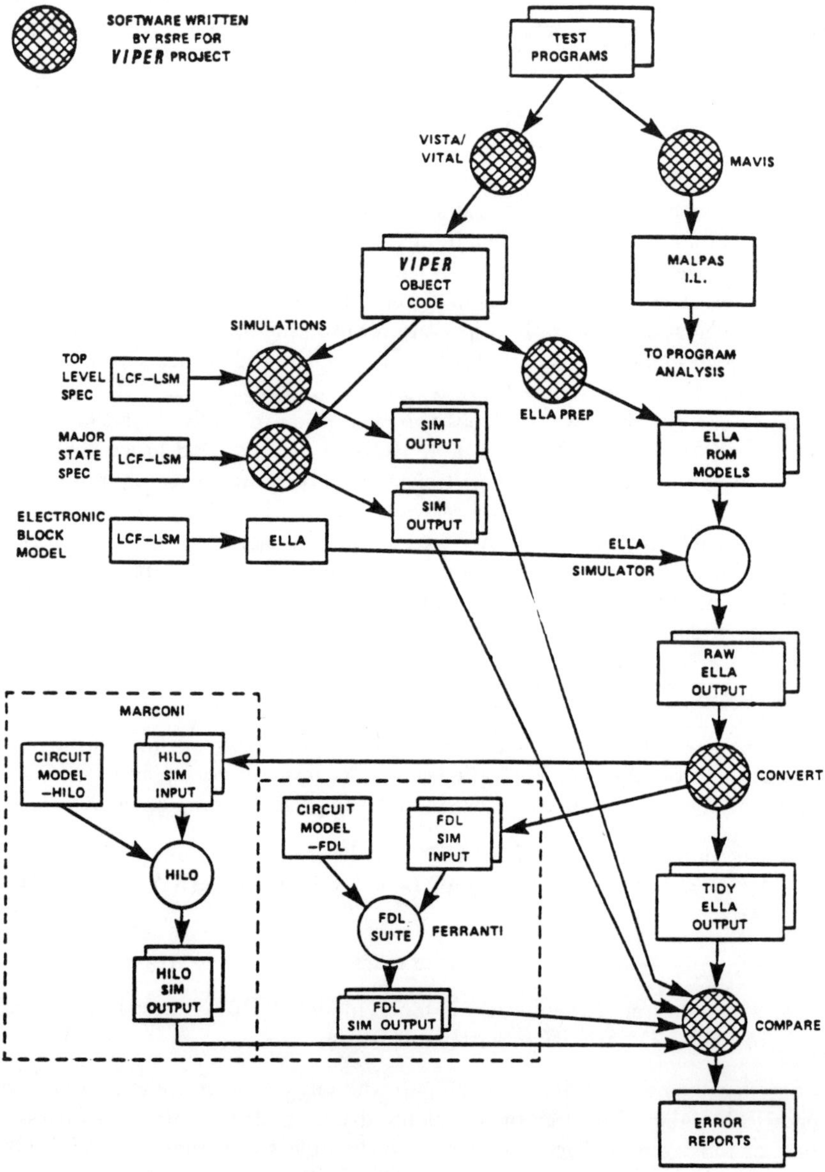

FIG 7 SIMULATIONS OF ***VIPER***

As might be expected, many hundreds of mill hours of VAX time was used at RSRE, most of this on ELLA simulations.

The so-called 'animations' of the LCF-LSM texts proved to be very valuable. When dealing with abstract definitions such simulators produce a great deal of engineering confidence and can be used for practical demonstrations to prospective customers for the ultimate device. It is not clear if this comment applies to work in HOL but based on experience of the RSRE team to date this is a topic worth pursuing. It should be straightforward to simulate higher level HOL documents but not so easy to animate HOL descriptions of circuits where existential quantifiers are used to hide internal wires.

3 Implementing VIPER

In principle, VIPER could be fabricated in any VLSI technology, including cell arrays, semi-custom and so on. The route chosen for the prototypes was the use of cell arrays, namely:

- The Ferranti Electronics Ltd Uncommitted Logic Arrays (ULA), based on a bipolar technology, and
- The UK5000 cell arrays, as fabricated by Marconi Electronic Devices Ltd, which is a CMOS technology.

The companies worked in parallel and independently to produce designs for the electronic building blocks which make up the VIPER processor. Neither company has ELLA in use and therefore schematic capture and initial simulations were done on their current CAD systems and based on existing libraries. Once these designs had been produced to the level of 'net lists' these descriptions were sent back to RSRE on magnetic tape. Using software written at RSRE, these texts were converted into ELLA so that the 'intelligent exhaustive testing' method described in Section 2.3 could be applied to detect any gate level mistakes.

Given the large amount of detail available at RSRE, it then proved possible to use the ELLA simulations of the two designs to generate test pattern tapes in Malvern, as illustrated in Figure 7. These tapes were used by the contractors to run acceptance tests at the ends of the respective production lines. To achieve this, simple programs were written which exercised all of the registers and facilities in VIPER. These were run through the ELLA simulator to generate the necessary waveforms on the external pins of the chip and these binary sequences were converted at RSRE into formats appropriate for the probe test machines in use by Marconi and Ferranti. This unusually close liason between a customer for a committed logic array and the silicon house undoubtedly contributed a great deal to the success of the project overall.

4 Programming VIPER

4.1 Choice of Language

Clearly assemblers and compilers could be written for VIPER to implement virtually any language. There is some point in sticking to existing languages, to gain commercial acceptance. However, for the most safety-critical systems the collateral work at RSRE on software verification suggests that a fresh language is needed. This seemingly radical decision has been taken after study of the options for writing software for potentially hazardous regimes. Each language was considered under the following headings:

- Existence of 'dangerous' constructs in the language;
- Knowledge of the operational semantics of the language, which affects the ability to carry out static code analysis and/or formal verification;
- Existence of a rigorous numerical model underlying the arithmetic, and
- Existence of any 'trusted' subsets of the language.

Using these criteria, assembler level languages were compared with PASCAL, CORAL 66 and Ada. Overall it was concluded that none of these languages was satisfactory, although a subset of PASCAL produced by Carre shows great promise and has been used on several critical UK projects [2]. Accordingly, it was decided to produce a new high order language for programming VIPER, which has been called 'NewSpeak', since it was designed in 1984 and is calculated to stop programmers thinking 'wrong thoughts'.

If it is worth producing a new language, then its definition and implementation must show major improvements over anything in the market place, otherwise industry will not be prepared to accept the overhead of training. Ian Currie of RSRE has designed the NewSpeak language to fulfill this requirement. As well as a compiler for VIPER, the language may well be valuable for machines such as the 68000 and it is hoped that other NewSpeak compilers will emerge over the next three to four years.

4.2 NewSpeak

The language is described in more detail in Reference [8]. In this section an outline of the chief features is presented, sufficient to show the interactions which have taken place between the design of the VIPER architecture and the design of the NewSpeak language. Among the main features of NewSpeak are:

- Range-monitoring of both real and integer variables with most of the work being done by the compiler, i.e. very little run-time overhead;

- Accuracy monitoring of real values, so that poor numerical significance in an algorithm is spotted at compile time;

- Assertions so that the programmer can incorporate collateral knowledge (e.g. the accuracy of an algorithm) in a form which the compiler can use;

- The 'flavouring' of types eg two real variables might be flavoured METRES and SECOND; the compiler has rules for combining flavours, some of which deliver a new flavour (e.g. METRES/SECOND);

- All NewSpeak constructs are bounded in time and storage allocation. Time critical programs can be written without the use of interrupts by polling the external lines at regular intervals, using the compiler diagnostics to check that the time limits are not exceeded.

It will be realised from the above list that NewSpeak is a far more rigorous language than PASCAL or Ada. Its very strong typing, static storage allocation and guarantees on run time will make it uniquely suited for safety-critical work. It should be available commercially at the start of 1990. By combining this new language with the VIPER microprocessor, it is believed that new standards of certainty can be reached in the development of high integrity equipment.

5 Conclusions

The VIPER project has shown that the methods of hardware verification developed by Michael Gordon at Cambridge, when combined with the use of a modern hardware description language, ELLA, are powerful enough to use on practical projects. It is probable that HOL/ELLA will replace LCF-LSM/ELLA as the working tools at RSRE within the next twelve months. Two companies in the UK VLSI industry have proven that they can interface to this formal regime using their standard logic array technologies.

On the programming side, RSRE has concluded that no existing language is safe enough for the highest integrity systems. A new high order language, NewSpeak, has been defined for use with VIPER. By rigorous application of typing, range checks on variables and other features it is believed that an ultra-safe language can be produced, with commercial applications post-1990.

Aside from verified processors and a high integrity language, great care must be taken with the overall system design when implementing safety-critical systems. Multi-channel implementations will be mandatory, even if VIPER processors are in use in the individual lanes and NewSpeak has been used for programming. It is believed that all of the steps advocated in this paper will have to be taken if the public is to be protected from the consequences of design errors in computer-based real-time control systems.

Acknowledgements

VIPER has been developed by a team at RSRE made up of Dr J. Kershaw, Dr C. H. Pygott and the author. The formal verification work at Cambridge is being conducted by Dr A. Cohn and Dr M. J. C. Gordon. Mr I. F. Currie of RSRE is responsible for the design and specification of NewSpeak.

References

[1] Bramson B. D., "Malvern's Program Analysers", RSRE Research Review 1985.

[2] Carre B. A. and Debney C., "SPADE Pascal", Program Validation Ltd., UK, July 1985.

[3] Cohn A. and Gordon M. J. C., "A Mechanized Proof of Correctness of a Simple Counter", University of Cambridge Computer Laboratory, Technical Report No. 94, July 1986.

[4] Cohn A., "A Proof of Correctness of the VIPER Microprocessor: The First Level", in: VLSI Specification, Verification and Synthesis, (this volume).

[5] Cullyer W. J., "Hardware Integrity", Aeronautical Journal, Royal Aeronautical Society, UK, September 1985.

[6] Cullyer W. J., "VIPER Microprocessor: Formal Specification", RSRE Report 85013, October 1985.

[7] Cullyer W. J., "VIPER: Correspondence Between Specification and Major State Machine", RSRE Report 86004, January 1986.

[8] Cullyer W. J. and Kershaw J., "Programming the VIPER Microprocessor" Proceedings of Conference on Safety and Security, Glasgow, UK, October 1986.

[9] Gordon M. J. C., "LCF-LSM", University of Cambridge Computer Laboratory, Technical Report No. 41, 1983.

[10] Morison J. et al., "ELLA: Hardware Description or Specification?", Proceedings IEEE International Conference, CAD-84, Santa Clara, November 1984.

[11] Pygott C. H., "Formal Proof of Correspondence Between a Hardware Module and its Gate Level Implementation", RSRE Report 85012, June 1985.

Annex A:
Top Level Specification of VIPER In LCF-LSM

```
VALUE : (word32 # bool # bool) → word32
VALUE (result, carry, overflow) == result

OFLO : (word32 # bool # bool) → bool
OFLO (result, carry, overflow) == overflow

CARRY : (word32 # bool # bool) → bool
CARRY (result, carry, overflow) == carry

SVAL : (word32 # bool # bool) → bool
SVAL (result, b, abort) == abort

BVAL : (word32 # bool # bool) → bool
BVAL (result, b, abort) == b

TRIM33TO20 : word33 → word20
TRIM33TO20 (w) == (WORD20 (V (SEG (0, 19) (BITS33 w))))

TRIM32TO20 : word32 → word20
TRIM32TO20 (w) == (WORD20 (V (SEG (0, 19) (BITS32 w))))

TRIM34TO32 : word34 → word32
TRIM34TO32 (w) == WORD32 (V (TL (TL (BITS34 w))))

PAD20TO32 : word20 → word32
PAD20TO32 (w) == WORD32 (VAL20 w)

SIGNEXT : word32 → word33
SIGNEXT (w) ==
    LET bitlist = BITS32 w IN
    (WORD33 (V (CONS (EL 31 bitlist) bitlist)))

RIGHT : (bool # word32) → word32
RIGHT (b, r) == (WORD32 (V (CONS b (SEG (1, 31) (BITS32 r)))))

LEFT : (word32 # bool) → word32
LEFT (r, b) ==
    LET twice = V (TL (BITS32 r)) IN
    (b ⇒ (WORD32 ((twice + twice) + 1) | (WORD32 (twice + twice))))

RIGHTARITH : word32 → word32
RIGHTARITH (r) ==
    LET sign = EL 31 (BITS32 r) IN
    (WORD32 (V (CONS sign (SEG (1, 31) (BITS32 r)))))
```

```
NEG : (word33) → num
NEG (m) == (((VAL33 m) = 0) ⇒ 0 | (VAL33 (NOT33 m) + 1))

ADD32 : (word32 # word32) → (word32 # bool # bool)
ADD32 (r, m) ==
    LET sum = WORD34 ((VAL33 (SIGNEXT r)) +
                      (VAL33 (SIGNEXT m))) IN
    LET opposite = (EL 31 (BITS32 r)) XOR (EL 31 (BITS32 m)) IN
    (TRIM34TO32 sum, (EL 32 (BITS34 sum)) XOR opposite,
     ((EL 32 (BITS34 sum)) XOR (EL 31 (BITS34 sum))))

SUB32 : (word32 # word32) → (word32 # bool # bool)
SUB32 (r, m) ==
    LET dif = WORD34 ((VAL33 (SIGNEXT r)) +
                      (NEG (SIGNEXT m))) IN
    LET opposite = (EL31 (BITS32 r)) XOR (EL 31 (BITS32 m)) IN
    (TRIM34TO32 dif, (EL 32 (BITS34 dif)) XOR opposite,
     ((EL 32 (BITS34 dif)) XOR (EL 31 (BITS34 dif))))

INCP32 : (word20) → word32
INCP32 (p) == (VALUE (ADD32 (PAD20TO32 p) (WORD32 1)))

COMPARE : (word4 # word32 # word32 # bool) → bool
COMPARE (fsf, r, m, b) ==
    LET op = VAL4 fsf IN
    LET dif = WORD34 ((VAL33 (SIGNEXT r)) +
                      (NEG (SIGNEXT m))) IN
    LET equal = (r = m) IN
    LET less = EL 32 (BITS34 dif) IN
    LET borrow = (EL 32 (BITS34 dif)) XOR
                 ((EL 31 (BITS32 r)) XOR
                  (EL 31 (BITS32 m))) IN
    (op = 0 ⇒ less |
     op = 1 ⇒ NOT less |
     op = 2 ⇒ equal |
     op = 3 ⇒ NOT equal |
     op = 4 ⇒ (less OR equal) |
     op = 5 ⇒ NOT (less OR equal) |
     op = 6 ⇒ borrow |
     op = 7 ⇒ NOT borrow
     op = 8 ⇒ (less OR b) |
     op = 9 ⇒ ((NOT less) OR b) |
     op = 10 ⇒ (equal OR b) |
     op = 11 ⇒ ((NOT equal) OR b) |
     op = 12 ⇒ ((less OR equal) OR b) |
```

```
    op = 13 ⇒ ((NOT (less OR equal)) OR b) |
    op = 14 ⇒ (borrow OR b) |
              ((NOT borrow) OR b))

REG : (word2 # word32 # word32 # word32 # word20) → word32
REG (rsf, a, x, y, p) ==
    LET r = VAL2 rsf IN
    (r = 0 ⇒ a |
     r = 1 ⇒ x |
     r = 2 ⇒ y |
           PAD20TO32 p)

OFFSET : (word2 # word20 # word32 # word32) → word32
OFFSET (msf, addr, x, y) ==
    LET mf = VAL2 msf IN
    LET addr32 = PAD20TO32 addr IN
    (mf = 0 ⇒ addr32 |
     mf = 1 ⇒ addr32 |
     mf = 2 ⇒ (VALUE (ADD32 addr32 x)) |
              (VALUE (ADD32 addr32 y)))

INVALID : (word32) → bool
INVALID (value) ==
    (NOT (value = (PAD20TO32 (TRIM32TO20 value))))

INSTFETCH : (mem21_32 # word20) → word32
INSTFETCH (ram, p) ==
    (FETCH21 (ram WORD21 (V (CONS F (BITS20 p)))))

MEMREAD : (mem21_32 # word2 # word20 # word32 #
           word32 # bool # bool) → word32
MEMREAD (ram, msf, addr, x, y, io, nil) ==
     LET m = VAL2 msf IN
     (nil ⇒ WORD32 0 |
      m = 0 ⇒ PAD32TO20 addr |
            (FETCH21 ram
                     (WORD21
                      (V
                       (CONS io
                             (BITS20
                              (TRIM32TO20
                               (OFFSET msf addr x y))))))))

MEMWRITE : (mem21_32 # word32 # word2 # word20 #
            word32 # word32 # bool) → mem21_32
```

```
MEMWRITE (ram, source, msf, addr, x, y, io) ==
      LET m = VAL2 msf IN
      (m = 0 ⇒ ram |
                   (STORE21
                    (WORD21
                     (V
                      (CONS io
                              (BITS20
                               (TRIM32TO20
                                (OFFSET msf addr x y))))))
                    source ram))

ALU : (word4 # word2 # word3 # word32 # word32 # bool) →
        (word32 # bool # bool)
ALU (fsf, msf, dsf, r, m, b) ==
      LET ff = VAL4 fsf IN
      LET mf = VAL2 msf IN
      LET df = VAL3 dsf IN
      LET pwrite = (df = 3) OR (df = 4) OR (df = 5) IN
      (ff = 0 ⇒ ((NOT32 m), b, pwrite) |
       ff = 1 ⇒ (m, b, ((NOT pwrite) OR (INVALID m))) |
       ff = 2 ⇒ (m, b, pwrite) |
       ff = 3 ⇒ (m, b, (pwrite AND (INVALID m))) |
       ff = 4 ⇒ (LET sum = ADD32 (r, m) IN
                 ((VALUE sum), (CARRY sum), pwrite)) |
       ff = 5 ⇒ (LET sum = ADD32 (r, m) IN
                 ((VALUE sum), b,
                  (OFLO sum) OR
                   (pwrite AND (INVALID (VALUE sum))))) |
       ff = 6 ⇒ (LET dif = SUB32 (r, m) IN
                 ((VALUE dif), (CARRY dif), pwrite)) |
       ff = 7 ⇒ (LET dif = SUB32 (r, m) IN
                 ((VALUE dif), b,
                  (OFLO dif) OR (pwrite AND (INVALID (VALUE dif))))) |
       ff = 8 ⇒ (((r OR32 m) AND32 (NOT32 (r AND32 m))),
                 b, pwrite) |
       ff = 9 ⇒ ((r AND32 m), b, pwrite) |
       ff = 10 ⇒ ((NOT32 (r OR32 m)), b, pwrite) |
       ff = 11 ⇒ ((r AND32 (NOT32 m)), b, pwrite) |
       ff = 12 ⇒ (mf = 0 ⇒ ((RIGHTARITH r), b, pwrite) |
                  mf = 1 ⇒ ((RIGHT (b, r)), (EL 0 (BITS32 r)), pwrite) |
                  mf = 2 ⇒ (LET double = ADD32 (r, r) IN
                            ((VALUE double), b,
                             (OFLO double) OR pwrite)) |
```

```
                              (LEFT (r, b), (EL 31 (BITS32 r)), pwrite)) |
     ff = 13 ⇒ (r, b, T) |
     ff = 14 ⇒ (r, b, T) |
               (r, b, T))

WRITE : (word3 # word1) → bool
WRITE (dsf, csf) ==
     LET df = VAL3 dsf IN
     LET cf = VAL1 csf IN
     ((cf = 0) AND ((df = 7) OR (df = 6)))

NILM : (word3 # word1 # word4) → bool
NILM (dsf, csf, fsf) ==
     LET df = VAL3 dsf IN
     LET cf = VAL1 csf IN
     LET ff = VAL4 fsf IN
     ((cf = 0) AND NOT ((df = 7) OR (df = 6)) AND (ff = 12))

NOOP : (word3 # word1 # bool) → bool
NOOP (dsf, csf, b) ==
     LET df = VAL3 dsf IN
     LET cf = VAL1 csf IN
     ((cf = 0) AND (((df = 5) AND b) OR ((df = 4) and (NOT b))))

SPAREFUNC : (word3 # word1 # word4) → bool
SPAREFUNC (dsf, csf, fsf) ==
     LET df = VAL3 dsf IN
     LET cf = VAL1 csf IN
     LET ff = VAL4 fsf IN
     ((cf = 0) AND NOT ((df = 6) OR (df = 7))
              AND ((ff = 13) OR (ff = 14) OR (ff = 15)))

ILLEGALCALL : (word3 # word1 # word4) → bool
ILLEGALCALL (dsf, csf, fsf) ==
     LET df = VAL3 dsf IN
     LET cf = VAL1 csf IN
     LET ff = VAL4 fsf IN
     ((cf = 0) AND (ff = 1) AND ((df = 0) OR (df = 1) OR (df = 2)))

ILLEGALPDEST : (word3 # word1 # word4) → bool
ILLEGALPDEST (dsf, csf, fsf) ==
     LET df = VAL3 dsf IN
     LET cf = VAL1 csf IN
     LET ff = VAL4 fsf IN
     ((cf = 0) AND ((df=3) OR (df=4) OR (df=5))
              AND NOT ((ff=1) OR (ff=3) OR (ff=5) OR (ff=7)))
```

```
ILLEGALWRITE : (word3 # word1 # word2) → bool
ILLEGALWRITE (dsf, csf, msf) ==
   LET mf = VAL2 msf IN
   ((WRITE dsf csf) AND (mf = 0))

OUTPUT : (word3 # word1) → bool
OUTPUT (dsf, csf) ==
    LET df = VAL3 dsf IN
    LET cf = VAL1 csf IN
    ((cf = 0) AND (df = 6))

INPUT : (word3 # word1 # word4) → bool
INPUT (dsf, csf, fsf) ==
   LET df = VAL3 dsf IN
   LET cf = VAL1 csf IN
   LET ff = VAL4 fsf IN
   ((cf = 0) AND NOT ((df = 7) OR (df = 6)) AND (ff = 2))

NEXT : (mem21_32 # word20 # word32 #
         word32 # word32 # bool # bool) →
        (mem21_32 # word20 # word32 #
         word32 # word32 # bool # bool)
NEXT (ram, p, a, x, y, b, stop) ==
    LET instbits = BITS32 (INSTFETCH ram p) IN
    LET newp = TRIM32TO20 (INCP32 p) IN
    LET rsf = WORD2 (V (SEG (30, 31) instbits)) IN
    LET msf = WORD2 (V (SEG (28, 29) instbits)) IN
    LET dsf = WORD3 (V (SEG (25, 27) instbits)) IN
    LET csf = WORD1 (V (SEG (24, 24) instbits)) IN
    LET fsf = WORD4 (V (SEG (20, 23) instbits)) IN
    LET addr = WORD20 (V (SEG (0, 19) instbits)) IN
    LET df = VAL3 dsf IN
    LET cf = VAL1 csf IN
    LET ff = VAL4 fsf IN
    LET comp = (cf = 1) IN
    LET call = (cf = 0) AND (ff = 1) IN
    LET output = (OUTPUT dsf csf) IN
    LET input = (INPUT dsf csf fsf) IN
    LET io = (output OR input) IN
    LET writeop = (WRITE dsf csf) IN
    LET skip = (NOOP dsf csf b) IN
    LET notinc = INVALID (INCP32 p) IN
    LET illegaladdr = (NOT (NILM dsf csf fsf)) AND
                      NOT skip AND
                      (INVALID (OFFSET msf addr x y)) IN
```

```
LET illegalcl = (ILLEGALCALL dsf csf fsf) IN
LET illegalsp = (SPAREFUNC dsf csf fsf) IN
LET illegalonp = (ILLEGALPDEST dsf csf fsf) IN
LET illegalwr = (ILLEGALWRITE dsf csf msf) IN
LET source = (REG rsf a x y newp)) IN
(stop ⇒ (ram, p , a, x, y, b, T) | {machine has stopped}
 ((notinc OR illegaladdr) OR
  (illegalcl OR illegalsp) OR
  (illegalonp OR illegalwr)) ⇒ (ram, newp, a, x, y, b, T) |
 comp ⇒ (ram, newp, a, x, y,
           COMPARE
             (fsf source (MEMREAD ram msf addr x y io F) b),
           F) |
 writeop ⇒ ((MEMWRITE ram source msf addr x y io),
             newp, a, x, y, b, F) |
 skip ⇒ (ram, newp, a, x, y, b, F) |
 (LET m = (MEMREAD ram msf addr x y io (NILM dsf csf fsf)) IN
  LET aluout = (ALU fsf msf dsf source m b) IN
  (df = 0 ⇒ (ram, newp, VALUE aluout, x, y,
              BVAL aluout, SVAL aluout) |
   df = 1 ⇒ (ram, newp, a, VALUE aluout, y,
              BVAL aluout, SVAL aluout) |
   df = 2 ⇒ (ram, newp, a, x, VALUE aluout,
              BVAL aluout, SVAL aluout) |
   call ⇒ (ram, (TRIM32TO20 (VALUE aluout)), a,
           x, (INCP32 p), BVAL aluout, SVAL aluout) |
   (ram, (TRIM32TO20 (VALUE aluout)),
    a, x, y, BVAL aluout, SVAL aluout)))
```

2

A Proof of Correctness of the Viper Microprocessor: The First Level

Avra Cohn

University of Cambridge
Computer Laboratory
Corn Exchange Street
Cambridge, CB2 3QG
England.

Abstract: The Viper microprocessor designed at the Royal Signals and Radar Establishment (RSRE) is one of the first commercially produced computers to have been developed using modern formal methods. Viper is specified in a sequence of decreasingly abstract levels. In this paper a mechanical proof of the equivalence of the first two of these levels is described. The proof was generated using a version of Robin Milner's LCF system.

1 Introduction

The Viper microprocessor designed at the Royal Signals and Radar Establishment (RSRE) is one of the first commercially produced computers to have been developed using modern formal methods. Viper is specified in a sequence of decreasingly abstract levels. In this paper a mechanical proof of the equivalence of the first two of these levels is described. The approach used for this proof is based on HOL, a version of Robin Milner's LCF proof generating system.

There are two reasons for verifying at successive levels of abstraction. First, though it is ultimately a circuit that is being built, and the correctness of this circuit is the most important concern, successive levels of description represent successive stages in the development of the implementation, and one would like to know if these levels are correct. It is possible for the circuit to be correct even if a more abstract specification is incorrect, but the error is still worth knowing about. In fact, some minor errors were discovered, during the machine-checked proof, in the two specifications of Viper. As far as we know, the actual implementation does not reflect these difficulties. (See Section 7 for details.)

Second, working along a sequence of levels makes the verification of the circuit a more tractable problem; there is a logical transformation and an introduction of detail at each stage, and these can then be treated separately. We intend to continue the machine-checked proof of Viper down to more circuit-like levels of abstraction in the future.

This paper is intended to be self contained, but is necessarily brief on background material. A description of Viper can be found in Kershaw [13]; the two specifications can be found in Cullyer [4] and Cullyer [5]. (See also Cullyer [6].) The machine proof is based on an informal proof outline given in [5]. A description of the HOL system is given in Gordon [10]. More generally, the LCF approach to proof is described in the successive manuals [7] and [15]. [2] is a study by Cohn and Gordon of a similar but very much simpler machine-checked proof (of the correctness of a counter); it is based on Cullyer and Pygott [3].

We know of three other machine-checked proofs of computers: Gordon [8] verified a PDP-8 style machine in a precursor of HOL called LCF_LSM. This was redone and improved in HOL by Joyce [12]. In the Boyer-Moore paradigm, Hunt [11] has verified a PDP-11 like machine. Neither of these machines was intended for serious use, as is Viper.

The design of Viper, the high-level specification and host machine are due to Cullyer, Kershaw and Pygott. The only differences in the specifications as they appear in this paper are that they have been put into HOL format, and have been corrected for errors. The idea of viewing the host machine as a transition graph is also due to Cullyer [5], as is the informal proof outline [5]. We have formalised the notion of traversing the graph, made the notion of time explicit, formulated the correctness statements, and generated a machine-checked proof.

1.1 The HOL System

1.1.1 Proof in HOL

HOL, like LCF, is a system for generating formal proofs. In HOL, a logic in which problems can be expressed is interfaced to a programming language called ML in which proof strategies can be encoded. The logic is conventional higher-logic (Church, [1]). It is oriented towards hardware verification only in that it provides types, constants and axioms for representing bit strings. New types, constants and axioms can be introduced by the user, and organised into hierarchies of logical *theories*.

The type discipline of the programming language ensures that the only way to create theorems is by performing a proof; theorems have the ML type *thm*, objects of which type can only be constructed by the application of inference rules to other theorems or axioms. (Theorems are written with a turnstile, |-, in front of them.)

Formal proofs (sequences of elements each of which is either an axiom or a theorem which follows from earlier elements of the sequence by a rule of inference) are *generated* in HOL in the sense that for each element of the sequence which is not an axiom, an ML procedure representing the rule of inference is *executed* to produce that element. That the final element, the fact proved, is actually a theorem is guaranteed by the type discipline of the programming language. The sequences themselves are not retained. LCF-style proof is discussed further in Sections 5.1 and 5.4.

1.1.2 The Logic

The HOL system uses the ASCII characters `~`, `\/` and `/\`, `==>`, `!` and `\` to represent the logical symbols $\neg$, $\vee$, $\wedge$, $\supset$, $\forall$ and λ respectively.

For the purposes of this paper, a *term* of higher-order logic can be one of eleven kinds.

- A *variable*
- A *constant* such as `T` or `F` (which represent the truth-values *true* and *false* respectively)
- A *function application* of the form $\mathtt{t}_1\mathtt{t}_2$ where the term $\mathtt{t}_1$ is called the *operator* and the term $\mathtt{t}_2$ the *operand*
- An *abstraction* of the form `\x.t` where the variable `x` is called the *bound variable* and the term `t` the *body*
- A *negation* of the form `~t` where `t` is a term
- A *conjunction* of the form $\mathtt{t}_1$`/\`$\mathtt{t}_2$ where $\mathtt{t}_1$ and $\mathtt{t}_2$ are terms
- A *disjunction* of the form $\mathtt{t}_1$`\/`$\mathtt{t}_2$ where $\mathtt{t}_1$ and $\mathtt{t}_2$ are terms
- An *implication* of the form $\mathtt{t}_1$`==>`$\mathtt{t}_2$ where $\mathtt{t}_1$ and $\mathtt{t}_2$ are terms
- A *universal quantification* of the form `!x.t` where the variable `x` is the bound variable and the term `t` is the body
- A *conditional* of the form `t=>`$\mathtt{t}_1$`|`$\mathtt{t}_2$ where `t`, $\mathtt{t}_1$ and $\mathtt{t}_2$ are terms; this has *if-part* `t`, *then-part* $\mathtt{t}_1$ and *else-part* $\mathtt{t}_2$
- A *"local declaration"* of the form `let x=`$\mathtt{t}_1$ `in` $\mathtt{t}_2$, where `x` is a variable and $\mathtt{t}_1$ and $\mathtt{t}_2$ are terms; this is provably equivalent to `(\x.`$\mathtt{t}_2$`)`$\mathtt{t}_1$ (see Section 5.1)

All terms in HOL have a *type*. The expression `t:ty` means `t` has type `ty`; for example, the expressions `T:bool` and `F:bool` indicate that the truth-values `T` and `F` have type `bool` for boolean, and `3:num` indicates that `3` is a number.

If `ty` is a type then `(ty)list` (also written `ty list`) is the type of lists whose components have type `ty`. If $\mathtt{ty}_1$ and $\mathtt{ty}_2$ are types, then $\mathtt{ty}_1$`->`$\mathtt{ty}_2$ is the type of functions whose arguments have type $\mathtt{ty}_1$ and results of type $\mathtt{ty}_2$. The cartesian product operator is represented by `#`, so that $\mathtt{ty}_1$`#`$\mathtt{ty}_2$ is the type of pairs whose first components have type $\mathtt{ty}_1$ and second, $\mathtt{ty}_2$.

The HOL system provides a number of predefined types and constants for reasoning about hardware. The types include `word`n, the type of n-bit words and `mem`n_1`_`n_2 for memories of n_2-bit words addressed by n_1-bit words. `#`$\mathtt{b}_{n-1}\cdots\mathtt{b}_0$ (where $\mathtt{b}_i$ is either `0` or `1`) denotes an n-bit word in which $\mathtt{b}_0$ is the least significant bit.

The predefined constants used in this paper are shown below.

- `V:(bool)list->num` converts a list of truth-values to a number
- `VAL`n`:word`n`->num` converts an n-bit word to a number
- `BITS`n`:word`n`->(bool)list` converts an n-bit word to a list of booleans
- `WORD`n`:num->word`n converts a number to an n-bit word
- `FETCH`n`:mem`n_1`_`n_2 `->(word`n_1`->word`n_2`)` looks up a word in memory

List functions include `HD` for taking the head of a list and `CONS` for constructing lists. $[t_1;\cdots;t_n]$ denotes the list containing $t_1,\cdots,t_n$.

To make terms more readable, HOL uses certain conventions. One is that a term $t_1 t_2 \cdots t_n$ abbreviates $(\cdots(t_1 t_2)\cdots t_n)$; function application associates to the left. The product operator `#` associates to the right and binds more tightly than the operator `->`. For example, the function used to model Viper's `ALU` (see Section 3.1) has type

```
word4#word2#word3#word32#word32#bool->word32#bool#bool
```

which abbreviates

```
(word4#(word2#(word3#(word32#(word32#bool)))))->(word32#(bool#bool))
```

From the axioms defining the various constants we can prove **Theorem 0**:

```
|- !b. HD(BITS1(WORD1(V[b]))) = b
```

1.2 What is Proved

At the most abstract level, Viper can be viewed as a function from a state of the machine to a new state, where a state reflects the configuration of the memory, registers and program counter. This is called the high-level or functional specification. The functional specification contains an *operational* semantics of the Viper instruction set. It carries an implicit notion of time in the concepts of *current* and *next* state.

Each state transition at the top level is implemented by a sequence of events at the lower level (the host or major state machine), each of which may affect the *internal* state of the machine. The sequence is determined as the events are performed, according to the internal state and the current event. The possible sequences define a graph of events, each event pointing to one or more others, and one designated *initial*. The lower level can thus be viewed as a function from internal states and nodes in the graph to new internal states and new nodes in the graph.

An internal state is more detailed than a high-level state; it includes the high-level state and also the state of some internal registers. The latter are collectively called the *transient*. The graph is traversed by starting at a fixed node at a fixed time, and moving from node to node until the initial node is reached again. The host machine is modelled with an explicit notion of time in the sequencing and accumulation of the effects. (This level of description is a first step toward a description of the Viper hardware.)

The host and target machine time scales are different, but coincide at intervals. In fact, the host machine provides a specific event (at the beginning of each sequence) during which registers may be examined. Because the notions of time are related in this way, the proof that the two levels of description correspond is not logically difficult. It involves examining each possible sequence of events at the lower level and relating its cumulative effect (the visible part of it) to the corresponding state transition at the higher level. As it happens, there are twenty-four such sequences, hence twenty-four cases to consider in the proof.

To give an idea of what is being proved, and how the proof is generated mechanically, one of the twenty-four cases is examined in some detail: the execution of a procedure call to a literal address.

1.3 The Top Level Correctness Statement

We have formulated and proved the following statement of correctness for Viper. All concepts are explained in detail later in the paper.

Theorem 1: The Top Level Correctness of Viper

```
|- HOST_NEXT_SIG(state_sig,transient_sig,node_sig) /\
   AT node_sig #10000 n ==>
   let d = NUMBER_OF_STEPS(state_sig n) in
   NEXT_TIME(n,n+d)(AT node_sig #10000) /\
   (state_sig(n+d) = NEXT(state_sig n))
```

The terms ending with `_sig` are signals: functions from time (a number) to something else: `state_sig` is a function from time to states, and `state_sig n` means the state at time `n`; similarly for `transient_sig` and `node_sig`. `HOST_NEXT_SIG` is an abbreviation for:

```
!n.(state_sig(n+1),transient_sig(n+1)),node_sig(n+1) =
   HOST_NEXT((state_sig n,transient_sig n),node_sig n)
```

`HOST_NEXT` represents the host machine. `HOST_NEXT_SIG` holds of a state-, transient- and node-signal if the host machine always (for every `n`) takes the state, transient and node at the current time and returns the state, transient and node at the next time. `AT` is defined by:

```
AT f x n = (f n = x)
```

It means: `f` at time `n` is `x`. `NEXT` is the high-level or functional specification. `NEXT_TIME(n1,n2) P` means that `n2` is the next time after `n1` that the predicate

P is true. NUMBER_OF_STEPS takes a state s and returns the total number of node transitions required to return to the initial node starting in state s. The initial node happens to be called #10000.

Therefore the correctness statement means that if:

- HOST_NEXT applied to the state, transient and node at any time gives the state, transient and node at the next time, and
- the node at some particular time n is #10000

then

- d is the number of steps it takes to return to node #10000, and
- after d steps, the state attained by the host (HOST_NEXT) is the same as the state specified functionally (by NEXT).

(The transient does not matter in this comparison; it is just used by HOST_NEXT in its graph traversal.)

The correctness statement is deceptively compact. To prove it requires computing for each possible path through the graph the number of steps that path comprises and the final state accumulated. Each final state must be compared to the state specified at the higher level, under the conditions that caused that particular path to be chosen. The proof depends on the definitions of NEXT and HOST_NEXT, as well as on various properties of numbers.

In this paper, one such path (the execution of a literal *call* instruction) is examined in detail to illustrate the nature of the proof. Two main theorems are proved for this path; the first is discussed in Section 5.2.3, and the second in Sections 5.3 and 5.5.

In this case the first theorem has the form:

Theorem 2: The Number of Steps for the Call Path

```
C(state_sig n) /\
HOST_NEXT_SIG(state_sig,transient_sig,node_sig) /\
(node_sig n = #10000) ==>
NEXT_TIME(n,n+4)(AT node_sig #10000)
```

where C is some property of the state which causes the literal *call* path through the graph to be chosen. The theorem says: given that condition C applies to the state at some time n, and given that the state, transient and node at any time are computed by HOST_NEXT applied to the state, transient and node at the previous time, and given that the node at time n is #10000, then n+4 happens to be the next time at which the node is again #10000.

The second theorem has the form:

Theorem 3: Equivalence of Host and Specification for the Call Case

```
C(state_sig n) /\
HOST_NEXT_SIG(state_sig,transient_sig,node_sig) /\
(node_sig n = #10000) ==>
(state_sig(n+4) = NEXT(state_sig n))
```

This states the correctness of Viper for the one kind of instruction. It says that under condition `C`, and given that `HOST_NEXT` computes the successive states, transients and nodes, and starting at node `#10000` at time `n`, the state attained by the host machine at time `n+4` agrees with the high-level transformation of the time-`n` state.

A similar pair of theorems is proved for each of the twenty-four possible paths through the graph. To tie these together into the main theorem, it has to be shown that at each node, the conditions for choosing the next node cover all possibilities. Then the main theorem follows by analysis of all logical cases.

2 The Design of Viper

Viper has a 32-bit memory. Addresses are 20-bit words, but the memory is addressed by 21-bit words, where the most significant bit distinguishes main from peripheral memory; peripheral memory is for input-output operations.

Instructions are interpreted in the context of a memory (`ram`) of 32-bit words, a 20-bit program counter (`p`), three 32-bit registers (accumulator `a` and index registers `x` and `y`), a 1-bit register (`b`, to hold the results of comparisons, *etc*) and a flag for stopping the machine (`stop`) should an anomolous situation arise. These seven components comprise the configurations (or states) described by the functional specification.

Instructions are 32 bit words, segmented as follows:

2 bits	2 bits	3 bits	1 bit	4 bits	20 bits
source register field	memory address control field	destin-ation control field	compare flag	function field	address

Each field is a bit string. The first five are represented by the variables `rsf`, `msf`, `dsf`, `csf` and `fsf`, respectively.

The top twelve bits encode the register source, the memory source, the destination, and the function part of an instruction. The bottom twenty bits are the address. The source register field can hold the values 0, 1, 2 or 3. These values respectively indicate that the source register for the operation is the `a`, `x`, or `y` register, or the program counter. The memory address control field can also hold the values 0, 1, 2 or 3; these indicate that the memory source is the literal address of the instruction, the contents of the address, or the contents of the address offset by the value in the `x` or `y` register. The destination control field can hold the values 0,...,7, which indicate the destination of the operation. The `a`, `x` and `y` registers are indicated by values 0, 1 and 2, respectively; the program counter by 3; the program counter *if* the `b`-flag is set, by 4; the program counter *if* the `b`-flag is not set, by 5; and the location given by the 20-bit address field (in peripheral or main memory) by 6 and 7, respectively. The 1-bit compare flag indicates a compare instruction if it holds the value 1, and a non-compare if 0. The function field can hold values 0,...,15. A *call* instruction is indicated by the value 1; a peripheral memory operation by 2; various other functions which require a memory source by 0 and 3,...,11; and functions which do not need a memory source by 12. The values 13, 14 and 15 are spare instructions whose attempted use indicates an error.

3 The Specification in HOL

To specify the behaviour of Viper in HOL, a hierarchy of logical theories is constructed in which the new types, constants and definitions required can be neatly organized. All of the definitions in Sections 3.1, 3.2 and 4.1 are corrected HOL versions of those in [4] and [5].

3.1 Basic Definitions

A high-level state `(ram,p,a,x,y,b,stop)` is represented in HOL as an object with the following type:

```
mem21_32#word20#word32#word32#word32#bool#bool
```

The following logical constants are used in [4]:

- `AND:bool->(bool->bool)` for the logical $\wedge$
- `OR:bool->(bool->bool)` for the logical $\vee$
- `NOT:bool->bool` for the logical $\neg$

In this paper, we use HOL's /\ and \/ for `AND` and `OR`, respectively, to avoid having two equivalent symbols in different places; in the actual proof `AND`, `OR` and `NOT` are used in the places they occur in [4], and simple substitutions are made so that HOL's inference rules for /\, and \/ and ~ apply. We retain `NOT` in this paper, however, as HOL's ~ is unfortunately rather unreadable.

There are some function constants for acting on bit strings:

- `PAD20TO32:word20->word32` for extending a 20-bit word to 32 bits
- `TRIM32TO20:word32->word20` for truncating a 32-bit word to 20 bits

- o `INCP32:word20->word32` for padding a 20-bit word, then incrementing

The first two are constrained by the axiom

Axiom 1: Trim-Pad Axiom

```
|- !w. TRIM32TO20(PAD20TO32 w) = w
```

The function `REG` for selecting a source register according to the source register field `rsf` has the type

```
REG:word2#word32#word32#word32#word20->word32
```

and is defined by

```
|- REG(rsf,a,x,y,p) =
   let r = VAL2 rsf in
   ((r=0) => a | (r=1) => x | (r=2) => y | PAD20TO32 p)
```

The function `INSTFETCH` to fetch from main memory according to the program counter has the type

```
INSTFETCH:mem21_32#word20->word32
```

and is defined below. A 21-bit address is formed from `p` and `F`.

```
|- INSTFETCH(ram,p) = FETCH21 ram (WORD21(V(CONS F(BITS20 p))))
```

The boolean value (`F` in this case) distinguishes main from peripheral memory. There are functions to extract the various fields of an instruction:

- o `R:word32->word2` to extract the source register field
- o `M:word32->word2` to extract the memory address field
- o `D:word32->word3` to extract the destination field
- o `C:word32->word1` to extract the compare field
- o `FF:word32->word4` to extract the function field
- o `A:word32->word20` to extract the address.

(The name `FF` is used instead of `F`, as in [4], to avoid confusion with the truth value `F`.) There are constants for recognizing certain illegal instructions:

- o `INVALID:word32->bool` for detecting invalid addresses
- o `ILLEGALCALL:word3#word1#word4->bool` for detecting illegal call instructions
- o `ILLEGALPDEST:word3#word1#word4->bool` for detecting illegal uses of the program counter as a destination
- o `ILLEGALWRITE:word3#word1#word2->bool` for detecting illegal write instructions

- SPAREFUNC:word3#word1#word4->bool for detecting attempted uses of the spare ALU functions fields.

defined, for example, by:

```
|- INVALID value = NOT(value = PAD20TO32(TRIM32TO20 value))

|- ILLEGALCALL(dsf,csf,fsf) =
   (let df = VAL3 dsf in
    let cf = VAL1 csf in
    let ff = VAL4 fsf in
    (cf=0) /\ ((ff=1) /\ ((df=0) \/ ((df=1) \/ (df=2)))))
```

INVALID tests whether the top twelve bits of a word are actually being used; a valid address can only use the bottom twenty bits. From the definition of INVALID and Axiom 1, it follows that

Theorem 4: Validity of Padded Addresses

```
|- !w. NOT(INVALID(PAD20TO32 w))
```

which means that a 20-bit address padded to 32 bits is always valid.

ILLEGALCALL tests whether an instruction is a *call* whose destination is the a, x or y register. If so it is illegal; the program counter must be the destination of the ALU result, so that the jump to the procedure can occur.

There is a function NOOP for recognizing instructions that are non-operations; it has the type

```
NOOP:word3#word1#bool->bool
```

and is defined by

```
|- NOOP(dsf,csf,b) =
   let df = VAL3 dsf in
   let cf = VAL1 csf in
   (cf=0) /\ (((df=5) /\ b) \/ ((df=4) /\ (NOT b)))
```

An instruction is a non-operation if it is not a comparison and if its destination field holds the value 5 while the b flag is set, or 4 while it's not. (This allows for conditional *call* and *jump* instructions whose conditions fail.)

Next, there is the important function that represents the behaviour of the ALU:

```
ALU:word4#word2#word3#word32#word32#bool->word32#bool#bool
```

ALU takes a function field, a memory address field and a destination field, a register source, a memory source and the b register, and returns a 32-bit result (a memory source), along with values for the b register and the stop flag. At the moment, we are only interested in the behaviour of ALU for *call* functions, in which case only the destination field and memory source matter. The computed result is the memory value given, the b-value is the value given, and the stop-value is true only if the destination is not the program counter, or if the memory source is invalid. (For *calls*, the memory source returned represents the location of the procedure being called.) "..." abbreviates parts of the definition not relevant to *call* instructions.

```
|- ALU(fsf,msf,dsf,r,m,b) =
   let ff = VAL4 fsf in
   let mf = VAL2 msf in
   let df = VAL3 dsf in
   let pwrite = (df=3) \/ ((df=4) \/ (df=5)) in
   ((ff=0) => ... |
    (ff=1) => (m,b,(NOT pwrite) \/ (INVALID m)) |
     (ff=2) => ... |
      (ff=3) => ... |
       (ff=4) => ... |
        (ff=5) => ... |
         (ff=6) => ... |
          (ff=7) => ... |
           (ff=8) => ... |
            (ff=9) => ... |
             (ff=10) => ... |
              (ff=11) => ... | (ff=12) => ... |
               (ff=13) => ... | (ff=14) => ... | ... )
```

The functions VALUE, BVAL and SVAL extract the respective components of the 3-tuple returned by ALU. They are defined simply by

```
|- VALUE(result,b,stop) = result
|- BVAL(result,b,stop) = b
|- SVAL(result,b,stop) = stop
```

It is easy to unfold the definition of ALU and prove:

Theorem 5: The ALU Result for the Call Case

```
|- (VAL4 fsf = 1) ==>
   (ALU(fsf,msf,dsf,r,m,b) =
    m,b,
    (NOT((VAL3 dsf = 3) \/ ((VAL3 dsf = 4) \/ (VAL3 dsf = 5)))) \/
    (INVALID m))
```

Finally, there are three more function constants:

- OFFSET:word2#word20#word32#word32->word32
- NILM:word3#word1#word4->bool
- MEMREAD:mem21_32#word2#word20#word32#word32#bool#bool->word32

defined below. (The definition of ADD32, which adds the contents of two 32-bit registers, is not given here; its importance for now is only that the addition may generate an invalid address. ADD32 returns a 32-bit result and two booleans, so VALUE can be used to extract the result.)

```
|- OFFSET(msf,addr,x,y) =
   let mf = VAL2 msf in
   let addr32 = PAD20TO32 addr in
   ((mf=0) => addr32 | (mf=1) => addr32 |
    (mf=2) => VALUE(ADD32(addr32,x)) | VALUE(ADD32(addr32,y)))

|- NILM(dsf,csf,fsf) =
   let df = VAL3 dsf in let cf = VAL1 csf in
   let ff = VAL4 fsf in
   (cf=0) /\ ((NOT((df=7) \/ (df=6))) /\ (ff=12))

|- MEMREAD(ram,msf,addr,x,y,io,nilm) =
   let m = VAL2 msf in
   (nilm => ... |
    (m=0) => PAD20TO32 addr |
     FETCH21 ram
     (WORD21(V(CONS io(BITS20(TRIM32TO20(OFFSET(msf,addr,x,y))))))))
```

OFFSET returns a memory value according to the memory address control field, an address, and the x and y registers. It either pads the address to 32 bits (if the memory address field indicates that the address is a literal or an indirect address) or adds the x or y register to the address (if it is an offset address).

NILM is a predicate which is true of the parts of an instruction if they indicate that no memory source is required to interpret the instruction. It holds of non-comparisons with non-memory destinations whose ALU operations do not require a memory source.

MEMREAD reads from memory; it takes a memory and a memory address control field, an address, two registers, a flag to distinguish main from peripheral memory, and a flag to indicate whether a memory source is required at all. For instructions which *do* require a memory source: if the memory address control field holds value 0, the literal address is returned (padded to 32 bits); otherwise the contents of the address of the (possibly offset) address is fetched from the appropriate part of memory. (For instructions which do not require a memory source, MEMREAD returns some arbitrary value.)

3.2 The Specification

The high-level specification of Viper can now be stated. It is called `NEXT` since it takes a state and returns the next state. The HOL definition (with parts not immediately relevant abbreviated as "...") is:

```
|- NEXT(ram,p,a,x,y,b,stop) =
   let fetched = INSTFETCH(ram,p) in
   let newp = TRIM32TO20(INCP32 p) in
   let rsf = R fetched in
   let msf = M fetched in
   let dsf = D fetched in
   let csf = C fetched in
   let fsf = FF fetched in
   let addr = A fetched in
   let df = VAL3 dsf in
   let cf = VAL1 csf in
   let ff = VAL4 fsf in
   let comp = (cf=1) in
   let call = ((cf=0) /\ (ff=1)) in
   let output = ((cf=0) /\ (df=6)) in
   let input = ((cf=0) /\ NOT((df=7) \/ (df=6)) /\ (ff=2)) in
   let io = (output \/ input) in
   let writeop = ((cf=0) /\ ((df=7) \/ (df=6))) in
   let skip = NOOP(dsf,csf,b) in
   let noinc = INVALID(INCP32 p) in
   let illegaladdr = (NOT(NILM(dsf,csf,fsf))) /\
                     ((INVALID(OFFSET(msf,addr,x,y))) /\
                      (NOT skip)) in
   let illegalcl = ILLEGALCALL(dsf,csf,fsf) in
   let illegalsp = SPAREFUNC(dsf,csf,fsf) in
   let illegalonp = ILLEGALPDEST(dsf,csf,fsf) in
   let illegalwr = ILLEGALWRITE(dsf,csf,msf) in
   let source = REG(rsf,a,x,y,newp) in
   (stop =>
    (ram,p,a,x,y,b,T) |
    (noinc \/ illegaladdr) \/
     ((illegalcl \/ illegalsp) \/ (illegalonp \/ illegalwr)) =>
     (ram,newp,a,x,y,b,T) |
     (comp => ... |
      (writeop => ... |
       (skip =>
        (ram,newp,a,x,y,b,F) |
        let m = MEMREAD(ram,msf,addr,x,y,io,NILM(dsf,csf,fsf)) in
        let aluout = ALU(fsf,msf,dsf,source,m,b) in
```

```
(df=0) =>
 (ram,newp,VALUE aluout,x,y,BVAL aluout,SVAL aluout) |
 (df=1) =>
  (ram,newp,a,VALUE aluout,y,BVAL aluout,SVAL aluout) |
  (df=2) =>
   (ram,newp,a,x,VALUE aluout,BVAL aluout,SVAL aluout) |
   (call =>
    (ram,TRIM32TO20(VALUE aluout),a,x,INCP32 p,BVAL aluout,
     SVAL aluout) | ... )))))
```

NEXT first tests whether the **stop** flag is set, and if so, returns the original state unchanged. Otherwise, it fetches a new instruction from memory according to the program counter, and examines its various fields. It decodes the instruction with a series of tests. The new instruction is either an illegal instruction, a comparison, a write instruction, a non-operation, an **ALU** operation with the **a**, **x** or **y** register as its destination, a call instruction, or a jump. In each of these nine cases, a new state is determined, representing the state *after* the new instruction has been executed. The new state may have the memory changed, the program counter incremented or otherwise changed, and so on.

Illegal instructions include the five sorts mentioned in 3.1, as well as instructions with illegal addresses. These latter are instructions for which the **ALU** requires a memory source, the address is invalid, and the operation indicated is *not* a non-operation.

Not all of the possible new states are of interest at the moment; we are really only interested in the conditional branch for call. For a *call*, the memory source for the **ALU** is provided by **MEMREAD** (and the register source is selected by **REG**). The value computed by the **ALU**, trimmed to 20 bits, is the program counter of the new state, and the incremented original program counter is the **y** register of the new state. In this way the new state "points to" the address of the procedure being called, and carries information for an eventual return to the original location (plus 1) in the **y** register. The **b** register and **stop** flag of the new state are also set according to the result of the **ALU** operation.

To arrive at the *call* branch of the conditional, certain conditions must obviously apply. For one thing, the **stop** flag must be false. Also, the six *illegal* situations must be avoided: first, the incremented program counter must be valid. Second, the address of the new fetched instruction must be legal. (For the example case, the address is literal, *i.e.* the value of the memory address control field is 0, so **OFFSET** just pads the address; by Theorem 4 a padded address is necessarily valid. Thus the address *is* legal.) Third, the instruction must be a legal *call*; the new destination cannot be the **a**, **x** or **y** register. This excludes the values 0, 1 and 2 for the new destination field. The other three *illegal* conditions must also be avoided. The instruction cannot be a comparison (the compare field does not hold 1) or a write operation (the values 6 and 7 are also excluded for the destination field) or a non-operation. Finally, for the *call* branch to be selected, the function field must hold the value 1. The complete list of conditions is as follows:

```
1. NOT stop
2. NOT(INVALID(INCP32 p))
3. VAL2(M(INSTFETCH(ram,p))) = 0
4. NOT(ILLEGALCALL(D(INSTFETCH(ram,p)),C(INSTFETCH(ram,p)),
                    FF(INSTFETCH(ram,p))))
5. NOT(SPAREFUNC(D(INSTFETCH(ram,p)),C(INSTFETCH(ram,p)),
                  FF(INSTFETCH(ram,p))))
6. NOT(ILLEGALPDEST(D(INSTFETCH(ram,p)),C(INSTFETCH(ram,p)),
                     FF(INSTFETCH(ram,p))))
7. NOT(ILLEGALWRITE(D(INSTFETCH(ram,p)),C(INSTFETCH(ram,p)),
                     M(INSTFETCH(ram,p))))
8. VAL1(C(INSTFETCH(ram,p))) = 0
9. NOT(VAL3(D(INSTFETCH(ram,p))) = 7) /\
   NOT(VAL3(D(INSTFETCH(ram,p))) = 6)
10. NOT(NOOP(D(INSTFETCH(ram,p)),C(INSTFETCH(ram,p)),b))
11. VAL4(FF(INSTFETCH(ram,p))) = 1
```

4 The Host Machine in HOL

4.1 Event Sequences

The functional specification of Viper gives the new state (after a new instruction is executed) directly from the current state. (The new instruction is implicit in the state because the state includes a memory and a program counter.) At the host machine level, however, there are several stages of computation for each state transition at the higher level. A new instruction is fetched and placed in internal registers. The state and these registers are then transformed in stages until the instruction is completely executed. Only then is a single transition at the time scale of functional specification completed.

For the interpretation of instructions by the host machine, five internal registers (of appropriate size) are used to hold the first five fields of an instruction; these (as mentioned in Section 2) are called `rsf`, `msf`, `dsf`, `csf` and `fsf`. In addition, a multi-purpose 32-bit register (`t`) is used to hold a (padded) address or various other information. These six registers together comprise what is called the *transient*, and are invisible to the functional specification. The type of the transient `(t,rsf,msf,dsf,csf,fsf)` is:

```
word32#word2#word2#word3#word1#word4
```

The host machine interprets an instruction by executing a sequence of (from three to seven) events. Each event in a sequence can affect the state and/or the transient. The event, in the context of the state and transient, determines whether there is a next event to be executed and if so what it is.

For example, a stop is a simple event during which the `stop` flag is set. It is always the last event in the sequence in which it occurs.

An instruction fetch is an event during which a new instruction is found in memory according to the program counter, and its various fields placed in the appropriate registers of the transient. The new address is placed in the `t` register. The state is affected only insofar as the program counter is imcremented, and possibly the `stop` flag set, depending on the new instruction and new program counter. Whether there is a next event in the sequence, and if so what it is, is determined by inspection of the new state and transient. The next event may be one of several, including the preparation for a call, or a stop event.

The preparation for a call causes the `y` register of the state to contain the program counter, and the `stop` flag to be set to false since nothing new can go wrong at this point; the transient is not affected. The next event must be an `ALU` operation. (The purpose is to save the program counter so that it can be restored after the called procedure is finished.)

Performing an operation can mean either performing a comparison or performing an `ALU` operation. Performing an `ALU` operation is an event during which either the a, x, or `y` register or the program counter receives a value computed by the `ALU`. The transient does not change. The next event is based on the result computed by the `ALU`; it is either a stop event, or the end of the sequence in which the `ALU` event occurred is signalled.

Thus the following sequence of events is possible:

- Fetch a new instruction
- Prepare for a call
- Perform an `ALU` operation

For allowing the `stop` flag to be noticed, a dummy event is placed at the beginning of the sequence, during which the state and transient remain unchanged. The event following a dummy event is an instruction fetch, unless the `stop` flag is set, so the sequence of events is:

- Dummy event
- Fetch a new instruction
- Prepare for a call
- Perform `ALU` operation

Viper is modelled at the host level as continuously running. Its infinite sequence of events consists of repetitions of twenty-four possible finite sequences each beginning with a dummy event and ending as soon as the end of the sequence is indicated (*i.e.* a dummy event is the necessary next event).

Events are associated with numbers corresponding to nodes in the transition graph. (The numbers look random here because not all the possible events have been revealed.) The numbers are represented by 5-bit strings.

```
16. DUMMY,   8. STOP,   1. FETCH,   3. PRECALL,   4. PERFORM
```

In HOL, the functions `DUMMY`, `STOP`, `FETCH`, `PRECALL` and `PERFORM` formalize the five events mentioned so far. The auxiliary functions `FMOVE` and `PMOVE` compute the next event after a `FETCH` and `PERFORM` event respectively. As usual, "..." abbreviates parts of the definition not relevant to the *call* sequence.

```
|- DUMMY((ram,p,a,x,y,b,stop),t,rsf,msf,dsf,csf,fsf) =
   let stopstate = WORD5 8 in
   let fetch = WORD5 1 in
   (stop =>
    (((ram,p,a,x,y,b,T),t,rsf,msf,dsf,csf,fsf),stopstate) |
    (((ram,p,a,x,y,b,F),t,rsf,msf,dsf,csf,fsf),fetch))
```

```
|- STOP((ram,p,a,x,y,b,stop),t,rsf,msf,dsf,csf,fsf) =
   let dummy = WORD5 16 in
   ((ram,p,a,x,y,b,T),t,rsf,msf,dsf,csf,fsf),dummy
```

```
|- FMOVE(msf,dsf,csf,fsf,b) =
   let mf = VAL2 msf in
   let df = VAL3 dsf in
   let cf = VAL1 csf in
   let ff = VAL4 fsf in
   let b' = HD(BITS1 b) in
   let noop = NOOP(dsf,csf,b') in
   let precall = WORD5 3 in
   let stopstate = WORD5 8 in
   let dummy = WORD5 16 in
   let ... = ... in let ... = ... in let ... = ... in
   let ... = ... in let ... = ... in let ... = ... in
   ((cf=1) =>  ((mf=0) => ... | ... )|
   ((df=7) =>
    ((mf=0) => ... | ((mf=1) => ... | ...)) |
    ((df=6) =>
     ((mf=0) => ... | ((mf=1) => ... | ... )) |
     (noop => dummy |
      ((mf=0) =>
       (((cf=0) /\ (ff=1)) => precall | ... ) |
       ((mf=1) =>
        (((cf=0) /\ (ff=2)) => ... |
         (((cf=0) /\ (ff=12)) => ... | ... )) |
         (((cf=0) /\ (ff=12)) => ... | ... )))))))
```

```
|- FETCH((ram,p,a,x,y,b,stop),t,rsf,msf,dsf,csf,fsf) =
   let fetched = INSTFETCH(ram,p) in
   let newp = TRIM32TO20(INCP32 p) in
   let newr = R fetched in let newm = M fetched in
   let newd = D fetched in let newc = C fetched in
   let newf = FF fetched in let newt = PAD20TO32(A fetched) in
   let notinc = INVALID(INCP32 p) in
   let illegalcl = ILLEGALCALL(newd,newc,newf) in
   let illegalsp = SPAREFUNC(newd,newc,newf) in
   let illegalonp = ILLEGALPDEST(newd,newc,newf) in
   let illegalwr = ILLEGALWRITE(newd,newc,newm) in
   let stopstate = WORD5 8 in let b' = WORD1(V[b]) in
   (notinc \/
    (illegalcl \/ (illegalsp \/ (illegalonp \/ illegalwr))) =>
    (((ram,newp,a,x,y,b,T),newt,newr,newm,newd,newc,newf),stopstate) |
    (((ram,newp,a,x,y,b,F),newt,newr,newm,newd,newc,newf),
     FMOVE(newm,newd,newc,newf,b')))

|- PRECALL((ram,p,a,x,y,b,stop),t,rsf,msf,dsf,csf,fsf) =
   let perform = WORD5 4 in
   ((ram,p,a,x,PAD20TO32 p,b,F),t,rsf,msf,dsf,csf,fsf),perform

|- PMOVE halt =
   let stopstate = WORD5 8 in
   let dummy = WORD5 16 in (halt => stopstate | dummy)

|- PERFORM((ram,p,a,x,y,b,stop),t,rsf,msf,dsf,csf,fsf) =
   let comp = (VAL1 csf = 1) in
   let df = VAL3 dsf in
   let source = REG(rsf,a,x,y,p) in
   let dummy = WORD5 16 in
   (comp => (((...,...,...,...,...,F),
              ...,...,...,...,...,...), ... ) |
    let result = ALU(fsf,msf,dsf,source,t,b) in
    let ans = VALUE result in let newb = BVAL result in
    let halt = SVAL result in
    ((df=0) =>
     (((ram,p,ans,x,y,newb,halt),t,rsf,msf,dsf,csf,fsf),PMOVE halt) |
     (df=1) =>
      (((ram,p,a,ans,y,newb,halt),t,rsf,msf,dsf,csf,fsf),PMOVE halt) |
      (df=2) =>
       (((ram,p,a,x,ans,newb,halt),t,rsf,msf,dsf,csf,fsf),PMOVE halt) |
       (((ram,TRIM32TO20 ans,a,x,y,newb,halt),t,rsf,msf,dsf,csf,fsf),
        PMOVE halt)))
```

4.2 The Transition Graph

Each of the events defined in Section 4.1 determines one or more subsequent events. Together they determine a transition graph of events, part of which is already known:

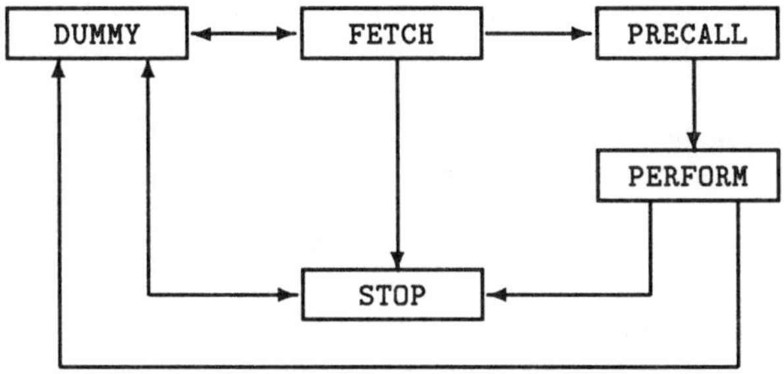

Figure 1: The Graph of Events

Each possible path through the graph, starting and ending at the DUMMY node, can be shown in a tree:

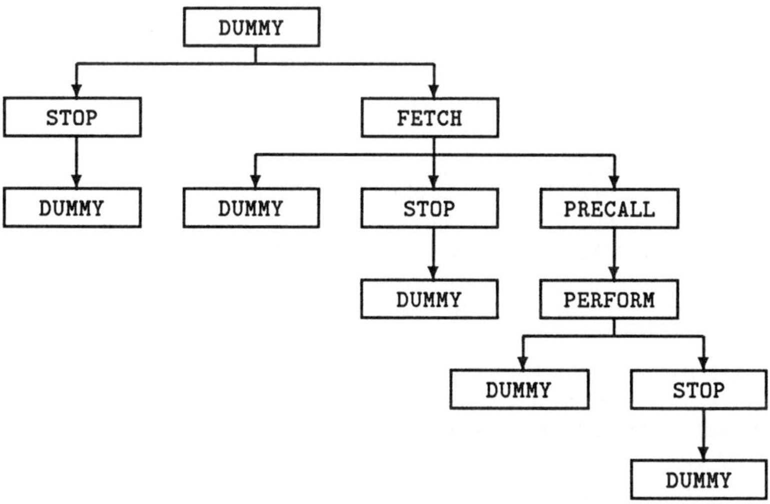

Figure 2: Paths through the Graph

Of the five possible sequences, we only consider in this paper the one containing PRECALL which does not stop. It is clear from the event definitions that where there is a choice of next event, certain conditions must hold in order that the desired next event be chosen. As DUMMY is executed, FETCH will be chosen if the stop flag is set to false. As FETCH is executed, STOP can be avoided if none of the five *illegal* conditions apply to the new fetched instruction. Furthermore, PRECALL will be chosen if the new value of the compare field is not 1 (*i.e.* is 0); the new value of the destination field neither 7 nor 6; NOOP is false of the destination field, compare field and b flag; the new value of the memory address control field is 0; and the new value of the function field is 1. As PERFORM is executed, STOP is again avoided if *either* the current value of the compare field is 1 *or* the stop value returned by performing the ALU operation is false; that is, if NOT(SVAL(ALU(fsf,msf,dsf,REG(rsf,a,x,y,p),t,b))) holds for the values in the various registers when PERFORM is executed. PRECALL gives no choice of next node.

To formalize these conditions, the predicates c1, c3 and c17, from states to boolean values, are introduced. (The conjunction of these was called C in Section 1.3. The twenty four paths determine in all thirty-four different conditions of which these are three.)

The examinations of the fields that take place must take into account that in the particular sequence DUMMY, FETCH, PRECALL, PERFORM, after FETCH is executed, it is the fields of the *new* instruction that occupy the transient registers. To make the desired choice of next node during DUMMY, we require:

```
c1(ram,p,a,x,y,b,stop) = NOT stop
```

To make the correct choice during FETCH:

```
c3(ram,p,a,x,y,b,stop) =
(NOT
 ((INVALID(INCP32 p)) \/
  ((ILLEGALCALL(D(INSTFETCH(ram,p)),C(INSTFETCH(ram,p)),
               FF(INSTFETCH(ram,p)))) \/
   ((SPAREFUNC(D(INSTFETCH(ram,p)),C(INSTFETCH(ram,p)),
              FF(INSTFETCH(ram,p)))) \/
    ((ILLEGALPDEST(D(INSTFETCH(ram,p)),C(INSTFETCH(ram,p)),
                  FF(INSTFETCH(ram,p)))) \/
     (ILLEGALWRITE(D(INSTFETCH(ram,p)),C(INSTFETCH(ram,p)),
                  M(INSTFETCH(ram,p))))))))) /\
((NOT(VAL1(C(INSTFETCH(ram,p))) = 1)) /\
 ((NOT((VAL3(D(INSTFETCH(ram,p))) = 7) \/
      (VAL3(D(INSTFETCH(ram,p))) = 6))) /\
  ((NOT(NOOP(D(INSTFETCH(ram,p)),C(INSTFETCH(ram,p)),b))) /\
   ((VAL2(M(INSTFETCH(ram,p))) = 0) /\
    ((VAL1(C(INSTFETCH(ram,p))) = 0) /\
     (VAL4(FF(INSTFETCH(ram,p))) = 1))))))
```

and to make the correct choice during PERFORM:

```
c17(ram,p,a,x,y,b,stop) =
(VAL1(C(INSTFETCH(ram,p))) = 1) \/
(NOT(SVAL
     (ALU
      (FF(INSTFETCH(ram,p)),M(INSTFETCH(ram,p)),D(INSTFETCH(ram,p)),
       REG(R(INSTFETCH(ram,p)),a,x,PAD20TO32(TRIM32TO20(INCP32 p)),
           TRIM32TO20(INCP32 p)),PAD20TO32(A(INSTFETCH(ram,p))),b))))
```

Some elementary inference can be done on the combined conditions: since the function field holds 1, the compare field holds 0 and the call is not illegal, it follows that the destination field holds neither 0, 1 nor 2. Indeed, since it does not hold 6 or 7 either, and its value has to be 0,...,7, it must hold 3, 4 or 5. That is:

```
|- NOT(ILLEGALCALL(dsf,csf,fsf)) /\
   (VAL1 csf = 0) /\ NOT(VAL3 dsf = 7) /\ NOT(VAL3 dsf = 6) /\
   (VAL4 fsf = 1)
   ==>
   ((VAL3 dsf = 3) \/ (VAL3 dsf = 4) \/ (VAL3 dsf = 5)) /\
   NOT(VAL3 dsf = 0) /\ NOT(VAL3 dsf = 1) /\ NOT(VAL3 dsf = 2)
```

In particular it follows that:

Theorem 6: Legal Call Destinations

```
|- NOT(ILLEGALCALL(D(INSTFETCH(ram,p)),C(INSTFETCH(ram,p)),
                   FF(INSTFETCH(ram,p)))) /\
   (VAL1(C(INSTFETCH(ram,p))) = 0) /\
   NOT(VAL3(D(INSTFETCH(ram,p))) = 7) /\
   NOT(VAL3(D(INSTFETCH(ram,p))) = 6) /\
   (VAL4(FF(INSTFETCH(ram,p))) = 1)
   ==>
   ((VAL3(D(INSTFETCH(ram,p))) = 3) \/
   (VAL3(D(INSTFETCH(ram,p))) = 4) \/
   (VAL3(D(INSTFETCH(ram,p))) = 5)) /\
   NOT(VAL3(D(INSTFETCH(ram,p))) = 0) /\
   NOT(VAL3(D(INSTFETCH(ram,p))) = 1) /\
   NOT(VAL3(D(INSTFETCH(ram,p))) = 2)
```

Because the value of the compare field is 0, it follows from `c3` and `c17` that:

```
NOT(SVAL
     (ALU
      (FF(INSTFETCH(ram,p)),M(INSTFETCH(ram,p)),D(INSTFETCH(ram,p)),
       REG(R(INSTFETCH(ram,p)),a,x,PAD20TO32(TRIM32TO20(INCP32 p)),
           TRIM32TO20(INCP32 p)),PAD20TO32(A(INSTFETCH(ram,p))),b)))
```

In fact, this last fact also follows from c3, Theorem 4, Theorem 5 and Theorem 6, so c17 is actually redundant for this path. We keep it as a reminder that there *is* a choice of next node during PERFORM.

The three conditions and their corollaries give all of the conditions for choosing the *call* branch of the conditional in the specification, NEXT, in Section 3.2. (The instruction does not have an illegal address for the same reason as before.)

5 The Equivalence Proof of Specification and Host Machine

What has to be proved is that *under the conditions* c1, c3 and c17, the functional specification agrees with the host machine on the visible state. The first thing to establish is what that state is; the second is whether NEXT gives that state.

The events comprising the host machine (Section 4.1) only say implicitly what the transformation is; they determine a sequence of events (a path through the graph shown in Section 4.2) which transforms the initial state and transient in stages. The transformation can be made *more* explicit by defining a "control" function (HOST_NEXT) linking the events. (Branches which are not needed at the moment are filled in by "...".) The type major abbreviates state#transient. HOST_NEXT has type major#node->major#node. Its partial definition is:

```
|- HOST_NEXT(maj,node) =
   (let nodenum = VAL5 node in
   (nodenum=0)  => ... |            (nodenum=1)  => FETCH maj |
   (nodenum=2)  => ... |            (nodenum=3)  => PRECALL maj |
   (nodenum=4)  => PERFORM maj |    (nodenum=5)  => ... |
   (nodenum=6)  => ... |            (nodenum=7)  => ... |
   (nodenum=8)  => STOP maj |       (nodenum=9)  => ... |
   (nodenum=10) => ... |            (nodenum=11) => ... |
   (nodenum=12) => ... |            (nodenum=13) => ... |
   (nodenum=14) => ... |            (nodenum=15) => ... |
   (nodenum=16) => DUMMY maj |      (nodenum=17) => ... |
   (nodenum=18) => ... |            (nodenum=19) => ... |
   (nodenum=20) => ... |            (nodenum=21) => ... |
   (nodenum=22) => ... |            (nodenum=23) => ... |
   (nodenum=24) => ... |            (nodenum=25) => ... |
   (nodenum=26) => ... |            (nodenum=27) => ... |
   (nodenum=28) => ... |            (nodenum=29) => ... |
   (nodenum=30) => ... |            (nodenum=31) => ... | ... )
```

HOST_NEXT says how to step through the graph. It defines the host machine, which takes a state, transient and node to a new state, transient and node. The new state, transient and node are still not fully explicit; they are computed by functions such as PRECALL which in turn may either call other functions or return a new major and node. In Sections 5.1 and 5.2 the transformation is made completely explicit.

5.1 A Digression on Forward Proof in HOL

First, each transition in the graph is characterized by saying exactly how HOST_NEXT changes a state and transient and selects a next node during that transition. It is clearly true, for example, by the definition of HOST_NEXT and the fact that VAL5 #00011 = 3 that

```
|- HOST_NEXT(((ram,p,a,x,y,b,stop),t,rsf,msf,dsf,csf,fsf),#00011) =
   PRECALL((ram,p,a,x,y,b,stop),t,rsf,msf,dsf,csf,fsf)
```

and that by the definition of PRECALL

```
|- PRECALL((ram,p,a,x,y,b,stop),t,rsf,msf,dsf,csf,fsf) =
   let perform = WORD5 4 in
   ((ram,p,a,x,PAD20TO32 p,b,F),t,rsf,msf,dsf,csf,fsf),perform
```

By definition of let and the fact that WORD5 4 = #00100, it follows that

```
|- (let perform = WORD5 4 in
    ((ram,p,a,x,PAD20TO32 p,b,F),t,rsf,msf,dsf,csf,fsf),perform) =
   ((ram,p,a,x,PAD20TO32 p,b,F),t,rsf,msf,dsf,csf,fsf),#00100
```

hence

Theorem 7: HOST_NEXT for Node 3

```
|- HOST_NEXT(((ram,p,a,x,y,b,stop),t,rsf,msf,dsf,csf,fsf),#00011) =
   ((ram,p,a,x,PAD20TO32 p,b,F),t,rsf,msf,dsf,csf,fsf),#00100
```

Simple as this reasoning appears, it represents a long chain of primitive inferences. To unfold HOST_NEXT the facts NOT(3=0), NOT(3=1), NOT(3=2) and 3=3, as well as (T=>t1|t2) = t1 and (F=>t1|t2) = t2 must be used to rewrite the definition.

Furthermore, each expression of the form let z = t1 in t2 is logically equivalent to (\z.t2)t1. Taking that inference step on the definition of PRECALL gives:

```
|- PRECALL((ram,p,a,x,y,b,stop),t,rsf,msf,dsf,csf,fsf) =
   (\z.
     (\perform.
       ((ram,p,a,x,PAD20TO32 p,b,F),t,rsf,msf,dsf,csf,fsf),perform)
     z)
   (WORD5 4)
```

The next inference step is beta-conversion; expressions of the form `(\z.t2)t1` can be reduced to `t2[t1/z]`, the result of substituting `t1` for free occurrences of `z` in `t2` (subject to restrictions on free variable capture). Taking that step once gives:

```
|- PRECALL((ram,p,a,x,y,b,stop),t,rsf,msf,dsf,csf,fsf) =
   (\perform.
     ((ram,p,a,x,PAD20TO32 p,b,F),t,rsf,msf,dsf,csf,fsf),perform)
   (WORD5 4)
```

and taking it again

```
|- PRECALL((ram,p,a,x,y,b,stop),t,rsf,msf,dsf,csf,fsf) =
   ((ram,p,a,x,PAD20TO32 p,b,F),t,rsf,msf,dsf,csf,fsf),WORD5 4.
```

Transitivity is needed to complete the chain: if `t1 = t2` and `t2 = t3` then `t1 = t3`. The fact that `WORD5 4 = #00100` is also required.

This chain of primitive inferences is the proof of Theorem 7. All of the proofs discussed in this paper are proofs in that sense: a chain of inferences justified by axioms and inference rules of the logic. Obviously, it is not practical to construct proofs of any size manually, nor would they be very interesting to look at, but one likes to know that they exist and *could* be displayed. (The whole correctness proof described in this paper comprises over a million primitive inferences, for example.) In HOL, the general purpose programming language (ML) interfaced to the logic allows the user to write procedures to generate the actual chains of inferences at some level of abstraction above primitive inference steps. (The level of abstraction depends on the cleverness of the procedure.)

A general ML procedure which proves Theorem 7 unfolds (rewrites) the definitions of all new constants (`HOST_NEXT`, `PRECALL`, *etc*), unfolds the definition of `let`, and does beta-conversion repeatedly until there are no lambda expressions left. Wherever possible, the procedure uses facts about numbers, substitutes equals for equals, use the transitivity of equality, and so on. By executing this general procedure, the proof is generated with a single high-level command. In fact, over 7,000 primitive inferences are performed in the course of generating the proof using this procedure. The number is as large as that partly *because* the procedure is so general; a great deal of inference is done in the course of each unfolding (rewriting) in order to find the location of the replacement and build up a new theorem by inference. This saves the user the trouble of specifying exactly each replacement to be made and its location in the structure of a term, or indeed of thinking very hard about how to prove the fact; the procedure is a general way of unfolding *any* definition phrased in terms of previous definitions and `let` expressions. This sort of proof by rewriting is central to HOL (and LCF) methodology; it is at once very powerful and quite expensive. For further discussion of this trade-off, see Section 8. For more on LCF-style rewriting see the LCF manual (Paulson, [15]) and also Paulson [14].

This is an example of *forward* proof: the user supplies a general ML procedure with the definitions it will need (and the bit string #00011 in this case), and HOL unfolds definitions and lets and does routine inferences until there is no more to do. The user does not necessarily have to know in advance what the final term will be; the procedure computes it, and proves it equal to the initial term. This is one mode in which HOL may be used; the other is discussed in Section 5.4.

5.2 Stepping through the Graph of the Host Machine

5.2.1 Node to Node

Using the same procedure, the new state, transient and node that HOST_NEXT gives for DUMMY, STOP, FETCH and PERFORM can also be produced.

For DUMMY all choice can be limited (by properties of conditionals) to the node component.

Theorem 8: HOST_NEXT for Node 16

```
|- HOST_NEXT(((ram,p,a,x,y,b,stop),t,rsf,msf,dsf,csf,fsf),#10000) =
   ((ram,p,a,x,y,b,stop),t,rsf,msf,dsf,csf,fsf),(stop => #01000|#00001)
```

For STOP there is no choice.

Theorem 9: HOST_NEXT for Node 8

```
|- HOST_NEXT(((ram,p,a,x,y,b,stop),t,rsf,msf,dsf,csf,fsf),#01000) =
   ((ram,p,a,x,y,b,T),t,rsf,msf,dsf,csf,fsf),#10000
```

For FETCH, the new stop flag is set if an illegal condition has arisen, and the next node (if not STOP) is selected by FMOVE. The transient receives parts of the new instruction. The choice can be limited to the stop and node components. The procedure uses Theorem 0 as a rewrite rule at the appropriate point. (Theorem 10 is easier to read with some of the let-expressions not unfolded.)

Theorem 10: HOST_NEXT for Node 1

```
|- HOST_NEXT(((ram,p,a,x,y,b,stop),t,rsf,msf,dsf,csf,fsf),#00001) =
   let fetched = INSTFETCH(ram,p) in
   let new_msf = M fetched in
   let new_dsf = D fetched in
   let new_csf = C fetched in
   let new_fsf = FF fetched in
   ((ram,TRIM32TO20(INCP32 p),a,x,y,b,
     (INVALID(INCP32 p)) \/
     ((ILLEGALCALL(new_dsf,new_csf,new_fsf)) \/
      ((SPAREFUNC(new_dsf,new_csf,new_fsf)) \/
       ((ILLEGALPDEST(new_dsf,new_csf,new_fsf)) \/
```

```
        (ILLEGALWRITE(new_dsf,new_csf,new_msf)))))),
   PAD20TO32(A(INSTFETCH(ram,p))),R(INSTFETCH(ram,p)),new_msf,new_dsf,
   new_csf,new_fsf),
  ((INVALID(INCP32 p)) \/
   ((ILLEGALCALL(new_dsf,new_csf,new_fsf)) \/
    ((SPAREFUNC(new_dsf,new_csf,new_fsf)) \/
     ((ILLEGALPDEST(new_dsf,new_csf,new_fsf)) \/
      (ILLEGALWRITE(new_dsf,new_csf,new_msf))))) =>
   #01000 |
   ((VAL1 new_csf = 1) =>
    ((VAL2 new_msf = 0) => ... |
                             ((VAL2 new_msf = 1) => ... | ... )) |
    ((VAL3 new_dsf = 7) =>
     ((VAL2 new_msf = 0) => ... |
                              ((VAL2 new_msf = 1) => ... | ... )) |
     ((VAL3 new_dsf = 6) =>
      ((VAL2 new_msf = 0) => ... |
                               ((VAL2 new_msf = 1) => ... | ... )) |
      ((VAL1 new_csf = 0) /\
       (((VAL3 new_dsf = 5) /\ b) \/
        ((VAL3 new_dsf = 4) /\ (NOT b))) =>  #10000 |
        ((VAL2 new_msf = 0) =>
        ((VAL1 new_csf = 0) /\ (VAL4 new_fsf = 1) => #00011 | ... ) |
        ((VAL2 new_msf = 1) =>
         ((VAL1 new_csf = 0) /\ (VAL4 new_fsf = 2) => ... |
          ((VAL1 new_csf = 0) /\ (VAL4 new_fsf = 12) =>
           ... | ... )) |
         ((VAL1 new_csf = 0) /\ (VAL4 new_fsf = 12) =>
          ... | ... ))))))))
```

For PERFORM, only the memory component and the transient involve no choice.

Theorem 11: HOST_NEXT for Node 4

```
|- HOST_NEXT(((ram,p,a,x,y,b,stop),t,rsf,msf,dsf,csf,fsf),#00100) =
   ((ram,
     ((VAL1 csf = 1) => ... |
      ((VAL3 dsf = 0) =>  p |
       ((VAL3 dsf = 1) =>  p |
        ((VAL3 dsf = 2) =>  p |
         TRIM32TO20(VALUE(ALU(fsf,msf,dsf,REG(rsf,a,x,y,p),t,b))))))),
     ((VAL1 csf = 1) =>  ... |
      ((VAL3 dsf = 0) =>
       VALUE(ALU(fsf,msf,dsf,REG(rsf,a,x,y,p),t,b)) | a)),
     ((VAL1 csf = 1) =>  ... |
```

```
       ((VAL3 dsf = 0) =>  x |
        ((VAL3 dsf = 1) =>
         VALUE(ALU(fsf,msf,dsf,REG(rsf,a,x,y,p),t,b)) | x))),
      ((VAL1 csf = 1) => ... |
       ((VAL3 dsf = 0) => y |
        ((VAL3 dsf = 1) => y |
         ((VAL3 dsf = 2) =>
          VALUE(ALU(fsf,msf,dsf,REG(rsf,a,x,y,p),t,b)) | y)))),
      ((VAL1 csf = 1) =>
       ... | BVAL(ALU(fsf,msf,dsf,REG(rsf,a,x,y,p),t,b))),
      ((VAL1 csf = 1) =>
       F | SVAL(ALU(fsf,msf,dsf,REG(rsf,a,x,y,p),t,b)))),
     t,rsf,msf,dsf,csf,fsf),
    ((VAL1 csf = 1) =>  ... |
     (SVAL(ALU(fsf,msf,dsf,REG(rsf,a,x,y,p),t,b)) => #01000 | #10000))
```

5.2.2 Making Time Explicit

We now consider making time explicit. Times are represented as numbers. *Signals* are functions from times to registers:

```
  ram_sig:num->mem21_32            p_sig:num->word20
    a_sig:num->word32              x_sig:num->word32
    y_sig:num->word32              b_sig:num->bool
 stop_sig:num->bool                t_sig:num->word32
  rsf_sig:num->word2             msf_sig:num->word2
  dsf_sig:num->word3             csf_sig:num->word1
  fsf_sig:num->word4
```

`ram_sig n` means the value of `ram` at time `n`, and so on. A theorem of the following form expresses the behaviour of `HOST_NEXT` as it steps through a node `#.....`; a time unit is the time it takes for one event to happen at the host level.

```
 |- (!n.
      ((ram_sig(n+1),p_sig(n+1),a_sig(n+1),x_sig(n+1),
        y_sig(n+1),b_sig(n+1),stop_sig(n+1)),t_sig(n+1),
       rsf_sig(n+1),msf_sig(n+1),dsf_sig(n+1),csf_sig(n+1),
       fsf_sig(n+1)),node_sig(n+1) =
      HOST_NEXT
      (((ram_sig n,p_sig n,a_sig n,x_sig n,y_sig n,b_sig n,stop_sig n),
        t_sig n,rsf_sig n,msf_sig n,dsf_sig n,csf_sig n,fsf_sig n),
       node_sig n))
      ==>
    (node_sig n = #.....)
      ==>
```

```
(((ram_sig(n+1) = ...n...) /\
  (p_sig(n+1) = ...n...) /\
  (a_sig(n+1) = ...n...) /\
  (x_sig(n+1) = ...n...) /\
  (y_sig(n+1) = ...n...) /\
  (b_sig(n+1) = ...n...) /\
  (stop_sig(n+1) = ...n...)) /\
 (t_sig(n+1) = ...n...) /\
 (rsf_sig(n+1) = ...n...) /\
 (msf_sig(n+1) = ...n...) /\
 (dsf_sig(n+1) = ...n...) /\
 (csf_sig(n+1) = ...n...) /\
 (fsf_sig(n+1) = ...n...)) /\
(node_sig(n+1) = ...)
```

The first antecedent could be abbreviated by use of `HOST_NEXT_SIG` as in Section 1.3. It gives a sequence of 14-tuples of signals such that `HOST_NEXT` applied to the fourteen signals at any time gives the fourteen signals at the next time. The second antecedent fixes a particular event (a node in the graph) at some time `n`. The fourteen equations then give the value of the signals at time `n+1` in terms of the signals at time `n`, after that node is traversed. For `PRECALL` (event 3, node `#00011`), the theorem is:

Theorem 12: Timed HOST_NEXT for Node 3

```
|- (!n.
    ((ram_sig(n+1),p_sig(n+1),a_sig(n+1),x_sig(n+1),
      y_sig(n+1),b_sig(n+1),stop_sig(n+1)),t_sig(n+1),
     rsf_sig(n+1),msf_sig(n+1),dsf_sig(n+1),csf_sig(n+1),
     fsf_sig(n+1)),node_sig(n+1) =
    HOST_NEXT
    (((ram_sig n,p_sig n,a_sig n,x_sig n,y_sig n,b_sig n,stop_sig n),
      t_sig n,rsf_sig n,msf_sig n,dsf_sig n,csf_sig n,fsf_sig n),
     node_sig n)) ==>
   (node_sig n = #00011) ==>
   (((ram_sig(n+1) = ram_sig n) /\
     (p_sig(n+1) = p_sig n) /\
     (a_sig(n+1) = a_sig n) /\
     (x_sig(n+1) = x_sig n) /\
     (y_sig(n+1) = PAD20T032(p_sig n)) /\
     (b_sig(n+1) = b_sig n) /\
     (stop_sig(n+1) = F)) /\
    (t_sig(n+1) = t_sig n) /\
    (rsf_sig(n+1) = rsf_sig n) /\
    (msf_sig(n+1) = msf_sig n) /\
```

```
    (dsf_sig(n+1) = dsf_sig n) /\
    (csf_sig(n+1) = csf_sig n) /\
    (fsf_sig(n+1) = fsf_sig n)) /\
   (node_sig(n+1) = #00100)
```

This gives the effect over one time unit of passing through the PRECALL node: if the node signal at time n is #00011, then the y register at time n+1 holds the (padded) value of the program counter at time n, and so on. The relation to Theorem 7 is clear. To prove Theorem 12, it is assumed for some n that

```
|- node_sig n = #00011
```

The first antecedent is assumed and instantiated to that n:

Assumption 1: Sequence Assumption

```
|- ((ram_sig(n+1),p_sig(n+1),a_sig(n+1),x_sig(n+1),
     y_sig(n+1),b_sig(n+1),stop_sig(n+1)),t_sig(n+1),
    rsf_sig(n+1),msf_sig(n+1),dsf_sig(n+1),csf_sig(n+1),
    fsf_sig(n+1)),node_sig(n+1) =
   HOST_NEXT
   (((ram_sig n,p_sig n,a_sig n,x_sig n,y_sig n,b_sig n,stop_sig n),
     t_sig n,rsf_sig n,msf_sig n,dsf_sig n,csf_sig n,fsf_sig n),
    node_sig n)
```

Assumption 1 is unfolded using the first assumption and then by using Theorem 7 (which says what HOST_NEXT does at node #00011). The result can be expressed (using properties of tuples) as fourteen pairwise equalities with all assumptions discharged. This gives Theorem 12.

The same ML procedure which generates the proof of Theorem 12 also generates theorems for DUMMY, STOP, FETCH and PERFORM, which are not shown here but bear a similar relation to Theorems 8-11 (respectively) as Theorem 12 does to Theorem 7.

This is another example of a general procedure being used to generate a forward proof (in the case of Theorem 12, a proof of about 2,200 inferences); one does not have to know in advance the theorem being proved, but only its form.

5.2.3 End Results of Sequences in the Host Machine

Finally, the result of the whole sequence DUMMY, FETCH, PRECALL, PERFORM can be produced. To do this, it is assumed again that for some n, node_sig n = #10000 (we start at DUMMY at time n). The assumption is made which asserts that the signals at any time are the result of HOST_NEXT at the previous time. The three conditions (c1, c3 and c17) are assumed to hold of the state signals at time n, to ensure that the correct choices are made at each node. Then the path is traversed by referring at each node to the corresponding theorem (Theorem 12 for node 3, *etc*) which describes the effects over the one time unit taken to traverse that node. That causes certain changes in the fourteen components, which are compounded with the changes at the previous node (if there was one). At each stage, the assumed conditions are used to help make decisions about the node signal, *i.e.* the next event. For each node, a theorem follows, describing the accumulated changes at that time.

Again, an ML procedure is written to generate the proof. It is a recursive procedure which refers initially to the theorem (analagous to Theorem 12) for node `#10000`, dismisses the antecedents because these have already been assumed, and then uses the pairwise equalities to rewrite the components and select the next node. Successive theorems are generated by examining the node, and referring to the theorem about `HOST_NEXT` for that node. The process continues until the first appearance of node `#10000` as the next node. At each stage, more complex simplifications must also be done; for example, where there is a choice of next node, the conditional expression denoting the next node has to be reduced to a particular bit string by using assumed facts such as the fact that `c17` holds of the state at time `n`. At some stages, the procedure must also use lemmas; for example, in the final stage of the literal *call* path, Theorem 6 is required to avoid choices involving a destination of 0, 1 or 2. Under the assumptions

```
1. !n.
    ((ram_sig(n+1),p_sig(n+1),a_sig(n+1),x_sig(n+1),
      y_sig(n+1),b_sig(n+1),stop_sig(n+1)),t_sig(n+1),
     rsf_sig(n+1),msf_sig(n+1),dsf_sig(n+1),csf_sig(n+1),
     fsf_sig(n+1)),node_sig(n+1) =
    HOST_NEXT
    (((ram_sig n,p_sig n,a_sig n,x_sig n,y_sig n,b_sig n,stop_sig n),
      t_sig n,rsf_sig n,msf_sig n,dsf_sig n,csf_sig n,fsf_sig n),
     node_sig n)
2. node_sig n = #10000
3. c1(ram_sig n,p_sig n,a_sig n,x_sig n,y_sig n,b_sig n,stop_sig n)
4. c3(ram_sig n,p_sig n,a_sig n,x_sig n,y_sig n,b_sig n,stop_sig n)
5. c17(ram_sig n,p_sig n,a_sig n,x_sig n,y_sig n,b_sig n,stop_sig n)
```

four theorems are proved. These, conjoined, are called **Theorem 13**.

At time `n+1`, `DUMMY` has been executed without having to stop.

```
|- ((((ram_sig(n+1) = ram_sig n) /\
      (p_sig(n+1) = p_sig n) /\
      (a_sig(n+1) = a_sig n) /\
      (x_sig(n+1) = x_sig n) /\
      (y_sig(n+1) = y_sig n) /\
      (b_sig(n+1) = b_sig n) /\
      (stop_sig(n+1) = F)) /\
     (t_sig(n+1) = t_sig n) /\
     (rsf_sig(n+1) = rsf_sig n) /\
```

```
      (msf_sig(n+1) = msf_sig n) /\
      (dsf_sig(n+1) = dsf_sig n) /\
      (csf_sig(n+1) = csf_sig n) /\
      (fsf_sig(n+1) = fsf_sig n)) /\
     (node_sig(n+1) = #00001))
```

At time n+2 FETCH has been executed, so the program counter is incremented, the transient fields contains the fields of the new instruction, and the t register holds the new address.

```
|- ((((ram_sig(n+2) = ram_sig n) /\
       (p_sig(n+2) = TRIM32TO20(INCP32(p_sig n))) /\
       (a_sig(n+2) = a_sig n) /\
       (x_sig(n+2) = x_sig n) /\
       (y_sig(n+2) = y_sig n) /\
       (b_sig(n+2) = b_sig n) /\
       (stop_sig(n+2) = F)) /\
      (t_sig(n+2) = PAD20TO32(A(INSTFETCH(ram_sig n,p_sig n)))) /\
      (rsf_sig(n+2) = R(INSTFETCH(ram_sig n,p_sig n))) /\
      (msf_sig(n+2) = M(INSTFETCH(ram_sig n,p_sig n))) /\
      (dsf_sig(n+2) = D(INSTFETCH(ram_sig n,p_sig n))) /\
      (csf_sig(n+2) = C(INSTFETCH(ram_sig n,p_sig n))) /\
      (fsf_sig(n+2) = FF(INSTFETCH(ram_sig n,p_sig n)))) /\
     (node_sig(n+2) = #00011))
```

At time n+3 PRECALL has been executed, so the y register holds the last value of the program counter.

```
|- ((((ram_sig(n+3) = ram_sig n) /\
       (p_sig(n+3) = TRIM32TO20(INCP32(p_sig n))) /\
       (a_sig(n+3) = a_sig n) /\
       (x_sig(n+3) = x_sig n) /\
       (y_sig(n+3) = INCP32(p_sig n)) /\
       (b_sig(n+3) = b_sig n) /\
       (stop_sig(n+3) = F)) /\
      (t_sig(n+3) =
       PAD20TO32(A(INSTFETCH(ram_sig n,p_sig n)))) /\
      (rsf_sig(n+3) = R(INSTFETCH(ram_sig n,p_sig n))) /\
      (msf_sig(n+3) = M(INSTFETCH(ram_sig n,p_sig n))) /\
      (dsf_sig(n+3) = D(INSTFETCH(ram_sig n,p_sig n))) /\
      (csf_sig(n+3) = C(INSTFETCH(ram_sig n,p_sig n))) /\
      (fsf_sig(n+3) = FF(INSTFETCH(ram_sig n,p_sig n)))) /\
     (node_sig(n+3) = #00100))
```

At time n+4 PERFORM has been executed. By the definition of PERFORM, the program counter now holds the result (the first value) computed by the ALU *if* the compare field of the new instruction does not hold 1 (which it doesn't) and *if* the destination field of the new instruction is not 0, 1 or 2 (which it is not). The b flag holds the b value (the second value) computed by the ALU, and the stop flag the stop value (the third). Based on Theorem 11, the memory source supplied to the ALU comes from the t register, which at time n+3 on this path held the (padded) new address. The other arguments are likewise dictated by Theorem 11 and the state and transient at time n+3. The whole ALU expression is:

```
ALU
(FF(INSTFETCH(ram_sig n,p_sig n)),
 M(INSTFETCH(ram_sig n,p_sig n)),
 D(INSTFETCH(ram_sig n,p_sig n)),
 REG
 (R(INSTFETCH(ram_sig n,p_sig n)),a_sig n,x_sig n,
  PAD20TO32(TRIM32TO20(INCP32(p_sig n))),
  TRIM32TO20(INCP32(p_sig n))),
 PAD20TO32(A(INSTFETCH(ram_sig n,p_sig n))),b_sig n)
```

For convenience this is abbreviated by a new constant, ALU_ABBR6 (one of eight needed for the twenty-four cases). For *any* ram, p, a, x, y and b:

```
|- ALU_ABBR6(ram,p,a,x,y,b) =
   ALU
   (FF(INSTFETCH(ram,p)),M(INSTFETCH(ram,p)),D(INSTFETCH(ram,p)),
    REG
    (R(INSTFETCH(ram,p)),a,x,PAD20TO32(TRIM32TO20(INCP32 p)),
     TRIM32TO20(INCP32 p)),PAD20TO32(A(INSTFETCH(ram,p))),b)
```

Since it has been assumed that VAL4(FF(INSTFETCH(ram_sig n,p_sig n))) = 1, Theorem 5 can be used to compute the result of the ALU operation:

```
PAD20TO32(A(INSTFETCH(ram_sig n,p_sig n))),b_sig n,
NOT
 ((VAL3(D(INSTFETCH(ram_sig n,p_sig n))) = 3) \/
  ((VAL3(D(INSTFETCH(ram_sig n,p_sig n))) = 4) \/
   (VAL3(D(INSTFETCH(ram_sig n,p_sig n))) = 5)))) \/
(INVALID(PAD20TO32(A(INSTFETCH(ram_sig n,p_sig n))))
```

The padded address cannot be invalid (Theorem 4), and the destination of the new instruction must be 3, 4 or 5 (Theorem 6), so the ALU returns:

```
PAD20TO32(A(INSTFETCH(ram_sig n,p_sig n))),b_sig n,F
```

Thus at time n+4 PERFORM has been executed:

```
|- (((ram_sig(n+4) = ram_sig n) /\
     (p_sig(n+4) = A(INSTFETCH(ram_sig n,p_sig n))) /\
     (a_sig(n+4) = a_sig n) /\
     (x_sig(n+4) = x_sig n) /\
     (y_sig(n+4) = INCP32(p_sig n)) /\
     (b_sig(n+4) = b_sig n) /\
     (stop_sig(n+4) = F)) /\
    (t_sig(n+4) =
     PAD20TO32(A(INSTFETCH(ram_sig n,p_sig n)))) /\
    (rsf_sig(n+4) = R(INSTFETCH(ram_sig n,p_sig n))) /\
    (msf_sig(n+4) = M(INSTFETCH(ram_sig n,p_sig n))) /\
    (dsf_sig(n+4) = D(INSTFETCH(ram_sig n,p_sig n))) /\
    (csf_sig(n+4) = C(INSTFETCH(ram_sig n,p_sig n))) /\
    (fsf_sig(n+4) =
     FF(INSTFETCH(ram_sig n,p_sig n)))) /\
   (node_sig(n+4) = #10000)
```

The next node is `#10000`, so the sequence is ended. The program counter now holds the address of the procedure to be called, and the `y` register stores the return address (the original program counter's value plus one).

These four theorems provide the basis for proving Theorem 2. The fourth is part of the proof of Theorem 3. (See Section 1.3). (Here, the theorems are stated in terms of the fourteen components. Various properties of n-tuples are required to formulate them as in Section 1.3.)

5.3 The Equivalence Proof

Now the high-level specification, `NEXT` (Section 3.2) is considered in relation to the results of the graph traversal. We wish to prove:

Theorem 14: Result of Specification for Call

```
c1(ram,p,a,x,y,b,stop) /\
c3(ram,p,a,x,y,b,stop) /\
c17(ram,p,a,x,y,b,stop) ==>
(NEXT(ram,p,a,x,y,b,stop) = ram,A(INSTFETCH(ram,p)),a,x,INCP32 p,b,F)
```

That is, for any state, `NEXT` specifies the same changes as have accumulated in the registers at the end of the sequence produced by the host machine (disregarding time).

The proof consists in reducing the conditional in the definition of `NEXT`. Obviously, `stop` is false (`c1` holds). The five *illegal* conditions mentioned in `c3` are all false. Furthermore, the new address cannot be illegal: the new memory address control field holds 0, so `OFFSET` gives a padded 20-bit address. Since the new compare field also holds 0, `comp` is false; `writeop` and `skip` are also obviously false. The new destination field is neither 0, 1 or 2 (by Theorem 6) and so the *call* branch is arrived at. The program counter, the `b` flag and the `stop` flag of the new state in this case depend on the result (`aluout`) computed by `ALU`. The new state is:

```
(ram,TRIM32TO20(VALUE aluout),a,x,INCP32 p,BVAL aluout,SVAL aluout)
```

In the fully unfolded definition of `NEXT`, `aluout` is

```
ALU
(FF(INSTFETCH(ram,p)),M(INSTFETCH(ram,p)),D(INSTFETCH(ram,p)),
 REG(R(INSTFETCH(ram,p)),a,x,y,TRIM32TO20(INCP32 p)),
 MEMREAD
 (ram,M(INSTFETCH(ram,p)),A(INSTFETCH(ram,p)),x,y,
  ((VAL1(C(INSTFETCH(ram,p))) = 0) /\
   (VAL3(D(INSTFETCH(ram,p))) = 6)) \/
  ((VAL1(C(INSTFETCH(ram,p))) = 0) /\
   NOT((VAL3(D(INSTFETCH(ram,p))) = 7) \/
     (VAL3(D(INSTFETCH(ram,p))) = 6)) /\
   (VAL4(FF(INSTFETCH(ram,p))) = 2)),
  ((VAL1(C(INSTFETCH(ram,p))) = 0) /\
   (NOT(VAL3(D(INSTFETCH(ram,p))) = 7) /\
    NOT(VAL3(D(INSTFETCH(ram,p))) = 6)) /\
   (VAL4(FF(INSTFETCH(ram,p))) = 12))),b)
```

which is abbreviated by a new constant `ALU_ABBR2`, so that the new state specified by `NEXT` is

```
(ram,TRIM32TO20(VALUE(ALU_ABBR2(ram,p,a,x,y,b))),a,x,
 INCP32 p,BVAL(ALU_ABBR2(ram,p,a,x,y,b)),
 SVAL(ALU_ABBR2(ram,p,a,x,y,b)))
```

Since `VAL4(FF(INSTFETCH(ram,p))) = 1`, the reasoning is as for `ALU_ABBR6` in Section 5.2.3: Theorem 5 is used to conclude that `ALU` in this case returns `MEMREAD(...)` as its result, `b` for the `b` flag, and

```
NOT((VAL3(D(INSTFETCH(ram,p))) = 3) \/
   ((VAL3(D(INSTFETCH(ram,p))) = 4) \/
    (VAL3(D(INSTFETCH(ram,p))) = 5))) \/
(INVALID(MEMREAD(...)))
```

for the `stop` flag, where "..." stands for the arguments to `MEMREAD`.

The value of the new destination field is 3 or 4 or 5, so the `stop` flag becomes `F \/ (INVALID(MEMREAD(...)))`. It remains to work out the value of `MEMREAD(...)`. Its sixth argument is a boolean value which distinguishes main from peripheral memory, and the seventh, a boolean which indicates whether no memory source is required. The seventh works out to be false, since the value of the function field is not 12 (but rather 1). The sixth also works out to be false, since the function field is not 2, nor is the destination field 6. Thus, since the memory address control field is 0, **MEMREAD** returns just `PAD20TO32(A(INSTFETCH(ram,p)))`. This cannot be invalid (Theorem 4). Therefore the **ALU** returns:

```
PAD20TO32(A(INSTFETCH(ram,p))),b,F
```

so the state returned by **NEXT** is

```
(ram,TRIM32TO20(PAD20TO32(A(INSTFETCH(ram,p)))),a,x,INCP32 p,b,F)
```

which by Axiom 1 is just

```
(ram,A(INSTFETCH(ram,p)),a,x,INCP32 p,b,F)
```

which is what was wanted.

In this case, the memory component (`ram`) and the `a`, `x` and `y` components of the state given by **NEXT** agree immediately with the end results of **HOST_NEXT**. That the other three registers agree depends on an argument about the **ALU**. The essence of the argument can be expressed by a theorem relating the two values computed by the **ALU**: the one referring to **MEMREAD**, in the case of the specification, and the one determined one by Theorem 11 for the **PERFORM** node (at the correct point in the sequence), in the case of the host machine.

Theorem 15: Equivalence of ALU in Specification and Host

```
|- ((VAL2(M(INSTFETCH(ram,p))) = 0) /\
    (VAL1(C(INSTFETCH(ram,p))) = 0) /\
    NOT(VAL3(D(INSTFETCH(ram,p))) = 7) /\
    NOT(VAL3(D(INSTFETCH(ram,p))) = 6) /\
    (VAL4(FF(INSTFETCH(ram,p))) = 1)) ==>
   (ALU_ABBR2(ram,p,a,x,y,b) = ALU_ABBR6(ram,p,a,x,y,b))
```

5.4 A Digression on Goal-Oriented Proof

For the HOL proof of Theorem 14, we are starting for a change with a *goal*. We know that **NEXT** should unfold to the state specified by the host machine, in the circumstances; but do not know whether it really does, nor how to attempt to prove that it does. Here, a *goal oriented proof* is called for. One starts with a goal (a term with an associated list of assumptions), and applies strategies or *tactics* to the goal to produce subgoals; then tactics are applied to the subgoals, and so on, until subgoals are reached which are known to be true (that is, which correspond to previously proved theorems or axioms). The tactics justify each decomposition into subgoals so that a proof can be assembled in the end. The proof is built up by inferences, starting with the theorems corresponding to the final set of subgoals. For example, if a tactic `T` is applied to a goal `(A,t)` (term `t` and assumptions `A`) to produce two subgoals

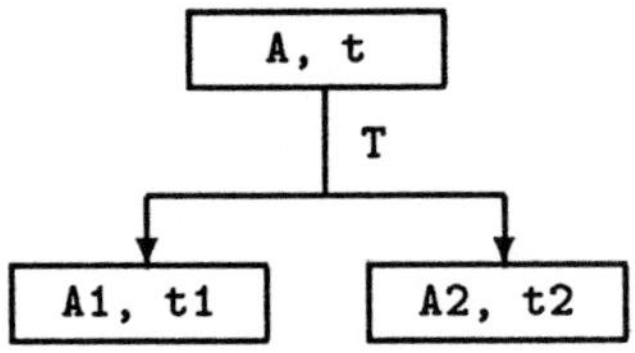

then the ML procedure which justifies that decomposition expects two theorems

- o `|-t1` with no assumptions beyond `A1`, and
- o `|-t2` with no assumptions beyond `A2`

and infers a theorem `|-t` (with no assumptions beyond `A`). Tactics can be sequenced and combined in various ways to form compound tactics. In this way, arbitrarily complex proof strategies can be expressed as procedures which generate proofs.

This style of proof revolves around the list of current assumptions in a natural way. Tactics use the same rewriting mechanism as forward proof. One frequently-used tactic generates a subgoal by rewriting the goal using all current assumptions of that goal as rewrite rules. Each rewrite is justified by a chain of inferences which is used used later in assembling the proof.

Sometimes groups of assumptions can logically imply a new assumption; a useful tactic finds new conclusions and adds them to the assumption list. This provides a simple kind of resolution in HOL. For example, it follows from the definition of invalidity (Section 3.1) that

```
|- !p1. NOT(INVALID p1) ==> (PAD20TO32(TRIM32TO20 p1) = p1)
```

This has an instance

```
|- NOT(INVALID(INCP32 p)) ==> (PAD20TO32(TRIM32TO20(INCP32 p))=INCP32 p)
```

which can be used to simplify the expression

```
PAD20TO32(TRIM32TO20(INCP32 p))
```

during the proof of Theorem 14. The instance can be identified and its conclusion added as an assumption as soon as its antecedent, `NOT(INVALID(INCP32 p))`, joins the assumptions. The conclusion is not a *new* assumption because it can always be dismissed by a series of simple inference steps.

5.5 The Equivalence Proof in HOL

To prove Theorem 14 in HOL, first the definitions of `c1`, `c3` and `c17` are unfolded. The definition of `NEXT` is unfolded, using a version of the definition without the `let` and lambda expressions (analagous to the version of the definition of `PRECALL` in Section 5.1). (Part of this unfolding has been seen in the form of the **ALU** expression, Section 5.3.) The assumptions are "resolved" with Theorem 6, all of whose antecedents are by this point among the assumptions. This adds new members to the list of current assumptions of the subgoal, including:

- `NOT(VAL3(D(INSTFETCH(ram,p))) = 0)`
- `NOT(VAL3(D(INSTFETCH(ram,p))) = 1)`
- `NOT(VAL3(D(INSTFETCH(ram,p))) = 2)`

The subgoal can now be rewritten using all of its current assumptions as well as facts such as Axiom 1 and Theorem 5 (to reduce the **ALU** expressions). By this process the **NEXT** expression (a conditional with ten branches) reduces to the correct case, and the state in that case to the desired state: both sides of the subgoal are equal.

A subgoal with a term `t=t` is known to correspond to an axiom: namely, reflexivity. From this point HOL starts with the axiom of reflexivity and builds up the proof by using the justification of each decomposition into subgoals. The book-keeping is all done by HOL, the user only supplying the tactics. The whole proof can be generated with one command, given the correct (compound) tactic. The point about forward proof is that it would be very difficult to structure a forward proof effort starting from `|-t=t` and applying inference rules to produce a theorem as complex as Theorem 14.

The only real tricks in this process are supplying the lemmas needed, and causing the rewriting to happen in an efficient way so the proof can be done in a reasonable amount of time. The former requires the proof effort to be planned and structured sensibly. The latter is discussed further in Section 8.

6 The Rest of the Proof

So far, the specification and host machine have only been revealed insofar as they concern the literal *call* instruction. This in fact gives a good idea of what is claimed and proved in the other cases, and how it is stated and proved in HOL. The literal *call* proofs are about average in length and difficulty among the twenty-four cases. The other four paths suggested in Figure 2 are for states in which the `stop` flag is set; *call* instructions for which the **ALU** sets the `stop` flag; states for which fetching produces an illegal instruction; and states for which the new fetched instruction is a non-operation. The other nineteen paths are for *call* instructions with indirect or offset addresses; **ALU** operations with the `a`, `x` or `y` register as destination; write operations to memory or output; read operations from memory or input; and variations of the above with differents sorts of addressing and in which illegal situations arise and the machine stops. The tree of all possible paths can be seen in [5].

In fact, the definitions of **NEXT** and the host machines in the example proofs have been unfolded to their maximum depth, but that is not necessary for doing the proof. As soon as they are unfolded sufficiently, the host and specification are seen to be equal (by means of Theorem 15 and so on), and the proof is completed.

A large number of other lemmas and intermediate results were proved *en route* to the main theorems and not mentioned here; no one is very complicated, but they add up to a certain amount of unseen effort. Other results have been mentioned only in passing (see tables in Section 8).

7 Errors Found in the Viper Specifications

Aside from certain HOL conventions, such as consistently currying or uncurrying a function, and some corrections of typographical errors, the HOL versions of the functional specification and the host differ from the definitions in [4] at a few places where we found errors in the host and high-level specification. There was a type error in the function `FMOVE` which was passed a boolean `b` by `FETCH` when `FMOVE` was expecting a 1-bit string. More seriously, there were some other errors in `FETCH`, and one in `NEXT`. The function for `FETCH` had read:

```
|- FETCH((ram,p,a,x,y,b,stop),t,rsf,msf,dsf,csf,fsf) =
   let fetched = INSTFETCH(ram,p) in
   let newp = TRIM32TO20(INCP32 p) in
   let newr = R fetched in
   let newm = M fetched in
   let newd = D fetched in
   let newc = C fetched in
   let newf = FF fetched in
   let newt = PAD20TO32(A fetched) in
   let notinc = INVALID(INCP32 newp) in
   let illegalcl = ILLEGALCALL(dsf,csf,fsf) in
   let illegalsp = SPAREFUNC(dsf,csf,fsf) in
   let illegalonp = ILLEGALPDEST(dsf,csf,fsf)in
   let illegalwr = ILLEGALWRITE(dsf,csf,msf) in
   let stopstate = WORD5 8 in
   let b' = WORD1(V[b]) in
   (notinc \/
    (illegalcl \/ (illegalsp \/ (illegalonp \/ illegalwr))) =>
    (((ram,newp,a,x,y,b,T),newt,newr,newm,newd,newc,newf),stopstate) |
    (((ram,newp,a,x,y,b,F),newt,newr,newm,newd,newc,newf),
     FMOVE(newm,newd,newc,newf,b')))
```

so that the *phase* of fetch cycle was confused: it is actually the incremented program counter one wants to check for validity, not the twice incremented counter, and it is the *new* instruction one wants to check for legality, not the instruction that rested in the transient before the `FETCH` operation. This error became apparent in the course of doing the proof because the host's results disagreed with the specification's results.

The original specification had an incomplete check for illegal instructions; the possibility of the instruction being a non-operation was neglected. The relevant part read:

```
... in let illegaladdr = (NOT(NILM(dsf,csf,fsf)) /\
                         (INVALID(OFFSET(msf,addr,x,y)))) in ...
```

This naturally caused a problem with interpreting non-operations whose fields met the two conditions above but which were meant to be legal instuctions.

Finally, we added definitions to form a complete problem statement: the simple definition of `HOST_NEXT` to traverse the graph and the formal definitions of the conditions `c1` and so on. There were more conditions than forseen in the informal proof ([5]), similar to `c17`; these conditions always decide whether to stop because of an abnormality after an `ALU` operation, or whether to end the sequence naturally, depending on the `stop` value returned by the `ALU`. However, the inputs to the `ALU` are different in different paths through the graph, depending on which nodes have come before. In the literal *call* path the `ALU` expression was:

```
ALU
(FF(INSTFETCH(ram_sig n,p_sig n)),
 M(INSTFETCH(ram_sig n,p_sig n)),
 D(INSTFETCH(ram_sig n,p_sig n)),
 REG
 (R(INSTFETCH(ram_sig n,p_sig n)),a_sig n,x_sig n,
  PAD20TO32(TRIM32TO20(INCP32(p_sig n))),
  TRIM32TO20(INCP32(p_sig n))),
 PAD20TO32(A(INSTFETCH(ram_sig n,p_sig n))),b_sig n)
```

but this instance was a result of having been through the `DUMMY`, `FETCH` and in particular the `PRECALL` node, and no others. In fact, there are seven other similar conditions for the twenty-four total paths.

Furthermore, the conditions suggested did not make a distinction between the contents of the transient before and after a `FETCH` event. For example, in [5], `c3` was defined to be something like

```
c3(p,dsf,csf,fsf,msf,b) =
(NOT
 ((INVALID(INCP32 p)) \/
  ((ILLEGALCALL(dsf,csf,fsf)) \/
   ((SPAREFUNC(dsf,csf,fsf)) \/
    ((ILLEGALPDEST(dsf,csf,fsf)) \/
     (ILLEGALWRITE(dsf,csf,msf))))))) /\
```

```
((NOT(VAL1 csf = 1)) /\
 ((NOT((VAL3 dsf = 7) \/ (VAL3 dsf = 6))) /\
  ((NOT(NOOP(dsf,csf,b))) /\ ((VAL2 msf = 0) /\
    ((VAL1 csf = 0) /\ (VAL4 fsf = 1))))))
```

This is not what is required for traversing the graph (see Section 4.2) and in fact does not make sense; it is not a function of the state and so has no meaning at the higher level.

8 Performing the Proof

We have tried to present the proof as simply as possible in this paper, but the actual proof effort deserves some discussion.

The process of typing the definitions in [4] and [5] into HOL took about a week; it was straightforward since the definitions were written in a notation called LCF_LSM, a predecessor of HOL.

The generation of the proof took about six person-months of work. (It was done on a Sun 3 workstation with eight megabytes of memory.) This may seem a long time for a proof which is not conceptually difficult, but it should be borne in mind what was actually generated: a sequence of over a million inferences which ensures that the correctness statement is really true (see Section 1.1). The hand proof done by Cullyer ([5]) took only about three weeks but did not detect the errors described in Section 7.

The generation of the theorems about `HOST_NEXT` (Sections 5.1, 5.2.1, 5.2.2 and 5.2.3) was relatively quick and easy, as they were done by forward proof, *i.e.* letting HOL continuously unfold and simplify a definition to compute an unforseen theorem by a known method.

Most of the time was spent in proving the theorems about `NEXT`, for which the desired result was known in advance but the way to prove it was not. The latter is a more typical sort of HOL or LCF proof effort, and it is a very interactive process. One typically tries a tactic, examines the subgoals, backs up, tries another, and so on, until a successful structure of tactics is discovered. The proof effort has to be planned carefully and the proof well understood in advance. Lemmas have to be anticipated and supplied in a useful form. The difficulty is partly because HOL (and LCF) are really just *frameworks* for performing proofs; the only proof tool which is in any sense automatic is rewriting (either forward or goal oriented) and the mechanism which justifies it. Any proof tool or strategy one thinks of can be expressed in the programming language ML and then applied, but designing these is a research problem. One hopes that large proof efforts such as the present one suggest new possible proof tools.

The proofs are difficult also because of the sheer size of the theorems (for example, the versions of Theorem 13 before it is fully simplified are several pages long when printed in the style shown). A certain amount of the proof effort was directed toward structuring theorems into smaller ones, and defining abbreviations to control the expansion of theorems (such as `ALU_ABBR6` and `ALU_ABBR2`).

The proofs are also large in terms of the computation time; rewriting in particular can be costly of time, and can be avoided in various ways by investing more of the user's time in doing explicit local replacements or other *ad hoc* methods. Large theorems are even time-consuming to fetch from files or to pretty-print. This problem can probably be attacked both by increased computing resources, and research into efficient proof strategies. Computation time can also be reduced by identifying large inference steps which are computed more than once and proving them as lemmas. For example, in computing the end results of the host machine sequences, certain conditional subexpressions recur for choosing the next node. The reductions of these are proved once as lemmas rather than being computed several times in the course of the proof. This is also often more convenient than doing a proof within a larger proof.

To give a very rough idea of the magnitude of by the proof, the following tables give the number of primitive inference steps and the CPU time in seconds for the main theorems of the proof. (Garbage collection time has been ignored.) The remarks in Section 5.1 about HOL proofs all apply here; in particular, the apparent sizes and times of proofs can be greatly (and rather misleadingly) inflated by reliance on powerful, general rewriting strategies where more specific and local strategies could be found. Theorem 1, for example, has a large proof which is generated by a strategy which simply rewrites according to previously proved theorems; Theorem 14 has a relatively small proof which is generated by laborious instantiations and carefully specified substitutions. We mention this so that these figures are not taken too seriously.

The last six theorems in the first table are referred to but not shown in this paper. "Unfolded `NEXT`" is the fully unfolded definition of `NEXT` with no `lets`; likewise for `ALU`. "Covering" refers to the theorem stating that at each node, the conditions for choosing the next node cover all possibilities. "Numbers" refers collectively to all the properties of numbers involved in reasoning about the numbers of steps in paths. "Tuples" refers collectively to properties of n-tuples required to transform theorems in phrased in terms of the fourteen components into theorems phrased in terms of a state, transient and node.

Theorems used for All 24 Paths

Theorem	Steps	CPU Seconds
Section 3.1 Theorem 4	215	3
Section 5.1 Theorem 7	7,664	230
Section 5.2.1 Theorem 8	8,474	315
Section 5.2.1 Theorem 9	7,649	280
Section 5.2.1 Theorem 10	19,587	456
Section 5.2.1 Theorem 11	24,428	649
Section 5.2.2 Theorem 12	2,195	64
Similar for `DUMMY`	2,259	66
Similar for `STOP`	2,179	57
Similar for `FETCH`	10,291	232
Similar for `PERFORM`	8,019	213
Section 5.5 Unfolded `NEXT`	7,486	292
Unfolded `ALU`	4,492	82
Total	104,938	2,939

Theorems Just for the Literal Call Path

Theorem	Steps	CPU Seconds
Section 3.1 Theorem 5	233	43
Section 4.2 Theorem 6	1,836	35
Section 5.3 Theorem 15	2,545	81
Section 5.2.3 Theorem 13	14,037	312
Section 5.3Theorem 14	8,626	237
Section 1.3 Theorem 2	8,586	213
Section 1.3 Theorem 3	8,620	230
Total	44,483	1,151

Theorems tying the 24 Cases Together

Theorem	Steps	CPU Seconds
Section 1.3 Covering	19,772	364
Section 1.3 Numbers	20,173	222
Section 5.2.3 Tuples	5,116	120
Section 1.3 Theorem 1	35,554	384
Total	80,615	1,090

9 Conclusions

As the reader can tell, the machine-checked proof described in this paper was a very large and somewhat tedious project. Proofs of this sort, while perfectly feasible, require experts in theorem-proving rather than electrical engineers, and such experts are at present scarce. It is the experts' time rather than computation time which makes verification expensive[1]. At present, formal verification is only appropriate in selected applications where the cost can be justified.

One clear conclusion of this work is that very much more basic research must be done before formal verification becomes practical and commonplace in real-life applications. Work to date shows that verification of real examples is still too time-consuming and tedious, and requires too much user guidance, for routine use. We believe that HOL is an excellent framework in which to research the problems: to design more automatic proof tactics, better ways to abstract and describe proofs, and more efficient and flexible rewriting strategies. Experimental proofs of real hardware like the Viper chip suggest many research areas which must be addressed before larger applications can be attempted.

A second conclusion is that for theorem-proving experts to undertake hardware proofs, documentation is vital; the experts will not be knowledgeable in engineering matters, and informal explanations of the function of the systems will save a lot of time spent working out connections which are perhaps obvious to the engineer.

Viper is to be manufactured and sold by two companies, for both civil and military applications. As we have shown, the HOL methodology can detect design errors *at a certain level of abstraction* from the actual hardware. We feel it necessary to warn against a false sense of security afforded by an HOL proof; there are many classes of errors not even *visible* in the models used. In particular, we would caution against a false sense of security in such hazardous applications as nuclear reactor control systems. The art of formal verification is in its early days. We believe that *years* of basic research are needed before the techniques can be reliably applied to life-critical systems; even then they can only be applied with an understanding of their limitations.

Further, a proof that a design meets a specification is only as good as the specification, which, for a complex system, can be very difficult to produce. A perfectly correct proof may relate to a specification which misses some essential behaviour of the design, and so may not give much security. For an example, see [12] regarding the problems of specifying a simple computer design.

Finally, for verification to be effective, we must first move to a situation in which the same sources are used by designers, verifiers and manufacturers. For example, as mentioned in Section 1, the errors we found in Viper's specification

[1] The first level of the Viper proof, for example, takes several hours of CPU time to run, and that is using an inefficient prototype system. It took six months, however, to organize the proof and carry it out. We expect that subsequent lower-level stages will be much more complex.

and host machine are apparently not present in the actual chip; hence the manufacturers cannot have used the specification which we have started to verify. Uniting these various communities is the aim of the research at RSRE, but there is a long way to go.

Acknowledgements

Theorems 1, 2 and 3, which tie together the body of the proof (the twenty-four cases), and the basic lemmas about numbers and tuples were generated in HOL by Mike Gordon. The author is grateful to Elsa Gunter for assistance with typesetting and suggestions about the paper. The work described here was supported by a grant from RSRE, Research Agreement 2029/205(RSRE), and has been placed under Crown Copyright © HMSO London 1987. This paper has also appeared under the same title as a University of Cambridge Computer Laboratory Technical Report, No. 104, 1987.

References

[1] A. Church, "A Formulation of the Simple Theory of Types", Journal of Symbolic Logic 5, 1940

[2] A. Cohn and M. Gordon, "A Mechanized Proof of Correctness of a Simple Counter", University of Cambridge, Computer Laboratory, Tech. Report No. 94, 1986

[3] W. J. Cullyer and C. H. Pygott, "Hardware Proofs using LCF_LSM and ELLA", RSRE Memo. 3832, Sept. 1985

[4] W. J. Cullyer, "Viper Microprocessor: Formal Specification", RSRE Report 85013, Oct. 1985

[5] W. J. Cullyer, "Viper — Correspondence between the Specification and the 'Major State Machine' ", RSRE report No. 86004, Jan. 1986

[6] W. J. Cullyer, "Implementing Safety-Critical Systems: The Viper Microprocessor", In: VLSI Specification, Verification and Synthesis, Edited by G. Birtwistle and P. A. Subrahmanyam, (this volume)

[7] M. Gordon, R. Milner and C. P. Wadsworth, "Edinburgh LCF, Lecture Notes in Computer Science", Springer-Verlag, 1979

[8] M. Gordon, "Proving a Computer Correct", University of Cambridge, Computer Laboratory, Tech. Report No. 42, 1983

[9] M. Gordon, "HOL: A Machine Oriented Formulation of Higher-Order Logic", University of Cambridge, Computer Laboratory, Tech. Report No. 68, 1985

[10] M. Gordon, "HOL: A Proof Generating System for Higher-Order Logic", In: VLSI Specification, Verification and Synthesis, Edited by G. Birtwistle and P. A. Subrahmanyam (this volume), Also: University of Cambridge, Computer Laboratory, Tech. Report No. 103, 1987

[11] W. A. Hunt Jr., "FM8501: A Verified Microprocessor", University of Texas, Austin, Tech. Report 47, 1985

[12] J. J. Joyce, "Verification and Implementation of a Microprocessor", In: VLSI Specification, Verification and Synthesis, Edited by G. Birtwistle and P. A. Subrahmanyam, (this volume)

[13] J. Kershaw, "Viper: A Microprocessor for Safety-Critical Applications", RSRE Memo. No. 3754, Dec. 1985

[14] L. Paulson, "A Higher-Order Implementation of Rewriting", Science of Computer Programming 3, 119-149, 1983

[15] L. Paulson, "Interactive Theorem Proving with Cambridge LCF", Cambridge University Press, To Appear 1987

3

HOL: A Proof Generating System for Higher-Order Logic

Michael J.C. Gordon

University of Cambridge
Computer Laboratory
Corn Exchange Street
Cambridge, CB2 3QG
England.

Abstract: HOL is a version of Robin Milner's LCF theorem proving system for higher-order logic. It is currently being used to investigate (1) how various levels of hardware behaviour can be rigorously modelled and (2) how the resulting behavioral representations can be the basis for verification by mechanized formal proof. This paper starts with a tutorial introduction to the meta-language ML. The version of higher-order logic implemented in the HOL system is then described. This is followed by an introduction to goal-directed proof with *tactics* and *tacticals*. Finally, there is a little example of the system in action which illustrates how HOL can be used for hardware verification.

1 Introduction to HOL

Higher-order logic is a promising language for specifying all aspects of hardware behaviour [8], [1]. It was originally developed as a foundation for mathematics [2]. Its use for hardware specification and verification was first advocated by Keith Hanna [9].

The approach to mechanising logic described in this paper is due to Robin Milner [6]. He originally developed the approach for a system called LCF designed for reasoning about higher-order recursively defined functions. The HOL system is a direct descendant of this work. The original LCF system was implemented at Edinburgh and is called "Edinburgh LCF". The code for it was ported from Stanford Lisp to Franz Lisp by Gerard Huet at INRIA and formed the basis for a French research project called "Formel". Huet's Franz Lisp version of LCF was further developed at Cambridge by Larry Paulson and eventually became known as "Cambridge LCF" [18]. The HOL system is implemented on top of Cambridge LCF and consequently many (good and bad) features of LCF are found in it. In particular, the axiomatization of higher-order logic used is not the classical one due to Church, but an equivalent formulation strongly influenced by LCF.

To make this paper self-contained, a brief introduction to ML (the LCF metalanguage) is included. The version of ML described here (and included as part of the HOL system) is not Standard ML [16] but the version of ML documented in the ML Handbook [4].

The acronym "HOL" refers to both the computer system and the version of higher-order logic that it supports; the former will be called the "HOL system" and the latter the "HOL logic".

Because this paper is about the HOL system, the machine-readable syntax for the logic will be presented. For example, instead of the conventional logical symbols $\wedge$, $\vee$, $\neg$, $\forall$, $\exists$ and λ, we use the following strings of ASCII characters `/\`, `\/`, `~`, `!`, `?` and `\` respectively. In various other papers on HOL (*e.g.* [8], [1]) conventional notation, rather than the ASCII notation, is employed. The two notations are, of course, equivalent.

2 Introduction to ML

This section is a very brief introduction to the metalanguage ML. The aim is just to give a feel for what it is like to interact with ML and to introduce some features of the HOL system (*e.g.* the representation in ML of the types and terms of the HOL logic[1]).

ML is an interactive programming language like Lisp. At top level one can evaluate expressions and perform declarations. The former results in the expression's value and type being printed, the latter in a value being bound to a name. The boxes below contain a little session with the system. The interactions in these boxes should be understood as occurring in sequence. For example, variable bindings made in earlier boxes are assumed to persist to later ones. To enter the HOL system one types `hol` to Unix[2]; the HOL system then prints a sign-on message and puts one into ML. The ML prompt is `#`, so lines beginning with `#` are typed by the user and other lines are the system's response.

---

[1] Types and terms are explained in detail in Section 3.2.

[2] The Unix prompt is `%`.

```
% hol

 _  _           __        _
|__|           |  |      |
|  | IGHER     |__| RDER |__ OGIC
===============================
(Built on Sept  31)

#1.[2;3;4;5];;
[1; 2; 3; 4; 5] : int list
```

The ML expression `1.[2;3;4;5]` has the form e_1 *op* e_2 where e_1 is the expression `1` (whose value is the integer 1), e_2 is the expression `[2;3;4;5]` (whose value is a list of four integers) and *op* is the infixed operator '`.`' which is like Lisp's *cons* function. Other list processing functions include `hd` (*car* in Lisp), `tl` (*cdr* in Lisp) and `null` (*null* in Lisp). The double semicolon '`;;`' terminates a top-level phrase. The system's response is shown on the line not starting with a prompt. It consists of the value of the expression followed, after a colon, by its type. The ML typechecker infers the type of expressions using methods invented by Robin Milner [15]. The type `int list` is the type of 'lists of integers'; `list` is a unary type operator. The type system of ML is very similar to the type system of the HOL logic which is explained in Section 3.1

The value of the last expression evaluated at top-level in ML is always remembered in a variable called `it`.

```
let l = it;;
l = [1; 2; 3; 4; 5] : int list

#tl l;;
[2; 3; 4; 5] : int list

#hd it;;
2 : int

#tl(tl(tl(tl(tl l))));;
[] : int list
```

Following standard λ-calculus usage, the application of a function f to an argument x can be written without brackets as $f x$ (although the more conventional $f(x)$ is also allowed). The expression $f x_1 x_2 \cdots x_n$ abbreviates the less intelligible expression $(\cdots((f x_1) x_2)\cdots) x_n$ (function application is left associative).

Declarations have the form let x_1=e_1 and $\cdots$ and x_n=e_n and result in the value of each expression e_i being bound to the name x_i.

```
#let l1 = [1;2;3] and l2 = [‘a‘;‘b‘;‘c‘];;
l1 = [1; 2; 3] : int list
l2 = [‘a‘; ‘b‘; ‘c‘] : string list
```

ML expressions like `‘a‘`, `‘b‘`, `‘foo‘` *etc.* are *strings* and have type `string`. Any sequence of ASCII characters can be written between the quotes. The function `words` splits a single string into a list of single character strings, using space as separator.

```
#words‘a b c‘;;
[‘a‘; ‘b‘; ‘c‘] : string list
```

An expression of the form (e_1,e_2) evaluates to a pair of the values of e_1 and e_2. If e_1 has type σ_1 and e_2 has type σ_2 then (e_1,e_2) has type σ_1#σ_2. A tuple (e_1,...,e_n) is equivalent to (e_1,(e_2,...,e_n)) (*i.e.* ‘,’ is right associative). The brackets around pairs and tuples are optional; the system doesn't print them. The first and second components of a pair can be extracted with the ML functions `fst` and `snd` respectively.

```
#(1,true,‘abc‘);;
1,true,‘abc‘ : (int # bool # string)

#snd it;;
true,‘abc‘ : (bool # string)


#fst it;;
true : bool
```

The ML expressions `true` and `false` denote the two truthvalues of type `bool`.

ML types can contain the *type variables* `*`, `**`, `***`, *etc.* Such types are called *polymorphic*. A function with a polymorphic type should be thought of as possessing all the types obtainable by replacing type variables by types. This is illustrated below with the function `zip`.

Functions are defined with declarations of the form let $f\ v_1\ \ldots\ v_n = e$ where each v_i is either a variable or a pattern build out of variables [3]. Recursive functions are declared with `letrec` instead of `let`.

[3] The ML Handbook [4] gives exact details.

The function zip, below, converts a pair of lists `([`x_1`;...;`x_n`], [`y_1`;...;`y_n`])` to a list of pairs `[(`x_1`,`y_1`);...;(`x_n`,`y_n`)]`.

```
#letrec zip(l1,l2) =
#if null l1 or null l2
# then []
# else (hd l1,hd l2).zip(tl l1,tl l2);;
zip = - : ((* list # ** list) -> (* # **) list)

#zip([1;2;3],['a';'b';'c']);;
[1,'a'; 2,'b'; 3,'c'] : (int # string) list
```

Functions may be *curried*, *i.e.* take their arguments 'one at a time' instead of as a tuple. This is illustrated with the function `curried_zip` below:

```
#let curried_zip l1 l2 = zip(l1,l2);;
curried_zip = - : (* list -> ** list -> (* # **) list)

#let zip_num = curried_zip [0;1;2;3;4;5;6;7;8;9];;
zip_num = - : (* list -> (int # *) list)

#zip_num ['a';'b';'c'];;
[0,'a'; 1,'b'; 2,'c'] : (int # string) list
```

Curried function are useful because they can be 'partially applied' as illustrated above by the partial application of `curried_zip` to `[0;1;2;3;4;5;6;7;8;9]` which results in the function `zip_num`.

The evaluation of an expression either *succeeds* or *fails*. In the former case, the evaluation returns a value; in the latter case the evaluation is aborted and a failure string (usually the name of the function that caused the failure) is passed to whatever invoked the evaluation. This context can either propagate the failure (this is the default) or it can *trap* it. These two possibilities are illustrated below. A failure trap is an expression of the form $e_1?e_2$. An expression of this form is evaluated by first evaluating e_1. If the evaluation succeeds (*i.e.* doesn't fail) then the value of $e_1?e_2$ is the value of e_1. If the evaluation of e_1 fails, then the value of $e_1?e_2$ is obtained by evaluating e_2.

```
hd[];;
evaluation failed     hd

#hd[] ? `hd applied to empty list`;;
`hd applied to empty list` : string
```

Terms of the HOL logic are represented in ML by a type called **term**. For example, the expression "x /\ y ==> z" evaluates in ML to a term representing $x \wedge y \supset z$. Anything between a pair of quotes is parsed as a logical term. The quotation mechanism is described in Section 3.3. Terms can be manipulated by various built-in ML functions. For example, the ML function

```
dest_imp : term -> term # term
```

splits an implication into a pair of terms consisting of the antecedant and consequent, and the ML function

```
dest_conj : term -> term # term
```

splits a conjunction into its conjuncts.

```
#"x /\ y ==> z";;
"x /\ y ==> z" : term

#dest_imp it;;
"x /\ y","z" : (term # term)

#dest_conj(fst it);;
"x","y" : (term # term)
```

Terms of the HOL logic are quite similar to ML expressions and this can at first be confusing. Indeed, terms of the logic have types similar to ML expressions. For example, "(1,2)" is an ML expression with ML type **term**. The HOL type of this term is **num#num**. By contrast, the ML expression ("1","2") has type **term#term**.

The types of HOL terms form an ML type called **type**. Expressions of the form ": ⋯ " evaluate to logical types. The built-in function **type_of** has ML type **term->type** and returns the logical type of a term.

```
#"(1,2)";;
"1,2" : term

#type_of it;;
":num # num" : type

#("1","2");;
"1","2" : (term # term)

#type_of(fst it);;
":num" : type
```

To try to minimise confusion between the logical types of HOL terms and the ML types of ML expressions, we will call the former *object language types* and the latter *meta-language types*. For example, `"(1,T)"` is an ML expression that has meta-language type `term` and evaluates to a term with object language type `":num#bool"`.

```
#"(1,T)";;
"1,T" : term

#type_of it;;
":num # bool" : type
```

HOL terms can be input using explicit *quotation*, as above, or they can be constructed using ML constructor functions. The function `mk_var` constructs a variable from a string and a type. In the example below, three variables of type `bool` are constructed. These are used later.

```
#let x = mk_var(`x`,":bool")
#and y = mk_var(`y`,":bool")
#and z = mk_var(`z`,":bool");;
x = "x" : term
y = "y" : term
z = "z" : term
```

The constructors `mk_conj` and `mk_imp` construct conjunctions and implications respectively.

```
#let t = mk_imp(mk_conj(x,y),z);;
t = "x /\ y ==> z" : term
```

Theorems are represented in HOL by values of type `thm`. The only way to create theorems is by *proof*. For a logician, a formal proof is a sequence each of whose elements is either an *axiom* or follows from earlier members of the sequence by a *rule of inference*. In HOL (following LCF) a proof is a sequence in just that sense. There are five axioms of the HOL logic and eight primitive inference rules. The axioms are bound to ML names. For example, the Law of Excluded Middle is bound to the ML name `BOOL_CASES_AX`:

```
#BOOL_CASES_AX;;
|- !t. (t = T) \/ (t = F)
```

Theorems are printed with a preceding turnstile `|-` as illustrated above; the symbol "`!`" is the universal quantifier "∀". Rules of inference are ML functions that return values of type `thm`. For example, the rule of *specialization* (or ∀-elimination) is an ML function, `SPEC`, which takes as arguments a term "a" and a theorem `|-` `!`$x.t[x]$ and returns the theorem `|-` $t[a]$, the result of substituting a for x in $t[x]$.

```
#SPEC;;
- : (term -> thm -> thm)

#SPEC "1=2" BOOL_CASES_AX;;
|- ((1 = 2) = T) \/ ((1 = 2) = F)
```

A proof in the HOL system is constructed by repeatedly applying inference rules to axioms or to previously proved theorems.

Since proofs may consist of millions of steps, it is necessary to provide tools to make proof construction easier for the user. The proof generating tools in the HOL system are just those of LCF, and are described in Section 7.1.

Theorems have *assumptions* and a *conclusion*. The inference rule `ASSUME` generates theorems of the form t `|-` t. The ML printer prints each assumption as a dot. The function `dest_thm` decomposes a theorem into a pair consisting of list of assumptions and the conclusion. The ML type goal abbreviates `(term)list # term`, this is motivated in Section 7.1.

```
#let th1 = ASSUME "t1==>t2";;
th1 = . |- t1 ==> t2

#dest_thm th1;;
["t1 ==> t2"],"t1 ==> t2" : goal
```

The inference rule UNDISCH moves the antecedant of an implication to the assumptions.

```
#let th2 = UNDISCH th1;;
th2 = .. |- t2

#dest_thm th2;;
["t1 ==> t2"; "t1"],"t2" : goal
```

The rule DISCH takes a term and a theorem and 'discharges' the given term from the assumptions of the theorem. The functions hyp and concl select the list of hypotheses and the conclusion of a theorem, respectively. They just return the components of the pair computed by dest_thm.

```
#DISCH "t1==>t2" (DISCH "t1" th2);;
|- (t1 ==> t2) ==> t1 ==> t2

#let th3 = DISCH "t1==>t2" (DISCH "t1" th2);;
th3 = |- (t1 ==> t2) ==> t1 ==> t2

#hyp th3;;
[] : term list

#concl th3;;
"(t1 ==> t2) ==> t1 ==> t2" : term
```

In the sections that follow we systematically describe HOL. We then conclude with another session illustrating the various theorem-proving tools in action.

3 Syntax of the HOL logic

In this section, the structure of HOL terms and types is explained. The particular sets of types, constants and axioms that are available to a user of the

HOL system is specified by the *theory* he or she is working in. Each theory $\mathcal{T}$ specifies sets $\mathsf{Tyops}_{\mathcal{T}}$ and $\mathsf{Consts}_{\mathcal{T}}$ of type operators and constants. These sets then determine the sets $\mathsf{Types}_{\mathcal{T}}$ and $\mathsf{Terms}_{\mathcal{T}}$ of types and terms. How to set up theories is described in Section 5.

We start by giving fairly abstract definitions of the sets $\mathsf{Types}_{\mathcal{T}}$ and $\mathsf{Terms}_{\mathcal{T}}$. Various constructor and destructor functions are then described. Finally, we explain the quotation mechanism built-in to the ML parser. This mechanism includes a type inference algorithm that enables the user to input terms without having to write an excessive amount of explicit type information.

In what follows, a *name* is a sequence of letters, digits or primes (') beginning with a letter. The set of names is denoted by Names. For example, `x`, `fred23`, `fred23'` and `fred23''` are all members of Names.

3.1 HOL Types

Object language types are expressions that denote sets. Following tradition, we use σ, σ', σ'', ... , σ_1, σ_2, *etc.* to stand for arbitrary types. HOL terms will be defined in such a way that all well-formed terms are type-consistent. Before describing the terms of the HOL logic we must first specify its types.

There are four kinds of object language types. These can be described informally as follows.

1. **Type variables:** These are sequences of asterisks, optionally followed by sequences of digits. The set of type variables is denote by Tyvars; for example, `*`, `**`, `***`, `*1`, `***24` are all members of Tyvars. Type variables provide the system with a limited amount of *polymorphism* (this is explained later). Small Greek letters, possibly with subscripts or primes, are used to stand for type variables.

2. **Type constants:** These are names like `bool`, `num` and `tok`. They denote fixed sets of values.

3. **Function types:** If σ_1 and σ_2 are types, then σ_1`->`σ_2 is the function type with *domain* σ_1 and *range* σ_2; it denotes the set of functions from the set denoted by its domain to the set denoted by its range.

4. **Compound types:** These have the form `(`σ_1`,` ... `,`σ_n`)`*op*, where the types σ_1, ... , σ_n are the argument types and *op* is a *type operator* of arity n. Type operators must be names; they denote operations for constructing sets. The type `(`σ_1`,` ... `,`σ_n`)`*op* denotes the set resulting from applying the operation denoted by *op* to the sets denoted by σ_1, ... , σ_n. For example, `list` is a type operator with arity 1. It denotes the operation of forming all finite lists of elements from a given set. Another example is the type operator `prod` of arity 2 which denotes the cartesian product operation.

Although these four kinds of types are logically distinct, the HOL system (following LCF) represents constant types as compound types built with 0-ary type operators and function types as compound types built with a 2-ary type operator called `fun`. Thus the constant type `num` is represented as the compound type `()num` and σ_1`->`σ_2 is represented as `(`σ_1`,`σ_2`)fun`.

In general, compound types must be written in the postfixed form described above, but there are two exceptions (in addition to `->`).

- *Cartesian product* types of the form `(`σ_1`,`σ_2`)prod` can be written as σ_1`#`σ_2.
- *Disjoint union* types of the form `(`σ_1`,`σ_2`)sum` can be written as σ_1`+`σ_2.

Products and unions are not primitive, but because they are so useful, the HOL parser treats them specially. These types are explained in Section 5.5 and Section 5.6 respectively.

It is useful to describe the set of types of a theory a bit more formally. Each theory $\mathcal{T}$ determines a set $\mathsf{Tyops}_{\mathcal{T}}$ of type operators. A type operator is a pair (op, n) where op is the name of the operator and n is its arity. The set $\mathsf{Types}_{\mathcal{T}}$ of types of the theory $\mathcal{T}$ is the smallest set such that:

- $\mathsf{Tyvars} \subseteq \mathsf{Types}_{\mathcal{T}}$.
- If $\sigma_1 \in \mathsf{Types}_{\mathcal{T}}$ and $\sigma_2 \in \mathsf{Types}_{\mathcal{T}}$ then σ_1`->`$\sigma_2 \in \mathsf{Types}_{\mathcal{T}}$.
- If $\sigma_i \in \mathsf{Types}_{\mathcal{T}}$ (for all i between 1 and n) and $(op, n) \in \mathsf{Tyops}_{\mathcal{T}}$ then `(`σ_1`, ... ,`σ_n`)`$op \in \mathsf{Types}_{\mathcal{T}}$.

Types containing type variables are called *polymorphic*; others are *monomorphic*.

An *instance* σ' of a type σ is obtained by replacing all occurrences of a type variable in σ by a type. The only instance of a monomorphic type is itself. Note that $\mathsf{Types}_{\mathcal{T}}$ is closed under the operation of taking instances.

3.2 HOL Terms

A theory $\mathcal{T}$ specifies a set $\mathsf{Consts}_{\mathcal{T}}$ of *constants*. Each constant is a pair (c, σ) where c is a name and σ is a type of $\mathcal{T}$ (*i.e.* $\sigma \in \mathsf{Types}_{\mathcal{T}}$); σ is called the *generic type* of c.

Distinct constants of a theory cannot have the same name; (*i.e.* if (c, σ) and (c, σ') are both members of $\mathsf{Consts}_{\mathcal{T}}$ then $\sigma = \sigma'$).

The set $\mathsf{Terms}_{\mathcal{T}}$ of *terms* of the theory $\mathcal{T}$ is the smallest set such that:

1. If x is a name which is not the name of a constant in $\mathcal{T}$ (*i.e.* $\mathsf{Consts}_{\mathcal{T}}$ does not contain a pair whose first component is x) and $\sigma \in \mathsf{Types}_{\mathcal{T}}$, then $(x, \sigma) \in \mathsf{Terms}_{\mathcal{T}}$. Terms formed in this way are called *variables*.
2. If $(c, \sigma) \in \mathsf{Consts}_{\mathcal{T}}$ and $\sigma' \in \mathsf{Types}_{\mathcal{T}}$ is an instance of σ, then $(c, \sigma') \in \mathsf{Terms}_{\mathcal{T}}$. Terms formed in this way are called *constants*.
3. If $(t, \sigma'$`->`$\sigma) \in \mathsf{Terms}_{\mathcal{T}}$ and $(t', \sigma') \in \mathsf{Terms}_{\mathcal{T}}$ then $(comb, t_1, t_2, \sigma) \in \mathsf{Terms}_{\mathcal{T}}$. Terms formed in this way are called *combinations* or *function applications*.

4. If $(x, \sigma) \in \mathsf{Terms}_{\mathcal{T}}$ (where x is a name that is not the name of a constant in $\mathcal{T}$) and $(t, \sigma') \in \mathsf{Terms}_{\mathcal{T}}$ and σ->$\sigma' \in \mathsf{Types}_{\mathcal{T}}$ then $(abs, x, t, \sigma$->$\sigma') \in \mathsf{Terms}_{\mathcal{T}}$. Terms formed in this way are called ***abstractions*** or λ-terms (\ is HOL's ASCII approximation to λ).

Members of the sets $\mathsf{Types}_{\mathcal{T}}$ and $\mathsf{Terms}_{\mathcal{T}}$ correspond closely to the internal representations of ML values of type `type` and `term` in the HOL system. The following constructor functions (listed below with their ML types) can be used to input types and terms; they are explained in Section 3.3 below.

```
mk_vartype : string -> type

mk_type    : (string # type list) -> type

mk_var     : (string # type) -> term

mk_const   : (string # type) -> term

mk_comb    : (term # term) -> term

mk_abs     : (term # term) -> term
```

The HOL logic consists of the built-in theories `bool` and `ind` (see Sections 5.3 and 5.7) together with eight rules of inference. In the theory `bool` the type `bool` is introduced (we often name theories after the type which they introduce). Terms of type `bool` are called *formulas*. There are two constant formulas: `T` and `F`; these represent *true* and *false* respectively. It follows from the Law of Excluded Middle (an axiom of HOL) that any formula is either equivalant to `T` or to `F`.

3.3 Quotation

It would be extremely tedious to always have to input types and terms using the constructor functions. The HOL system has a special parser and type-checker that enable them to be input using a fairly standard syntax. The HOL printer also outputs types and terms using this syntax.

For example, the ML expression `":bool->bool"` denotes exactly the same value (of ML type `type`) as

```
mk_type(`fun`,[mk_type(`bool`,[]);mk_type(`bool`,[])])
```

and `"\x.x+1"` can be used instead of

```
mk_abs
 (mk_var(`x`,mk_type(`num`,[])),
  mk_comb
  (mk_comb
   (mk_const
    (`+`,
     mk_type(`fun`,[mk_type(`num`,[]);
                    mk_type(`fun`,[mk_type(`num`,[]);
                                   mk_type(`num`,[])])])),
    mk_var(`x`, mk_type(`num`,[]))),
   mk_const(`1`, mk_type(`num`,[]))))
```

Notice that there is no explicit type information in "\x.x+1". The HOL type-checker knows that (1, num) $\in$ $\mathsf{Consts}_{\mathsf{num}}$ and (+, num->(num->num)) $\in$ $\mathsf{Consts}_{\mathsf{num}}$, *i.e.* that the constants 1 and + have type num and num->(num->num), respectively. From the types of 1 and + the type-checker infers that both occurrences of x in "\x.x+1" could have type num. This is not the only type assignment possible; one could, for example, make the first occurrence of x have type bool and the second one have type num. In that case there would be two *different* variables with name x, namely (x, bool) and (x, num), the second of which is free. In fact, in HOL, the only way to construct a term with this second type assignment would be by using constructors, since the type-checker uses the heuristic that variables with the same name have the same type. The type-checker was designed by Robin Milner. It uses heuristics like the one above to infer a sensible type for all variables occurring in a term. It uses the types of any constants in making this inference. If there are not enough clues, then the system will complain with the message

```
evaluation failed    types indeterminate in quotation
```

To give the system a hint, one can explicitly indicate types by following any subterm by a colon and then a type. For example, "f(x:num):bool" will type-check with f and x getting types num->bool and num respectively. If there are polymorphic constants in a term, there must be enough type information to uniquely identify a type instance for each such constant.

The type-checking algorithm used for the HOL logic differs from that used for ML. For example, the ML expression \x.x will get ML type *->*, but the HOL term "\x.x" will fail to type-check and will result in the message shown above. To get this term to type-check one must indicate explicity the type of the variable x by writing, for example, "\x:*.x". This treatment of types is inherited from LCF.

The table below shows ML expressions for various kinds of type quotations. The expressions in adjacent columns are equivalent.

Types		
Kind of type	ML *quotation*	*Constructor expression*
Type variable	`":*..."`	`mk_vartype(‘*...‘)`
Type constant	`":`op`"`	`mk_type(‘`op`‘,[])`
Function type	`":`σ_1`->`σ_2`"`	`mk_type(‘fun‘, [":`σ_1`";":`σ_2`"])`
Compound type	`":(`σ_1`, ... ,` σ_n`)`op`"`	`mk_type(‘`op`‘, [":`σ_1`"; ... ;":`σ_n`"])`

Equivalent ways of inputting the four primitive kinds of term are shown in the next table.

Primitive terms		
Kind of term	ML *quotation*	*Constructor expression*
Variable	`"`var`:`σ`"`	`mk_var(‘`var`‘,":`σ`")`
Constant	`"`$const$`:`σ`"`	`mk_const(‘`$const$`‘,":`σ`")`
Combination	`"`t_1 t_2`"`	`mk_comb("`t_1`","`t_2`")`
Abstraction	`"\`x`.`t`"`	`mk_abs("`x`","`t`")`

The HOL quotation mechanism can translate various standard logical notations into primitive terms. For example, if `+` has been declared an infix (as explained in Section 5), then `"x+1"` is translated to `"$+ 1 2"`. The escape character `$` suppresses the infix behaviour of `+` and prevents the quotation parser getting confused. In general, `$` can be used to suppress any special syntactic behaviour a constant name might have; this is illustrated in the table below, in which the terms in the column headed "ML *quotation*" are translated by the quotation parser to the corresponding terms in the column headed "*Primitive term*". Conversely, the terms in the latter column are always printed in the form shown in the former one. The ML constructor expressions in the rightmost column evaluate to the same values (of type `term`) as the other quotations in the same row.

Non-primitive terms			
Kind of term	ML *quotation*	*Primitive term*	*Constructor expression*
Negation	"~t"	"$~ t"	mk_neg("t")
Disjunction	"t_1\/t_2"	"$\/ t_1 t_2"	mk_disj("t_1","t_2")
Conjunction	"t_1/\t_2"	"$/\ t_1 t_2"	mk_conj("t_1","t_2")
Implication	"t_1==>t_2"	"$==> t_1 t_2"	mk_imp("t_1","t_2")
Equality	"t_1=t_2"	"$= t_1 t_2"	mk_eq("t_1","t_2")
∀-quantification	"!x.t"	"$!(\\$x$.$t$)"	mk_forall("x","t")
∃-quantification	"?x.t"	"$?(\\$x$.$t$)"	mk_exists("x","t")
ε-term	"@x.t"	"$@(\\$x$.$t$)"	mk_select("x","t")
Conditional	"(t=>$t_1 \mid t_2$)"	"COND t t_1 t_2"	mk_cond("t","t_1","t_2")
let-expression	"let x=t_1 in t_2"	"LET(\\x.t_2)t_1"	mk_let("x","t_1","t_2")

The constants COND and LET are explained in Section 5.3. The constants \/, /\, ==> and = are examples of *infixes*. If c is declared to be an infix, then the HOL parser will translate "t_1 c t_2" to "\$$c$ t_1 t_2". The constants !, ? and @ are examples of *binders* (see Section 5.3 also). If c is declared to be a binder, then the HOL parser will translate "c x.t" to the combination "\$$c$(\\$x$.$t$)".

In addition to the kinds of terms in the tables above, the parser also supports the following syntactic abbreviations.

Syntactic abbreviations		
Abbreviated term	*Meaning*	*Constructor expression*
"$t\ t_1 \cdots t_n$"	"$(\cdots(t\ t_1)\cdots t_n)$"	list_mk_comb("t",["t_1"; ... ;"t_n"])
"\\$x_1 \cdots x_n$.$t$"	"\\$x_1$. ⋯ \\$x_n$.$t$"	list_mk_abs(["x_1"; ... ;"x_n"],"t")
"!$x_1 \cdots x_n$.t"	"!x_1. ⋯ !x_n.t"	list_mk_forall(["x_1"; ... ;"x_n"],"t")
"?$x_1 \cdots x_n$.t"	"?x_1. ⋯ ?x_n.t"	list_mk_exists(["x_1"; ... ;"x_n"],"t")

The parser will also convert "\\(x_1,x_2).t" to "UNCURRY(\\x_1 x_2.t)" (the constant UNCURRY is described in Section 5.5). This transformation is done recursively so that, for example,

"\\(x_1,x_2,x_3).t"

is converted to

"UNCURRY(\\$x_1$.UNCURRY(\\$x_2$ x_3.t))"

More generally, if b is a binder, then "b(x_1,x_2).t" is parsed as "\$b(\\(x_1,x_2).t)". For example, "!(x,y).x>y" parses to "$!(UNCURRY(\x.\y.$> x y))" (> is an infixed constant of the theory num meaning “is greater than”).

3.4 Antiquotation

Within a quotation, expressions of the form `^(`*t*`)` (where *t* is an ML expression of type `term` or `type`) are called *antiquotations*. An antiquotation `^(`*t*`)` evaluates to the ML value of *t*. For example, `"x \/ ^(mk_conj("y:bool","z:bool"))"` evaluates to the same term as `"x \/ (y /\ z)"`. The most common occurrence of antiquotation is when the term *t* is just an ML variable *x*; in this case `^(`*x*`)` can be abbreviated by `^`*x*. The following session illustrates antiquotion.

```
#let y = "x+1";;
y = "x + 1" : term

#let z = "y = ^y";;
z = "y = x + 1" : term

#"!x.?y.^z";;
"!x. ?y. y = x + 1" : term
```

4 Sequents, Theorems and Proof

The HOL system supports proof by ***natural deduction***. Assertions in the HOL logic are not individual formulas (*i.e.* boolean terms), but are ***sequents*** of the form (Γ, t), where Γ is a set of formulas called the ***assumptions*** and t a formula called the ***conclusion***. A sequent(Γ, t) asserts that if all the formulas in Γ are true, then so is t. Using sequents, one can design a deductive system in which proofs are more 'natural' than in other (*e.g.* Hilbert-style) systems.

Sequents are represented in ML by pairs with ML type `(term list)#term`; the first component represents the assumptions and the second component the conclusion. For example, `(["x>y";"y>z"],"x>z")` is an ML expression representing the sequent with assumptions `"x>y"` and `"y>z"` and conclusion `"x>z"`.

A *theorem* is a sequent that has a *proof*. More precisely, a theorem is a sequent that is either an *axiom*, or follows from other theorems by a *rule of inference*.

To guarantee that the only way to get theorems is by proof, the HOL system (following LCF) distinguishes sequents from theorems by having a separate ML type `thm`[4]. There are five initial theorems corresponding to the five axioms of the HOL logic. The only way to generate other objects with ML type `thm` is by using one of the eight primitive inference rules of the logic (see Section 5.4). ML's type discipline ensures that only functions that represent inference rules are applicable to theorems [5].

---

[4]In ML jargon, `thm` is an abstract type whose representing type is `(term list)#term`.

Interesting theorems require large numbers of applications of the primitive inference rules for their proof. For example, a recent verification of the design of a simple microprocessor took over a million inference steps [3]. The HOL system provides tools to help the user generate such proofs. One such tool is a *derived inference rule*. This is an ML function which invokes sequences of applications of the eight primitive rules. There are about a hundred derived inference rules predefined in the system. A brief discussion of derived rules can be found in Section 6. Proofs can be generated in a goal-oriented style using *tactics*; these were invented (for LCF) by Robin Milner and are explained in Section 7.1. The goal-directed proof style starts with a formula (representing the goal one wants to prove) and then reduces this to sub-goals, sub-sub-goals *etc.* until the problem is decomposed into trivial goals (*e.g.* instances of axioms). This is in contrast to *forward proof* in which one works forward from the axioms, using the rules of inference, until the desired theorem is deduced.

4.1 Axioms and definitions

Each theory $\mathcal{T}$ has a set $\mathsf{Axioms}_{\mathcal{T}} \subseteq \mathsf{Terms}_{\mathcal{T}}$ of terms of type `bool`, called axioms. These axioms usually specify the meaning of the constants of $\mathcal{T}$; however, the HOL system enables the user to assert *any* formula as an axiom.

A *definition* is an axiom of the form c=t, where c is a constant and t is a closed term[5]. A constant c is *defined in* $\mathcal{T}$ if there is exactly one axiom in $\mathcal{T}$ containing c and that axiom is a definition. If all the axioms of $\mathcal{T}$ are definitions, then $\mathcal{T}$ is a *definitional* theory. Definitional theories have the important property that they cannot introduce any new inconsistency that wasn't already present; they are what logicians call *conservative extensions*.

Axioms of the form $f\ x_1 \ldots x_n\ = t$, where the free variables of t are included among $x_1, \ldots, x_n$, are also regarded as definitions. This is because the formula $f\ x_1 \ldots x_n = t$ is equivalent to the ordinary definition $f = \backslash x_1 \ldots x_n . t$. Definitions are also allowed to have the form $f(x_1,\ldots,x_n) = t$.

Another kind of conservative extension is a *type definition*.

4.2 Type definitions

There does not seem to be any standard notion of type definition in the literature on higher-order logic. The mechanism described here is experimental; it was suggested to me by Mike Fourman and may change soon.

Recall that a constant type denotes a set and an n-ary type operator denotes an operation for combining n sets to form a set. The intuitive idea behind type definitions in the HOL logic is as follows:

- A constant type is defined by specifying it to be in one-to-one correspondence with a non-empty subset of an existing type.

---

[5] A term is *closed* if it contains no free variables.

- An n-ary type operator *op* is defined by specifying $(\alpha_1, \ldots, \alpha_n)op$ to be in one-to-one correspondence with a non-empty subset of an existing type constructed from $\alpha_1, \ldots, \alpha_n$.

All types in the HOL logic must denote non-empty sets. This is because the term `@x:`σ`.T` denotes a member of the set denoted by σ (for arbitrary σ). The binder `@` is explained in Section 5.3.

The HOL system has a single mechanism for defining types. The idea behind this mechanism is to define an n-ary type operator[6] *op* by specifying the set denoted by $(\alpha_1, \ldots, \alpha_n)op$ for all denotations of the type variables $\alpha_1, \ldots, \alpha_n$. This set is specified by giving a (polymorphic) term denoting its characteristic function[7].

This idea can perhaps be made clearer with a concrete example. Suppose we want to define a unary type operator `iso` such that the type denoted by `":(*)iso"` is the subset of the set denoted by `":*->*"` consisting of those functions which have an inverse. The characteristic function of the set of functions with inverses is denoted by the term `"\f:*->*.?g.!x. g(f x) = x"`. To ensure that only non-empty sets are used to define types, the HOL system requires the user to supply a theorem asserting non-emptiness when making a type definition. If t_{op} is the term representing the characteristic function defining *op*, then the user is required to supply a theorem of the form `|- ?`x`.`t_{op} x. In the case of our example, *op* is `iso` and $t_{\texttt{iso}}$ is `"\f:*->*.?g.!x. g(f x) = x"`, so the required theorem is:

```
|- ?f. (\f:*->*.?g.!x. g(f x) = x) f
```

The ML function

```
new_type_definition : (string # term # thm) -> thm
```

is used to make a new type definition. Suppose *op* is a name, t_{op} is a term whose object language type has the form σ`->bool` (where σ contains n distinct type variables $\alpha_1, \ldots, \alpha_n$) and *th* is the theorem `|- ?`x`.`t_{op} x. Then

`new_type_definition(`op`,`t_{op}`,`th`)`

declares *op* to be a new n-ary type-operator such that $(\alpha_1, \ldots, \alpha_n)op$ denotes the set with characteristic function t_{op}. This set is a subset of the set denoted by σ. The theorem *th* ensures that this subset is non-empty for all values of the type variables $\alpha_1, \ldots, \alpha_n$. The type σ is called the *representing type* of *op*. The result of the type definition is an axiom asserting the existence of a one-to-one function from $(\alpha_1, \ldots, \alpha_n)op$ onto the subset of σ characterized by t_{op}.

Suppose we have proved `|- ?f. (\f:*->*.?g.!x. g(f x) = x) f` and bound this theorem to the ML name `th`. Then the new type `iso` is defined by executing the ML expression

---

[6] Constant types are defined by taking $n = 0$.

[7] The *characteristic function* of a set is a boolean-valued function that maps an element to *true* if and only if the element is a member of the set.

```
new_type_definition(`iso`, "(\f:*->*.?g.!x. g(f x) = x)", th)
```

The HOL system then automatically declares a new 1-ary type operator `iso`, a new constant `REP_iso` (whose name is generated by prefixing the name of the type operator being defined with the string `` `REP_` ``), and finally, HOL automatically generates the axiom

```
|- ONE_ONE REP_iso /\
   (!f. (\f. ?g. !x. g(f x) = x)f = (?f'. f = REP_iso f'))
```

This axiom has the form

$$\texttt{|- ONE\_ONE REP\_}op\ \texttt{/\textbackslash}\ \texttt{!f.}t_{op}\ \texttt{f = (?f'. f = REP\_}op\ \texttt{f')}$$

where the function `ONE_ONE` is defined by

```
|- ONE_ONE f = (!x1 x2. (f x1 = f x2) ==> (x1 = x2))
```

The constant `ONE_ONE` is introduced in the theory `bool` (see Section 5.3) and so is built into the HOL system.

5 Theories

The set of types, constants and axioms that are available in HOL depends on the *theory* in which one is working (see Section 3). Theories form a hierarchy in which a theory $\mathcal{T}_1$ can be a *parent* of another theory $\mathcal{T}_2$, meaning that the types, constants and axioms of $\mathcal{T}_1$ will be available in $\mathcal{T}_2$. If $\mathcal{T}_1$ is a parent of $\mathcal{T}_2$, then $\mathcal{T}_2$ is an *immediate descendant* of $\mathcal{T}_1$. $\mathcal{T}_2$ is a *descendant* of $\mathcal{T}_1$ if the two theories are in the transitive closure of the immediate descendant relation. In this case we also say that $\mathcal{T}_1$ is an *ancestor* of $\mathcal{T}_2$. When first entering the HOL system one enters a theory called `HOL`; this is a descendant of several theories including `bool` and theories of numbers, lists and pairs. In the subsections below, we describe the ML functions provided for manipulating theories and the various theories built in to the HOL system. First we describe a bit more abstractly exactly what a theory is.

5.1 Abstract Definition of a Theory

A theory $\mathcal{T}$ is characterized by a 4-tuple

$$(\mathsf{Parents}_{\mathcal{T}},\ \mathsf{Tyops}_{\mathcal{T}},\ \mathsf{Consts}_{\mathcal{T}},\ \mathsf{Axioms}_{\mathcal{T}})$$

where:

- $\mathsf{Parents}_{\mathcal{T}}$ is a (finite) set of all the parent theories of $\mathcal{T}$.
- $\mathsf{Tyops}_{\mathcal{T}}$ is a set of type operators (see Section 4.2).
- $\mathsf{Consts}_{\mathcal{T}}$ is a set of constants (see Section 3.2).
- $\mathsf{Axioms}_{\mathcal{T}}$ is a set of theorems which are the axioms of $\mathcal{T}$ (see Section 4.1).

5.2 ML Functions for Manipulating Theories

In the HOL system, the four components of a theory are stored on disk in files with names of the form *name*`.th`, where *name* is the name given to the theory (with `new_theory`) when it is created. Various additional pieces of information are stored in the *name*`.th` files, including the parsing status of the contants (*i.e.* whether they are infixes or binders), which axioms are definitions (see Section 4.1) and the theorems that have been proved and saved by the user.

The ML functions for creating theories are listed below.

```
new_theory             : string -> void
new_parent             : string -> void
new_type               : int -> string -> void
new_constant           : (string # type) -> void
new_infix              : (string # type) -> void
new_binder             : (string # type) -> void
new_axiom              : (string # term) -> thm
new_definition         : (string # term) -> thm
new_infix_definition   : (string # term) -> thm
new_binder_definition  : (string # term) -> thm
new_type_definition    : (string # term # thm) -> thm
```

The effect of these functions should be reasonably clear from their names and types. The first argument of `new_type` is the arity of the type operator being declared; the second argument is its name. The arguments of type `string` to `new_axiom`, `new_definition` *etc.* are the names of the corresponding axioms and definitions. These names are used when accessing theories with the functions `axiom`, `definition`, *etc.*, described below. The various functions for setting up theories are illustrated in the example session in Section 8.

A theory with no descendants can be extended by adding new parents, types, constants, axioms and definitions. Theories that are already the parents of other theories cannot be extended because it would be messy (though not impossible) to implement the necessary checks to ensure that added types, constants *etc.* did not invalidate declarations in the descendant theories. When one is creating a new theory or extending an existing theory one is said to be in *draft mode*. When one is working in a completed theory, one is said to be in *working mode*; in this mode the functions with prefix "`new_`" listed above are not available. In draft mode there is the danger that one is not prevented from asserting inconsistent axioms such as `|- T=F`. See the discussion of definitional axioms in Section 4.1.

The functions for entering an already existing theory in either draft mode or working mode are, respectively:

```
extend_theory : string -> void
load_theory   : string -> void
```

To finish a session and write all new declarations to the theory file there is the function:

```
close_theory : void -> void
```

There are various functions for loading the contents of theory files:

```
parents     : string -> string list
types       : string -> (int # string) list
constants   : string -> term list
infixes     : string -> term list
binders     : string -> string list
axioms      : string -> (string # thm) list
definitions : string -> (string # thm) list
theorems    : string -> (string # thm) list
```

Once a theorem has been proved, it can be saved with the function

```
save_thm : string # thm -> thm
```

Evaluating `save_thm('Th1',th)` will save the theorem `th` with name `Th1` on the current theory. Individual axioms, definitions and theorems can be read (from ancestor theories) using the following ML functions:

```
axiom      : string -> string -> thm
definition : string -> string -> thm
theorem    : string -> string -> thm
```

Theories can be printed using the function `print_theory`, which takes a theory name (a value of type `string`) and then prints out the named theory in a readable format.

In the remaining subsections of this section we describe all the theories built into the HOL system[8]. The theory `HOL` has all of these as parents.

5.3 The Theory `bool`

The theory `bool` introduces the type `bool` and contains four of the five axioms for higher-order logic (the fifth axiom is in the theory `ind`). These axioms, together with the rules of inference described in Section 5.4, constitute the core of the HOL logic. Because of the way the HOL system evolved from LCF[9], the

---

[8]The actual theory structure in the HOL system is slightly different from that described in this paper. Specifically, the types and definitions for pairs are included in the theory `bool` (rather than being in a separate descendant theory called `prod`). It is hoped that future versions of the system will correct this anomaly.

[9]In order to simplify the porting of the LCF theorem-proving tools to the HOL system, the HOL logic was made as like PPLAMBDA (the logic built-in to LCF) as possible.

particular axiomatization of higher-order logic it uses is superficially different from the classical axiomatization due to Church [2]. The biggest difference is that in Church's formulation, type variables are in the meta-language, whereas in the HOL logic they are part of the object language.

There are three primitive constants in the theory `bool`: `=` (equality, an infix), `==>` (implication, an infix) and `@` (choice, a binder). Equality and implication are standard predicate calculus notions, but choice is more exotic: if t is a term having type σ`->bool`, then `@x.`t `x` (or, equivalently, `$@`t) denotes *some* member of the set whose characteristic function is t. If the set is empty, then `@x.`t `x` denotes an arbitrary member of the set denoted by σ. The constant `@` is a higher-order version of Hilbert's ε-operator; it is related to the constant ι in Church's formulation of higher-order logic. For more details, see Leisenring's book [12] and Church's original paper [2].

The logical constants `T` (truth), `F` (falsity), `~` (negation), `/\` (conjunction), `\/` (disjunction), `!` (universal quantification) and `?` (existential quantification) can all be defined in terms of equality, implication and choice. The definitions listed below are fairly standard; each one is preceded by its ML name. (Later definitions sometimes use earlier ones.)

```
T_DEF           |- T   = ((\x:*. x)=(\x. x))

FORALL_DEF      |- $!  = \P:*->bool. P=(\x. T)

EXISTS_DEF      |- $?  = \P:*->bool. P($@ P)

AND_DEF         |- $/\ = \t1 t2. !t. (t1 ==> t2 ==> t) ==> t

OR_DEF          |- $\/ = \t1 t2. !t. (t1 ==> t) ==> (t2 ==> t) ==> t

F_DEF           |- F   = !t. t

NOT_DEF         |- $~  = \t. t ==> F
```

There are four axioms in the theory `bool`:

```
BOOL_CASES_AX    |- !t. (t = T) \/ (t = F)

IMP_ANTISYM_AX   |- !t1 t2. (t1 ==> t2) ==> (t2 ==> t1) ==> (t1 = t2)

ETA_AX           |- !t. (\x. t x) = t

SELECT_AX        |- !P:*->bool x. P x ==> P($@ P)
```

The fifth and last axiom of the HOL logic is the Axiom of Infinity; this is in the theory `ind` described in Section 5.7. The theory `bool` supplies the definitions of a number of useful constants:

```
LET_DEF      |- LET       = \f x. f x

COND_DEF     |- COND      = \t t1 t2.@x.((t=T)==>(x=t1))/\((t=F)==>(x=t2))

ONE_ONE_DEF |- ONE_ONE f = (!x1 x2. (f x1 = f x2) ==> (x1 = x2))

ONTO_DEF     |- ONTO f    = (!y. ?x. y = f x)
```

The constant `LET` is used in representing terms containing local variable bindings (*i.e.* `let`-terms, as in Section 3.3). The constant `COND` is used in representing conditionals. The constant `ONE_ONE` is used in type definitions (see Section 3.3). The constant `ONTO` is used in stating the Axiom of Infinity (see Section 5.7).

5.4 Primitive Rules of Inference of the HOL Logic

There are eight primitive rules of inference of the HOL logic. We will specify these using standard natural deduction notation. The metavariables t, t_1, t_2, *etc.* stand for arbitrary terms. The theorems above the horizontal line are called the *hypotheses* of the rule and the theorem below the line is called the *result*. Each rule asserts that its result can be deduced from its hypotheses, provided any restrictions (listed after the bullets) hold.

The first three rules have no hypotheses; their results can always be deduced. The square brackets contain the ML names of the rules followed by their ML types.

5.4.1 Assumption introduction

[`ASSUME : term -> thm`]

$$\frac{}{t \;|\!-\; t}$$

`ASSUME "`t`"` evaluates to `|-` t.

5.4.2 Reflexivity

[`REFL : term -> thm`]

$$\frac{}{|\!-\; t = t}$$

`REFL "`t`"` evaluates to `|-` t `=` t.

5.4.3 Beta-conversion

[`BETA_CONV : term -> thm`]

$$\frac{}{\texttt{|-}\ (\backslash x.t_1)t_2\ \texttt{=}\ t_1[t_2/x]}$$

- where $t_1[t_2/x]$ denotes the result of substituting t_2 for x in t_1, with the restriction that no free variables in t_2 become bound after substitution into t_1.

`BETA_CONV` "$(\backslash x.t_1)t_2$" evaluates to the theorem $\texttt{|-}\ (\backslash x.t_1)t_2\ \texttt{=}\ t_1[t_2/x]$.

5.4.4 Substitution

[`SUBST : (thm#term)list -> term -> thm -> thm`]

$$\frac{\Gamma_1\ \texttt{|-}\ t_1\texttt{=}t'_1 \quad \cdots \quad \Gamma_n\ \texttt{|-}\ t_n\texttt{=}t'_n \quad \Gamma\ \texttt{|-}\ t[t_1,\ldots,t_n/x_1,\ldots,x_n]}{\Gamma_1\cup\cdots\cup\Gamma_n\cup\Gamma\ \texttt{|-}\ t[t'_1,\ldots,t'_n]}$$

- where $t[t_1,\ldots,t_n]$ denotes a term t with some free occurrences of the terms $t_1, \ldots, t_n$ singled out and $t[t'_1,\ldots,t'_n]$ denotes the result of simultaneously replacing each such occurrences of t_i by t'_i (for $1 \leq i \leq n$), with the restriction that the context $t[\,]$ must not bind any variable occurring free in either t_i or t'_i (for $1 \leq i \leq n$).
- $\cup$ denotes set union.

The first argument to `SUBST` is a list $[(\texttt{|-}t_1\texttt{=}t'_1,\ x_1);\ldots(\texttt{|-}t_n\texttt{=}\ t'_n,\ x_n)]$. The second argument is a template term $t[x_1,\ldots,x_n]$ in which occurrences of the variable x_i (where $1 \leq i \leq n$) are used to mark the places where substitutions with $\texttt{|-}\ t_i\texttt{=}t'_i$ are to be done.

5.4.5 Abstraction

[`ABS : term -> thm -> thm`]

$$\frac{\Gamma\ \texttt{|-}\ t_1\ \texttt{=}\ t_2}{\Gamma\ \texttt{|-}\ (\backslash x.t_1)\ \texttt{=}\ (\backslash x.t_2)}$$

- where x is not free in Γ.

If `th` is an ML name for the theorem $\Gamma\ \texttt{|-}\ t_1\texttt{=}t_2$, then `ABS` "$x$" `th` returns the theorem $\Gamma\ \texttt{|-}\ (\backslash x.t_1)\ \texttt{=}\ (\backslash x.t_2)$.

5.4.6 Type instantiation

[`INST_TYPE : (type#type) list -> thm -> thm)`]

$$\frac{\Gamma \ \texttt{|-}\ t}{\Gamma \ \texttt{|-}\ t[\sigma_1, \ \dots\ ,\sigma_n/\alpha_1, \ \dots\ ,\alpha_n]}$$

- $t[\sigma_1, \ \dots\ ,\sigma_n/\alpha_1, \ \dots\ ,\alpha_n]$ denotes the result of substituting (in parallel) the types $\sigma_1, \dots, \sigma_n$ for type variables $\alpha_1, \dots, \alpha_n$ in t, with the restriction that none of $\alpha_1, \dots, \alpha_n$ occur in Γ.

`INST_TYPE[(`σ_1`,`α_1`);...;(`σ_n`,`α_n`)]th` returns the result of instantiating each occurrence of α_i in the theorem `th` to σ_i (for $1 \leq i \leq n$).

5.4.7 Discharging an assumption

[`DISCH : term -> thm -> thm`]

$$\frac{\Gamma \ \texttt{|-}\ t_2}{\Gamma-\{t_1\} \ \texttt{|-}\ t_1 \ \texttt{==>}\ t_2}$$

- $\Gamma-\{t_1\}$ denotes the set obtained by removing t_1 from Γ.

If `th` is the theorem Γ `|-` t_2, then the ML expression `DISCH "`t_1`" th` evaluates to the theorem $\Gamma-\{t_1\}$ `|-` t_1 `==>` t_2. Note that if t_1 is not a member of Γ, then $\Gamma-\{t_1\} = \Gamma$.

5.4.8 Modus Ponens

[`MP : thm -> thm -> thm`]

$$\frac{\Gamma_1 \ \texttt{|-}\ t_1 \ \texttt{==>}\ t_2 \qquad \Gamma_2 \ \texttt{|-}\ t_1}{\Gamma_1 \cup \Gamma_2 \ \texttt{|-}\ t_2}$$

`MP` takes two theorems (in the order shown above) and returns the result of applying Modus Ponens.

5.5 The Theory `prod`

The binary type operator `prod` denotes the Cartesian product operator (see Section 3.1). Values of type `(`σ_1`,`σ_2`)prod` are ordered pairs whose first component has type σ_1 and whose second σ_2. The HOL parser recognises σ_1`#`σ_2 as an abbreviation for `(`σ_1`,`σ_2`)prod`.

Cartesian products can be defined by representing a pair `(`t_1`,`t_2`)` as the function `\x y.(x=`t_1`)/\(y=`t_2`)`. The representing type of σ_1`#`σ_2 is thus σ_1`->`σ_2`->bool`. To define pairs this way, we first evaluate the ML expressions:

```
new_definition
 (`MK_PAIR_DEF`, "MK_PAIR(x:*)(y:**) = \a b.(a=x)/\(b=y)")

new_definition
 (`IS_PAIR_DEF`, "IS_PAIR p = ?x:*.?y:**. p = MK_PAIR x y")
```

We then prove that:

```
|- ?p:*->**->bool. IS_PAIR p
```

which is easily done (since `|- IS_PAIR(MK_PAIR x y)` is easily proved). This theorem is called `PAIR_EXISTS`. Next we define the type operator `prod` by evaluating

```
new_type_definition(`prod`, "IS_PAIR:(*->**->bool)->bool", PAIR_EXISTS)
```

This results in the constant `REP_prod` being declared and the following axiom (called `DEF_prod`) being asserted.

```
|- ONE_ONE REP_prod /\ (!p. IS_PAIR p = (?p'. p = REP_prod p'))
```

The infix constructor "," and the selectors `FST:*#**->*` and `SND:*#**->**` can then be defined by:

```
new_infix_definition
 (`COMMA_DEF`, "$, (x:*) (y:**) = @p. REP_prod p = MK_PAIR x y")

new_definition
 (`FST_DEF`, "FST(p:(*,**)prod) = @x.?y. MK_PAIR x y = REP_prod p")

new_definition
 (`SND_DEF`, "SND(p:(*,**)prod) = @y.?x. MK_PAIR x y = REP_prod p")
```

From these definitions and the axiom `DEF_prod`, the following theorems are proved and stored in the theory `prod`.

```
PAIR      |- !x. (FST x,SND x) = x

FST       |- !x y. FST(x,y)    = x

SND       |- !x y. SND(x,y)    = y
```

In addition to the constants just described, the theory `prod` also contains the definitions of `CURRY` and `UNCURRY`.

```
|- CURRY f x y     = f(x,y)

|- UNCURRY f (x,y) = f x y
```

These constants are used for representing generalized the abstractions of the form `\(`x_1`,...,`x_n`).`t described in Section 3.3.

5.6 The Theory sum

A type `(`σ_1`,`σ_2`)sum` (which may be abbreviated as σ_1`+`σ_2) denotes the disjoint union of types σ_1 and σ_2. The type operator `sum` can be defined just as `prod` was, but we omit the details here[10]. From the user's point of view, all that is needed are the functions

```
INL  : * -> *+**
INR  : ** -> *+**
OUTL : *+** -> *
OUTR : *+** -> **
ISL  : *+** -> bool
ISR  : *+** -> bool
```

`INL` and `INR` are the *injections* for inserting elements into the sum; `OUTL` and `OUTR` are the corresponding *projections* out of the sum. The predicates `ISL` and `ISR` test whether an element of a sum is in the left or right summand respectively. The following theorems are pre-proved in the system.

```
ISL         |- (!e. ISL(INL e)) /\ (!e. ~ISL(INR e))
ISR         |- (!e. ISR(INR e)) /\ (!e. ~ISR(INL e))
ISL_OR_ISR  |- !s. ISL s \/ ISR s
OUTL        |- !e. OUTL(INL e) = e
OUTR        |- !e. OUTR(INR e) = e
INL         |- !s. ISL s ==> (INL(OUTL s) = s)
INR         |- !s. ISR s ==> (INR(OUTR s) = s)
```

These theorems follow from the definitions.

5.7 The Theory ind

The theory `ind` introduces the type `ind` of *individuals* and the *Axiom of Infinity*. This axiom states that the set denoted by `ind` is infinite. The four axioms of the theory `bool`, the rules of inference in Section 5.4 and the Axiom of Infinity, are together sufficient for developing all of standard mathematics. Thus, in principle, the user of the HOL system should never need to make a non-definitional theory. In practice, it is often very tempting to take the risk of introducing new axioms because deriving them from definitions can be tedious; proving that 'axioms' follow from definitions amounts to giving a consistency proof of them.

The Axiom of Infinity is called `INFINITY_AX` and states:

---

[10] The definition of disjoint unions in the HOL system is due to Tom Melham.

```
|- ?f:ind->ind. ONE_ONE f /\ ~(ONTO f)
```

This asserts that there exists a one-to-one map from `ind` to itself that is not onto. This implies that the type `ind` denotes an infinite set.

5.8 The Theory `num`

The type `num` of natural numbers can be defined as equivalent to a countable subset of `ind`. Peano's axioms can then be proved as theorems. However, this has not yet been done in the HOL system. The type `num` is declared as a new primitive type and the constants `0` and `SUC` are taken as primitive constants with types `num` and `num->num` respectively. Peano's axioms are then asserted using `new_axiom`. The resulting theorems are:

```
NOT_SUC      |- !n. ~(SUC n = 0)

INV_SUC      |- !m n. (SUC m = SUC n) ==> (m = n)

INDUCTION    |- !P. P 0 /\ (!n. P n ==> P(SUC n)) ==> (!n. P n)
```

In higher-order logic, Peano's axioms are sufficient for developing number theory because addition and multiplication can be defined. In first order logic these must be taken as primitive. Note also that `INDUCTION` could not be stated as a single axiom in first-order logic because predicates (*e.g.* `P`) cannot be quantified.

5.9 The Theory `prim_rec`

In classical logic, unlike domain theory logics such as PPLAMBDA, arbitrary recursive definitions are not allowed. For example, there is no function `f` (of type `num->num`) such that

```
!x. f x  =  (f x) + 1
```

Certain restricted forms of recursive definition do, however, uniquely define functions. An important example are the *primitive recursive* functions. For any `x` and `f` the *Primitive Recursion Theorem* tells us that there is a unique function `fun` such that:

```
(fun 0 = x) /\
(!m. fun(SUC m) = f(fun m)m)
```

The Primitive Recursion Theorem follows from Peano's axioms. When the HOL system is built, the following theorem is proved and stored in the theory `prim_rec`:

```
|- !x f. ?fun.
     (fun 0 = x) /\
     (!m. fun(SUC m) = f(fun m)m)
```

From this it follows that there exists a function PRIM_REC such that:

```
|- !m x f.
     (PRIM_REC x f 0  = x) /\
     (PRIM_REC x f (SUC m) = f(PRIM_REC x f m)m)
```

The function PRIM_REC can be used to justify any primitive recursive definition. In higher-order logic a recursion of the form

```
fun 0 x1...xn        = f1(x1,...,xn)

fun (SUC m) x1...xn = f2(fun m x1...xn, m, x1,...,xn)
```

is equivalent to:

```
fun 0       = \x1...xn. f1(x1,...,xn)

fun (SUC m) = \x1...xn. f2(fun m x1...xn, m, x1,...,xn)
            = (\f m x1...xn. f2(f x1...xn, m, x1,...,xn))(fun m)m
```

which is equivalent to the non-recursive definition:

```
fun = PRIM_REC
        (\x1...xn. f1(x1,...,xn))
        (\f m x1...xn. f2(f x1...xn, m, x1,...,xn))
```

For example, we can define addition and multiplication by:

```
|- + = PRIM_REC(\n. n)(\f m n. SUC(f n))

|- * = PRIM_REC(\n. 0)(\f m n. (f n) + n)
```

To automate the use of the Primitive Recursion Theorem, HOL provides two functions:

```
new_prim_rec_definition       : (string # term) -> thm
new_infix_prim_rec_definition : (string # term) -> thm
```

Evaluating

```
new_prim_rec_definition
 (`fun`,
  "(!x1...xn. fun 0 x1...xn = t1[x1,...,xn]) /\
   (!m x1...xn.
     fun (SUC m) x1...xn = t2[fun m x1...xn, m, x1,...,xn])")
```

automatically makes the (non-recursive) definition:

```
new_definition
 (`fun_DEF`,
  "fun = PRIM_REC
          (\x1...xn. t1[x1,...,xn])
          (\fun m x1...xn. t2[fun x1...xn, m, x1,...,xn])")
```

and then proves the theorem:

```
|-  fun 0       x1...xn = t1[x1,...,xn] /\
    fun (SUC m) x1...xn = t2[fun m x1...xn, m, x1,...,xn]
```

which is saved as the theorem with name `fun`.

The ML function `new_infix_prim_rec_definition` declares an infixed function by primitive recursion. For example, here is the definition of `+` used by the system:

```
new_infix_prim_rec_definition
 (`ADD`,
  "($+ 0 n = n) /\
   ($+ (SUC m) n = SUC($+ m n))")
```

The `$`'s are there to indicate that `+` is being declared an infix. Evaluating this ML expression will create a definition of the following form in the current theory:

```
ADD_DEF   |- $+ = PRIM_REC(\n. n)(\f m n. SUC(f n))
```

The `$` is now necessary since the call to `new_infix_prim_rec_definition` declares `+` as an infix. It also automatically proves the following theorem:

```
ADD   |- (0 + n = n) /\ ((SUC m) + n = SUC(m + n))
```

which is saved in the current theory with name `ADD`.

We might hope to define the less-than relation "<" by primitive recursion, but it is needed for the proof of the Primitive Recursion Theorem, so we must somehow define it before we have primitive recursion available. A definition that works is:

```
LESS  |- $< m  n = (?P. (!n. P(SUC n) ==> P n) /\ P m /\ ~P n)
```

Intuitively, this says that `m<n` if there exists a set (with characteristic function `P`) that is downward closed[11] and contains `m` but not `n`.

---

[11] A set of numbers is *downward closed* if whenever it contains the successor of a number, it also contains the number.

5.10 The Theory arithmetic

The theory `arithmetic` contains primitive recursive definitions of standard arithmetic operators. For example:

```
|- (0 + n = n) /\ ((SUC m) + n = SUC(m + n))

|- (0 - m = 0) /\ ((SUC m) - n = (m < n => 0 | SUC(m - n)))

|- (0 * n = 0) /\ ((SUC m) * n = (m * n) + n)
```

It also contains various non-recursive definitions.

```
|- m > n   = n < m

|- m <= n  = m < n \/ (m = n)

|- m >= n  = m > n \/ (m = n)

|- m DIV n = (@x. (n * x) <= m /\ m < ((n * x) + n))

|- m REM n = m - (n * (m DIV n))
```

An *ad hoc* collection of elementary arithmetic theorems are pre-proved in the theory `arithmetic`. For example:

```
|- !m n. n < m ==> (?p. p + n = m)

|- !m n p. m < n /\ n < p ==> m < p

|- !m n. ~(m < n /\ n < m)

|- !m n. m * n = n * m

|- !m n p. (m + n) * p = (m * p) + (n * p)
```

At the time of writing there are about seventy such pre-proved theorems.

5.11 The Theory list

The theory `list` introduces the unary type operator `list` and the list processing functions:

```
NIL  : (*)list
CONS : * -> (*)list -> (*)list
HD   : (*)list -> *
TL   : (*)list -> (*)list
NULL : (*)list -> bool
```

These satisfy the usual axioms:

```
|- NULL NIL /\ !(x:*) l. ~(NULL(CONS x l))

|- !(x:*) l. HD(CONS x l) = x

|- !(x:*) l. TL(CONS x l) = l

|- !l:(*)list. CONS(HD l)(TL l) = l
```

These theorems could be proved from suitable definitions (which we don't give); the rule of structural induction can also be derived from these definitions.

The HOL parser allows one to write `[]` instead of `NIL` and `[`t_1`;`t_2`;...;`t_n`]` instead of `CONS` t_1 `(CONS` $t_2 \cdots$ `(CONS` t_n `NIL)` $\cdots$ `)`.

5.12 The Theory `tok`

The theory `tok` (for token) introduces the types `char` and `tok` to represent characters and strings of characters respectively. The type `char` could be defined equivalent to a suitable subset of `num` (for example, representing ASCII codes) and the type `tok` could then be defined as equivalent to `(char)list`. In fact, characters and tokens are currently axiomatized rather than defined (but we plan to eventually change this). The HOL parser converts any expression of the form `‘`*string*`‘` into a constant of type `tok` (where *string* can be any string of characters not containing the quote `‘`).

There are two constants in the theory `tok`:

```
EXPLODE : tok -> (char)list
IMPLODE : (char)list -> tok
```

`EXPLODE` converts a token to the list of characters it contains; `IMPLODE` is its inverse.

6 Derived Inference Rules

Derived inference rules are rules that can be justified on the basis of the axioms and primitive inference rules. For example, consider the following rule for 'undischarging' assumptions:

$$\frac{\Gamma \ \texttt{|-}\ t_1 \texttt{==>} t_2}{\Gamma \cup \{t_1\} \ \texttt{|-}\ t_2}$$

This is valid because if we have a theorem Γ |- t_1==>t_2 then we can derive the theorem $\Gamma \cup \{t_1\}$ |- t_2 by Modus Ponens from the theorem t_1 |- t_1 (which follows from the primitive rule `ASSUME`).

Derived inference rules enable proofs to be done using bigger and more natural steps. They can be defined in ML simply as functions that call the primitive inference rules. For example, the undischarging rule just described can be implemented by:

```
let UNDISCH th = MP th (ASSUME(fst(dest_imp(concl th))))
```

Each application of `UNDISCH` will invoke an application of `ASSUME` followed by an application of `MP`. Some of the predefined derived rules in HOL can invoke many thousands of primitive inference steps.

The HOL system has all the standard introduction and elimination rules of Predicate Calculus predefined as derived inference rules. It is these, rather than the primitive rules, that one normally uses in practice. In addition, there are some special rules that do a limited amount of automatic theorem-proving. The most generally useful examples of these are a collection of *rewriting* rules developed by Larry Paulson [17]. Rewriting rules use equations of the form |- t_1=t_2 to repeatedly replace subterms matching t_1 by the corresponding instances of t_2. The Cambridge LCF Manual [18] documents the rewriting rules in the HOL system, so we do not describe them here.

7 Goal Directed Proof: Tactics and Tacticals

A *tactic* is an ML function which is applied to a goal to reduce it to subgoals. A *tactical* is a (higher-order) ML function for combining tactics to build new tactics[12].

For example, if T_1 and T_2 are tactics, then the ML expression T_1 `THEN` T_2 evaluates to a tactic which first applies T_1 to a goal and then applies T_2 to each subgoal produced by T_1. The tactical `THEN` is an infixed ML function.

The tactics and tacticals in the HOL system are derived from those in the Cambridge LCF system [18] (which evolved from the ones in Edinburgh LCF [6]).

7.1 Tactics

It simplifies the description of tactics if we use various *type abbreviations*. A type abbreviation is just a name given to a type . A type and its abbreviation

[12]The terms "tactic" and "tactical" are due to Robin Milner, who invented the concepts.

can be used interchangeably. The system prints types using any abbreviations that are in force. A type abbreviation is introduced by executing a declaration of the form

```
lettype name = type
```

For example:

```
lettype goal = term list # term
```

The following ML type abbreviations are also used in connection with tactics (they will be explained later).

```
proof      = thm list -> thm
subgoals   = goal list # proof
tactic     = goal -> subgoals
thm_tactic = thm -> tactic
conv       = term -> thm
```

If T is a tactic and g is a goal, then applying T to g (*i.e.* evaluating the ML expression T g) will result in an object of ML type `subgoals`, *i.e.* a pair whose first component is a list of goals and whose second component has ML type `proof`.

Suppose T g = `([`g_1`;...;`g_n`],`p`)`. The idea is that g_1 , ... , g_n are subgoals and p is a 'justification' of the reduction of goal g to subgoals g_1 , ... , g_n. Suppose further that we have solved the subgoals g_1 , ... , g_n. This would mean that we had somehow proved theorems th_1 , ... , th_n such that each th_i $(1 \leq i \leq n)$ 'achieves' the goal g_i. The justification p (produced by applying T to g) is an ML function which when applied to the list `[`th_1`;...;`th_n`]` returns a theorem, th, which 'achieves' the original goal g. Thus p is a function for converting a solution of the subgoals to a solution of the original goal. If p does this successfully, then the tactic T is called *valid*. Invalid tactics cannot result in the proof of invalid theorems; the worst they can do is result in insolvable goals or unintended theorems being proved. If T were invalid and were used to reduce goal g to subgoals g_1 , ... , g_n, then one might prove theorems th_1 , ... , th_n achieving g_1 , ... , g_n, only to find that these theorems are useless because p`[`th_1`;...;`th_n`]` doesn't achieve g (*i.e.* it fails, or else it achieves some other goal).

A theorem *achieves* a goal if the assumptions of the theorem are included in the assumptions of the goal *and* if the conclusion of the theorems is equal (up to renaming of bound variables) to the conclusion of the goal. More precisely, we say a theorem t_1, ... , t_m `|-` t achieves a goal `([`u_1`;...;`u_n`],`u`)` if and only if `{`t_1`,...,`t_m`}` is a subset of `{`u_1`,...,`u_n`}` and t is equal to u (up to renaming of bound variables). For example, the goal `(["x=y";"y=z";"z=w"],"x=z")` is achieved by the theorem `x=y, y=z |- x=z` (the assumption `"z=w"` is not needed).

We say that a tactic *solves* a goal if it reduces the goal to the empty list of subgoals. Thus T solves g if $T\ g$ = ([],p). If this is the case and if T is valid, then p[] will evaluate to a theorem achieving g. Thus if T solves g then the ML expression snd($T\ g$)[] evaluates to a theorem achieving g.

Tactics are specified using the following notation:

$$\frac{goal}{goal_1 \quad goal_2 \quad \ldots \quad goal_n}$$

For example, a tactic called CONJ_TAC is described by

$$\frac{t_1 \ /\backslash\ t_2}{t_1 \qquad t_2}$$

Thus CONJ_TAC reduces a goal of the form (Γ,"t_1/\t_2") to subgoals (Γ,"t_1") and (Γ,"t_2"). The fact that the assumptions of the top-level goal are propagated unchanged to the two subgoals is indicated by the absence of assumptions in the notation.

Another example is INDUCT_TAC, the tactic for doing mathematical induction on the natural numbers:

$$\frac{!n.t[n]}{t[0] \qquad \{t[n]\}\ t[\mathtt{SUC}\ n]}$$

INDUCT_TAC reduces a goal (Γ,"!$n.t[n]$") to a basis subgoal (Γ,"$t[0]$") and an induction step subgoal ($\Gamma \cup \{$"$t[n]$"$\}$,"$t[\mathtt{SUC}\ n]$"). The extra induction assumption "$t[\mathtt{SUC}\ n]$" is indicated in our tactic notation with set brackets.

Tactics generally fail (in the ML sense) if they are applied to inappropriate goals. For example, CONJ_TAC will fail if it is applied to a goal whose conclusion is not a conjunction. Some tactics never fail, for example ALL_TAC

$$\frac{t}{t}$$

is the 'identity tactic'; it reduces a goal (Γ,t) to the single subgoal (Γ,t) — *i.e.* it has no effect. ALL_TAC is useful for writing complex tactic using tacticals (*e.g.* see the definition of REPEAT in Section 7.3).

7.2 Using Tactics to Prove Theorems

Suppose one has a goal g to solve. If g is simple one might be able to think up a tactic, T say, which reduces it to the empty list of subgoals. If this is the case then executing

`let` gl,p `=` $T\ g$

will bind p to a function which when applied to the empty list of theorems yields a theorem th achieving g. (The declaration above will also bind gl to the empty list of goals.) Thus a theorem achieving g can be computed by executing

`let` th `=` p`[]`

After proving a theorem one usually wants to store it in the current theory. To do this one chooses an unused name, `Thm1` say, and then executes:

`save_thm(‘Thm1‘,` th`)`

This saves the theorem th with name `Thm1`. If the current theory is called $\mathcal{T}$ then to retrieve th (in a later session, in $\mathcal{T}$ or its descendants) one does:

`let` th `= theorem ‘`$\mathcal{T}$`‘ ‘Thm1‘`

The ML function `theorem` has type `string->string->thm`. Its first argument should be the name of a theory and its second argument the name of a theorem on that theory. The named theorem is returned.

To simplify the use of tactics there is a standard function called `prove_thm`:

```
prove_thm : (string # term # tactic) -> thm
```

`prove_thm(‘foo‘,`t`,`T`)` proves the goal `([],`t`)` (*i.e.* the goal with no assumptions and conclusion t) using tactic T and saves the resulting theorem with name `foo` on the current theory.

7.3 Tacticals

A *tactical* is an ML function that returns a tactic (or tactics) as result. Tacticals may take various parameters; this is reflected in the various ML types that the built-in tacticals have. Some important tactics in the HOL system are listed below (these are all in LCF also).

7.3.1 `ORELSE : tactic -> tactic -> tactic`

The tactical `ORELSE` is an ML infix. If T_1 and T_2 are tactics, then the ML expression T_1 `ORELSE` T_2 evaluates to a tactic which first tries T_1 and then if T_1 fails it tries T_2. It is defined in ML by

```
let (T1 ORELSE T2) g =  T1 g ? T2 g
```

7.3.2 `THEN : tactic -> tactic -> tactic`

The tactical `THEN` is an ML infix. If T_1 and T_2 are tactics, then the ML expression T_1 `THEN` T_2 evaluates to a tactic which first applies T_1

and then applies T_2 to all the subgoals produced by T_1. Its definition in ML is tricky and not given here.

7.3.3 `THENL : tactic -> tactic list -> tactic`

If T is a tactic which produces n subgoals and $T_1, \ldots, T_n$ are tactics then T `THENL` [T_1;...;T_n] is a tactic which first applies T and then applies T_i to the ith subgoal produced by T. The tactical `THENL` is useful if one wants to do different things to different subgoals.

7.3.4 `REPEAT : tactic -> tactic`

If T is a tactic then `REPEAT` T is a tactic which repeatedly applies T until it fails. It is defined in ML by;

```
letrec REPEAT T g = ((T THEN REPEAT T) ORELSE ALL_TAC) g
```

The extra argument `g` is needed because ML does not use lazy evaluation.

7.3.5 `EVERY_ASSUM : (thm -> tactic) -> tactic`

Applying `EVERY_ASSUM` f to a goal ([t_1;...;t_n],t) is equivalent to applying the tactic:

f(`ASSUME` t_1) `THEN` ... `THEN` f(`ASSUME` t_n)

7.3.6 `FIRST_ASSUM : (thm -> tactic) -> tactic`

Applying `FIRST_ASSUM` f to a goal ([t_1;...;t_n],t) is equivalent to applying the tactic:

f(`ASSUME` t_1) `ORELSE` ... `ORELSE` f(`ASSUME` t_n)

7.4 Tactics Built into HOL

We list below some of the tactics built into the HOL system, including those that are used in the example proof in Section 8.

Recall that the ML type `thm_tactic` abbreviates `theorem->tactic`, and the type `conv`[13] abbreviates `term->thm`.

[13] The type `conv` comes from Larry Paulson's theory of conversions [17].

7.4.1 ACCEPT_TAC : thm_tactic

- **Summary:** `ACCEPT_TAC` *th* is a tactic that solves any goal that is achieved by *th*.
- **Use:** Interfacing forward and backward proofs. For example, one might reduce a goal g to subgoals $g_1, \ldots, g_n$ using a tactic T and then prove theorems $th_1, \ldots, th_n$ achieving these goals by forward proof. The tactic

  `T THENL[ACCEPT_TAC` th_1`;...;ACCEPT_TAC` th_n`]`

  would then solve g.

7.4.2 DISJ_CASES_TAC : thm_tactic

- **Summary:** `DISJ_CASES_TAC |-` *u*`\/`*v* splits a goal into two cases: one with "*u*" as an assumption and the other with "*v*" as an assumption.

$$\frac{t}{\{u\}t \qquad \{v\}t}$$

- **Uses:** Case analysis. The tactic `ASM_CASES_TAC` (see below) is defined in ML by

```
let ASM_CASES_TAC t = DISJ_CASES_TAC(SPEC t EXCLUDED_MIDDLE)
```

  where `EXCLUDED_MIDDLE` is the theorem `|- !t. t \/ ~t`.

7.4.3 ASM_CASES_TAC : term -> tactic

- **Summary:** `ASM_CASES_TAC` "*u*" does case analysis on the boolean term "*u*".

$$\frac{t}{\{u\}t \qquad \{\sim u\}t}$$

- **Uses:** Case analysis.

7.4.4 COND_CASES_TAC : tactic

- **Summary:** Does a case split on the condition of a conditional term.

$$\frac{t[p\texttt{=>}u\,|\,v]}{\{p\texttt{=T}\}t[u] \qquad \{p\texttt{=F}\}t[v]}$$

  `COND_CASES_TAC` searches for a conditional term and then does cases on its if-part. It fails if the context $t[]$ captures any free variables in the if-part p.

- **Uses:** Most useful when there is only one conditional term in the goal (otherwise it is difficult to predict which condition will be chosen for the case split). Successive case analysis on all conditions can be done using `REPEAT COND_CASES_TAC`.

7.4.5 `REWRITE_TAC : thm list -> tactic`

- **Summary:** `REWRITE_TAC[`th_1`;...;`th_n`]` simplifies the goal by rewriting it with the explicitly given theorems $th_1, \ldots, th_n$, and various built-in rewriting rules.

$$\frac{\{t_1, \ldots, t_m\}t}{\{t_1, \ldots, t_m\}t'}$$

where t' is obtained from t by rewriting with

1. $th_1, \ldots, th_n$ and
2. the standard rewrites held in the ML variable `basic_rewrites`.

- **Uses:** Simplifying goals using previously proved theorems.
- **Other rewriting tactics** (based on `REWRITE_TAC`):

1. `ASM_REWRITE_TAC` adds the assumptions of the goal to the list of theorems used for rewriting.
2. `FILTER_ASM_REWRITE_TAC` p `[`th_1`;...;`th_n`]` simplifies the goal by rewriting it with the explicitly given theorems th_1 , ... , th_n , together with those assumptions of the goal which satisfy the predicate p and also the built-in rewrites in the ML variable `basic_rewrites`.
3. `PURE_ASM_REWRITE_TAC` is like `ASM_REWRITE_TAC`, but it doesn't use any built-in rewrites.
4. `PURE_REWRITE_TAC` uses neither the assumptions nor the built-in rewrites

7.4.6 `ASSUME_TAC : thm_tactic`

- **Summary:** `ASSUME_TAC |-`u adds u as an assumption.

$$\frac{t}{\{u\}t}$$

- **Uses:** Enriching the assumptions of a goal with previously proved theorems.

7.4.7 `CONJ_TAC : tactic`

- **Summary:** Splits a goal "t_1/\t_2" into two subgoals "t_1" and "t_2".

$$\frac{t_1 \ \texttt{/\textbackslash}\ t_2}{t_1 \qquad t_2}$$

- **Uses:** Solving conjunctive goals. `CONJ_TAC` is invoked by `STRIP_TAC` (see below).

7.4.8 `DISCH_TAC : tactic`

- **Summary:** Moves the antecedant of an implicative goal into the assumptions.

$$\frac{u \ \texttt{==>}\ v}{\{u\}v}$$

- **Uses:** Solving goals of the form "u ==> v" by assuming "u" and then solving "v". `STRIP_TAC` (see below) will invoke `DISCH_TAC` on implicative goals.

7.4.9 `GEN_TAC : tactic`

- **Summary:** Strips off one universal quantifier.

$$\frac{\texttt{!}x\texttt{.}t[x]}{t[x']}$$

  Where x' is a variant of x not free in the goal or the assumptions.

- **Uses:** Solving universally quantified goals. `REPEAT GEN_TAC` strips off all universal quantifiers and is often the first thing one does in a proof. `STRIP_TAC` (see below) applies `GEN_TAC` to universally quantified goals.

7.4.10 `IMP_RES_TAC : tactic`

- **Summary:** `IMP_RES_TAC` th 'resolves' (see below) th with the assumptions of the goal and then adds the results to the assumptions.

$$\frac{\{t_1, \ldots, t_m\}t}{\{t_1, \ldots, t_m, u_1, \ldots, u_n\}t}$$

where $u_1, \ldots, u_n$ are derived by 'resolving' *th* with $t_1, \ldots, t_m$. Resolution in HOL is not classical resolution, but just Modus Ponens with a bit of one-way pattern matching (not unification). The usual case is where *th* is of the form

`|- !`$x_1 \ldots x_p$`.`v_1`==>`v_2`==>...==>`v_q`==>`v.

`IMP_RES_TAC` *th* then tries to specialize $x_1, \ldots, x_p$ so that $v_1, \ldots, v_q$ match members of $\{t_1, \ldots, t_m\}$. If such a match is found then the appropriate instance of v is added to the assumptions, together with all appropriate instances of v_i`==>...`v_n`==>`v $(2 \leq i \leq n)$. `IMP_RES_TAC` can also be given a conjunction of implications in which case it will do 'resolution' with each of the conjuncts. In fact, it applies a canonicalization rule to its argument to split it into a list of theorems. Each theorem produced by this canonicalization process is resolved with the assumptions. Full details can be found in the Cambridge LCF Manual [18].

- **Uses:** Deriving new assumptions from existing ones and previously proved theorems so that subsequent tactics (*e.g.* `ASM_REWRITE_TAC`) have more to work with.

7.4.11 `STRIP_TAC : tactic`

- **Summary:** Breaks a goal apart. `STRIP_TAC` removes one outer connective from the goal, using `CONJ_TAC`, `DISCH_TAC`, `GEN_TAC`, *etc.* If the goal has the form t_1`/\`$\cdots$`/\`t_n `==>` t then `DISCH_TAC` makes each t_i into a separate assumption.
- **Uses:** Useful for spliting a goal up into manageable pieces. Often the best thing to do first is `REPEAT STRIP_TAC`.

7.4.12 `SUBST_TAC : thm list -> thm`

- **Summary:** `SUBST_TAC[|-`u_1`=`v_1`;...;|-`u_n`=`v_n`]` converts a goal of the form $t[u_1, \ldots, u_n]$ to a subgoal of the form $t[v_1, \ldots, v_n]$.
- **Uses:** To make replacements for terms in situation in which `REWRITE_TAC` is too general or would loop.

7.4.13 `ALL_TAC : tactic`

- **Summary:** Identity tactic for the tactical `THEN` (see end of Section 7.1).
- **Uses:**
  1. Writing tacticals (see description of `REPEAT` in Section 7.3).
  2. With `THENL`; for example, if tactic T produces two subgoals and we want to apply T_1 to the first one but to do nothing to the second, then the tactic to use is T `THENL[`T_1`;ALL_TAC]`.

7.4.14 NO_TAC : tactic

- **Summary:** Tactic that always fails.
- **Uses:** Writing tacticals (see the example in Section 8).

8 An Example of HOL in Action

The example in this section has been chosen to give a flavour of what it is like to use the HOL system. Although the theorems proved are simple, the way we prove them illustrates the kind of intricate 'proof engineering' that is typical. The proofs below could be done more elegantly, but presenting them that way would defeat our purpose of illustrating various features of HOL. We have tried to use a small example to give the reader a feel for what it is like to do a big one. Readers who are not interested in hardware verification should be able to learn something about the HOL system even if they do not wish to penetrate the details of the parity-checking example we use.

As in Section 2, the boxed interactions below should be understood as occurring in sequence. These interactions comprise the specification and verification of a device that computes the parity of a sequence of bits. More specifically, we verify the implementation of a device with an input `in` and an output `out` with the specification that the nth output on `out` is `T` if and only if there have been an even number of `T`'s input on `in`. We shall construct a new theory called `Parity` in which we specify and verify the device. The first thing we do is start up the HOL system and then enter draft mode for this theory.

```
% hol


 _  _          __        _
|__|          |  |      |
|  | IGHER    |__| RDER |__ OGIC
===============================
(Built on Sept  31)

#new_theory`Parity`;;
() : void
```

To specify the device we define a primitive recursive function `PARITY` such that `PARITY n f` is true if the number of `T`'s in the sequence `f(0)`, ... , `f(n)` is even.

```
#let PARITY_DEF =
# new_prim_rec_definition
#  (`PARITY_DEF`,
#   "(PARITY 0 f = T) /\
#    (PARITY(SUC n)f = (f(SUC n) => ~(PARITY n f) | PARITY n f))");;
PARITY_DEF =
|- (PARITY 0 f = T) /\
   (PARITY(SUC n)f = (f(SUC n) => ~PARITY n f | PARITY n f))
```

The effect of `new_prim_rec_definition` is to store the definition of the constant `PARITY` on the theory `Parity` and to bind the defining theorem to the ML variable `PARITY`. The specification of the device can now be given as:

```
!t. out t = PARITY t in
```

It is *intuitively* clear that this specification will be satisfied if the signal[14] functions `in` and `out` satisfy:

```
out(0) = T
```

and

```
!t. out(t+1)  =  (in(t+1) => ~(out t) | out t)
```

We can verify this *formally* in HOL by proving the following lemma:

```
!in out.
 (out 0 = T) /\ (!t. out(SUC t) = (in(SUC t) => ~(out t) | out t))
 ==>
 (!t. out t = PARITY t in)
```

The proof of this is by Mathematical Induction and, although trivial, is a good illustration of how such proofs are done. We will prove this lemma interactively using HOL's subgoal package[15]. We start the proof by putting the goal we want to prove on a goal stack using the function `set_goal` which takes a goal as argument.

---

[14] Signals are modelled as functions from numbers (representing times) to booleans.

[15] The subgoal package is part of Cambridge LCF, but not Edinburgh LCF. It was implemented by Larry Paulson and provides proof building tools similar to those found in Stanford LCF [14]. We do not describe the subgoal package in this paper, but hope that the simple uses of it that we make will be clear.

```
#set_goal
#([],
# "!in out.
#    (out 0 = T) /\ (!t. out(SUC t) = (in(SUC t) => ~(out t) | out t))
#    ==>
#    (!t. out t = PARITY t in)");;
"!in out.
   (out 0 = T) /\ (!t. out(SUC t) = (in(SUC t) => ~out t | out t)) ==>
   (!t. out t = PARITY t in)"
```

The subgoal package prints out the goal on the top of the goal stack. We now expand the top goal by stripping off the universal quantifier (with GEN_TAC) and then making the two conjuncts of the antecedant of the implication into assumptions of the goal (with STRIP_TAC). The ML function **expand** takes a tactic and applies it to the top goal; the resulting subgoals are pushed on to the goal stack. The message "OK.." is printed out just before the tactic is applied.

```
#expand(REPEAT GEN_TAC THEN STRIP_TAC);;
OK..
"!t. out t = PARITY t in"
    [ "out 0 = T" ]
    [ "!t. out(SUC t) = (in(SUC t) => ~out t | out t)" ]
```

Next we do induction on t using INDUCT_TAC, which does induction on the outermost universally quantified variable.

```
#expand INDUCT_TAC;;
OK..
"out(SUC t) = PARITY(SUC t)in"
    [ "out 0 = T" ]
    [ "!t. out(SUC t) = (in(SUC t) => ~out t | out t)" ]
    [ "out t = PARITY t in" ]

"out 0 = PARITY 0 in"
    [ "out 0 = T" ]
    [ "!t. out(SUC t) = (in(SUC t) => ~out t | out t)" ]
```

The assumptions of the two subgoals are shown in square brackets. The last goal printed is the one on the top of the stack, which is the basis case. This is solved by rewriting with its assumptions and the definition of PARITY.

```
#expand(ASM_REWRITE_TAC[PARITY_DEF]);;
OK..
goal proved
. |- out 0 = PARITY 0 in

Previous subproof:
"out(SUC t) = PARITY(SUC t)in"
    [ "out 0 = T" ]
    [ "!t. out(SUC t) = (in(SUC t) => ~out t | out t)" ]
    [ "out t = PARITY t in" ]
```

The top goal is proved, so the system pops it from the goal stack (and puts the proved theorem on a stack of theorems). The new top goal is the step case of the induction. This goal is also solved by rewriting.

```
#expand(ASM_REWRITE_TAC[PARITY_DEF]);;
OK..
goal proved
.. |- out(SUC t) = PARITY(SUC t)in
.. |- !t. out t = PARITY t in
|- !in out.
    (out 0 = T) /\ (!t. out(SUC t) = (in(SUC t) => ~out t | out t))
    ==>
    (!t. out t = PARITY t in)

Previous subproof:
goal proved
```

The goal is proved, *i.e.* the empty list of subgoals is produced. The system now applies the justification functions (see Section 7.1) produced by the tactics to the lists of theorems achieving the subgoals (starting with the empty list). These theorems are printed out in the order they are generated (note that assumptions of theorems are printed as dots).

The ML function

```
save_top_thm : string -> thm
```

saves the theorem just proved (*i.e.* on the top of the theorem stack) in the current theory with a given name.

```
#let UNIQUENESS_LEMMA = save_top_thm `UNIQUENESS_LEMMA`;;
UNIQUENESS_LEMMA =
|- !in out.
    (out 0 = T) /\ (!t. out(SUC t) = (in(SUC t) => ~out t | out t))
    ==>
    (!t. out t = PARITY t in)
```

this saves the theorem in the theory `Parity` with name `UNIQUENESS_LEMMA` and also binds it to an ML variable with the same name.

The lemma just proved suggests that the parity checker can be implemented by holding the parity value in a register and then complementing the contents of the register whenever `T` is input. To make the implementation more interesting, we will assume that registers 'power up' storing `F`. Thus the output at time `0` cannot be taken directly from a register, because the output of the parity checker at time `0` is specified to be `T`. Another tricky thing to notice is that if `t>0`, then the output of the parity checker at time `t` is a function of the input at time `t`. Thus there must be a combinational path from the input to the output.

The schematic diagram below shows the design of a device that is intended to implement this specification.

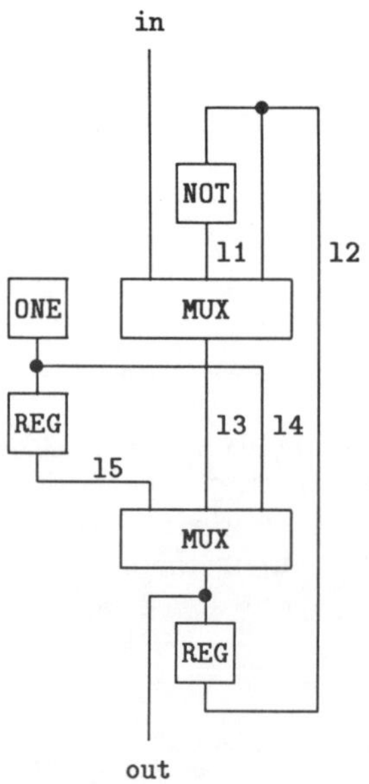

This works by storing the parity of the sequence input so far in the lower of the two registers. Each time `T` is input at `in`, this stored value is complemented.

Registers are assumed to 'power up' in a state in which they are storing `F`. The second register (connected to `ONE`) initially outputs `F` and then outputs `T` forever. Its role is just to ensure that the device works during the first cycle by connecting the output `out` to the device `ONE` via the lower multiplexer. For all subsequent cycles `out` is connected to `l3` and so either carries the stored parity value (if the current input is `F`) or the complement of this value (if the current input is `T`).

The devices making up this schematic will be modelled with predicates [8]. For example, the predicate `ONE` is true of a signal `out` if for all times `t`, the value of `out` is `T`.

```
#let ONE_DEF =
# new_definition
#  (`ONE_DEF`, "ONE(out:num->bool) = !t. out t = T");;
ONE_DEF = |- ONE out = (!t. out t = T)
```

Note that "`ONE_DEF`" is used both as an ML variable and as the name of the definition in the theory `Parity`.

The binary predicate `NOT` is true of a pair of signals `(in,out)` if the value of `out` is always the negation of the value of `in`. We thus model inverters as having no delay. This is appropriate for a register-transfer level model, but would be wrong at a lower level.

```
#let NOT_DEF =
# new_definition
#  (`NOT_DEF`, "NOT(in,out:num->bool) = !t. out t = ~(in t)");;
NOT_DEF = |- NOT(in,out) = (!t. out t = ~in t)
```

The final combinational device we need is a multiplexer. This is a 'hardware conditional'; the input `sw` selects which of the other two inputs are to be connected to the output `out`.

```
#let MUX_DEF =
# new_definition
#  (`MUX_DEF`,
#   "MUX(sw,in1,in2,out:num->bool) =
#     !t. out t = (sw t => in1 t | in2 t)");;
MUX_DEF =
 |- MUX(sw,in1,in2,out) = (!t. out t = (sw t => in1 t | in2 t))
```

The remaining devices in the schematic are registers. These are unit-delay elements; the values output at time `t+1` are the values input at the preceding time `t`, except at time `0` when the register outputs `F`[16] .

```
#let REG_DEF =
# new_definition
# (`REG_DEF`, "REG(in,out:num->bool) =
#              !t. out t = ((t=0) => F | in(t-1))");;
REG_DEF = |- REG(in,out) = (!t. out t = ((t = 0) => F | in(t - 1)))
```

The schematic diagram above can be represented as a predicate by conjoining the relations holding between the various signals and then existentially quantifying the internal lines. This techniques is explained elsewhere (*e.g.* see [8], [1]).

```
#let PARITY_IMP_DEF =
# new_definition
#  (`PARITY_IMP_DEF`,
#   "PARITY_IMP(in,out) =
#    ?l1 l2 l3 l4 l5.
#     NOT(l2,l1) /\ MUX(in,l1,l2,l3) /\ REG(out,l2) /\
#     ONE l4     /\ REG(l4,l5)       /\ MUX(l5,l3,l4,out)");;
PARITY_IMP_DEF =
|- PARITY_IMP(in,out) =
   (?l1 l2 l3 l4 l5.
     NOT(l2,l1) /\
     MUX(in,l1,l2,l3) /\
     REG(out,l2) /\
     ONE l4 /\
     REG(l4,l5) /\
     MUX(l5,l3,l4,out))
```

What we shall eventually prove is

```
|- !in out. PARITY_IMP(in,out) ==> (!t. out t = PARITY t in)
```

This states that *if* `in` and `out` are related as in the schematic diagram (*i.e.* as in the definition of `PARITY_IMP`), *then* the pair of signals `(in,out)` satisfies the specification.

We shall start by proving the following lemma:

[16] Time 0 represents when the device is switched on.

```
|- !in out.
      PARITY_IMP(in,out) ==>
       (out 0 = T) /\ !t. out(SUC t) = (in(SUC t) => ~(out t) | out t)
```

The correctness of the parity checker follows from this and UNIQUENESS_LEMMA by the transitivity of ==>.

```
#set_goal
# ([], "!in out.
#          PARITY_IMP(in,out) ==>
#          (out 0 = T) /\
#          !t. out(SUC t) = (in(SUC t) => ~(out t) | out t)");;
"!in out.
  PARITY_IMP(in,out) ==>
  (out 0 = T) /\ (!t. out(SUC t) = (in(SUC t) => ~out t | out t))"
```

We start proving this by rewriting with definitions followed by a decomposition of the resulting goal using STRIP_TAC. We use the rewriting tactic PURE_REWRITE_TAC which does no built-in simplifications, only the ones explicitly given in the list of theorems supplied as an argument. One of the built-in simplifications that REWRITE_TAC uses is |- (x = T) = T. We use PURE_REWRITE_TAC to prevent rewriting with this being done.

```
#expand(PURE_REWRITE_TAC
#         [PARITY_IMP_DEF;ONE_DEF;NOT_DEF;MUX_DEF;REG_DEF]
#        THEN REPEAT STRIP_TAC);;
OK..
2 subgoals
"out(SUC t) = (in(SUC t) => ~out t | out t)"
    [ "!t. l1 t = ~l2 t" ]
    [ "!t. l3 t = (in t => l1 t | l2 t)" ]
    [ "!t. l2 t = ((t = 0) => F | out(t - 1))" ]
    [ "!t. l4 t = T" ]
    [ "!t. l5 t = ((t = 0) => F | l4(t - 1))" ]
    [ "!t. out t = (l5 t => l3 t | l4 t)" ]

"out 0 = T"
    [ "!t. l1 t = ~l2 t" ]
    [ "!t. l3 t = (in t => l1 t | l2 t)" ]
    [ "!t. l2 t = ((t = 0) => F | out(t - 1))" ]
    [ "!t. l4 t = T" ]
    [ "!t. l5 t = ((t = 0) => F | l4(t - 1))" ]
    [ "!t. out t = (l5 t => l3 t | l4 t)" ]
```

The top goal is the one printed last; its conclusion is `out 0 = T` and its assumptions are equations relating the values on the lines in the circuit. We would like to expand the top goal by rewriting with the assumptions. However, if we do this the system will go into an infinite loop because the equations for `out`, `12` and `13` are mutually recursive. To prevent looping, we must be more delicate and only rewrite with a non-looping subset of the assumptions.

To enable the assumptions corresponding to particular lines to be selected for rewriting, we define an ML function `lines` such that `lines` '$l_1 \ldots l_n$' t is true if t has the form "`!t.` l_i `t =` ⋯ " for some l_i in the set specified by the string '$l_1 \ldots l_n$'. The functions `words` and `rator` used below are explained in the examples in Section 2.

```
#let lines tok t =
# (let x = fst(dest_var(rator(lhs(snd(dest_forall t)))))
#  in
#  mem x (words tok)) ? false;;
lines = - : (string -> term -> bool)
```

`FILTER_ASM_REWRITE_TAC(lines'out 14 15')[]` is a tactic which rewrites with only those assumptions that involve `out`, `14` and `15` (see Section 7.4.5).

```
#expand(FILTER_ASM_REWRITE_TAC(lines'out 14 15')[]);;
OK..
goal proved
... |- out 0 = T

Previous subproof:
"out(SUC t) = (in(SUC t) => ~out t | out t)"
    [ "!t. 11 t = ~12 t" ]
    [ "!t. 13 t = (in t => 11 t | 12 t)" ]
    [ "!t. 12 t = ((t = 0) => F | out(t - 1))" ]
    [ "!t. 14 t = T" ]
    [ "!t. 15 t = ((t = 0) => F | 14(t - 1))" ]
    [ "!t. out t = (15 t => 13 t | 14 t)" ]
```

The first of the two subgoals is proved. Inspecting the remaining goal we see that it will be solved if we expand `out(SUC t)` using only the last assumption (*i.e.* `!t. out t = (15 t => 13 t | 14 t)`). If we just rewrite with this assumption, then all the subterms of the form `out t` will also be expanded. To prevent this we are forced to use the messy and *ad hoc* tactic shown below.

```
FIRST_ASSUM
 (\th. if lines‘out‘(concl th)
        then SUBST_TAC[SPEC "SUC t" th]
        else NO_TAC)
```

`FIRST_ASSUM` is explained in Section 7.3.6. The function

```
(\th. if lines‘out‘(concl th)
       then SUBST_TAC[SPEC "SUC t" th]
       else NO_TAC)
```

converts a theorem of the form

```
|- !t. out t = tm[t]
```

to a tactic that replaces occurrences of `out(SUC t)` by *tm*`[SUC t]`. All other theorems are mapped to `NO_TAC`, a tactic that always fails.

```
#expand
# (FIRST_ASSUM
#   (\th. if lines‘out‘(concl th)
#          then SUBST_TAC[SPEC "SUC t" th]
#          else NO_TAC));;
OK..
"(15(SUC t) => 13(SUC t) | 14(SUC t)) = (in(SUC t) => ~out t | out t)"
    [ "!t. 11 t = ~12 t" ]
    [ "!t. 13 t = (in t => 11 t | 12 t)" ]
    [ "!t. 12 t = ((t = 0) => F | out(t - 1))" ]
    [ "!t. 14 t = T" ]
    [ "!t. 15 t = ((t = 0) => F | 14(t - 1))" ]
    [ "!t. out t = (15 t => 13 t | 14 t)" ]
```

The fact that we had to resort to this messy use of `FIRST_ASSUM` illustrates both the strengths and weaknesses of the HOL system. Trivial deductions sometimes require elaborate tactics, but on the other hand one never reaches an impasse. HOL experts can prove arbitrarily complicated theorems if they are willing to use sufficient ingenuity. Furthermore, the type discipline ensures that no matter how complicated and *ad hoc* are the tactics, it is impossible to prove an invalid theorem.

Inspecting the goal above, we see that the next step is to unwind the equations for lines `11`, `13`, `14` and `15` and then, when this is done, unwind with the equation for line `12`.

```
#expand(FILTER_ASM_REWRITE_TAC(lines`l1 l3 l4 l5`)[]
#       THEN FILTER_ASM_REWRITE_TAC(lines`l2`)[]);;
OK..
"(((SUC t = 0) => F | T) =>
  (in(SUC t) =>
   ~((SUC t = 0) => F | out((SUC t) - 1)) |
   ((SUC t = 0) => F | out((SUC t) - 1))) |
  T) =
 (in(SUC t) => ~out t | out t)"
    [ "!t. l1 t = ~l2 t" ]
    [ "!t. l3 t = (in t => l1 t | l2 t)" ]
    [ "!t. l2 t = ((t = 0) => F | out(t - 1))" ]
    [ "!t. l4 t = T" ]
    [ "!t. l5 t = ((t = 0) => F | l4(t - 1))" ]
    [ "!t. out t = (l5 t => l3 t | l4 t)" ]
```

This goal can now be solved by rewriting with two standard theorems:

```
NOT_SUC      |- !n. ~(SUC n = 0)

SUC_SUB1     |- !m. (SUC m) - 1 = m
```

```
#expand(REWRITE_TAC[NOT_SUC;SUC_SUB1]);;
OK..
goal proved
|- (((SUC t = 0) => F | T) =>
    (in(SUC t) =>
     ~((SUC t = 0) => F | out((SUC t) - 1)) |
     ((SUC t = 0) => F | out((SUC t) - 1))) |
    T) =
   (in(SUC t) => ~out t | out t)
..... |- (l5(SUC t) => l3(SUC t) | l4(SUC t)) =
         (in(SUC t) => ~out t | out t)
...... |- out(SUC t) = (in(SUC t) => ~out t | out t)
|- !in out.
    PARITY_IMP(in,out) ==>
    (out 0 = T) /\ (!t. out(SUC t) = (in(SUC t) => ~out t | out t))

Previous subproof:
goal proved
```

We name the theorem just proved `PARITY_LEMMA` and save it in the current theory.

```
#let PARITY_LEMMA = save_top_thm`PARITY_LEMMA`;;
PARITY_LEMMA =
|- !in out.
    PARITY_IMP(in,out) ==>
    (out 0 = T) /\ (!t. out(SUC t) = (in(SUC t) => ~out t | out t))
```

We could have proved `PARITY_LEMMA` in one step with a single compound tactic. To illustrate this we set up the goal again.

```
#set_goal
# ([], "!in out.
#         PARITY_IMP(in,out) ==>
#         (out 0 = T) /\
#         !t. out(SUC t) = (in(SUC t) => ~(out t) | out t)");;
"!in out.
  PARITY_IMP(in,out) ==>
  (out 0 = T) /\ (!t. out(SUC t) = (in(SUC t) => ~out t | out t))"
```

We then expand with a single tactic corresponding to the sequence of tactics that we used interactively.

```
#expand
# (PURE_REWRITE_TAC[PARITY_IMP_DEF;ONE_DEF;NOT_DEF;MUX_DEF;REG_DEF]
#  THEN REPEAT STRIP_TAC
#  THENL
#   [FILTER_ASM_REWRITE_TAC(lines`out 14 15`)[];ALL_TAC]
#  THEN FIRST_ASSUM
#        (\th. if lines`out`(concl th)
#               then SUBST_TAC[SPEC "SUC t" th]
#               else NO_TAC)
#  THEN FILTER_ASM_REWRITE_TAC(lines`11 13 14 15`)[]
#  THEN FILTER_ASM_REWRITE_TAC(lines`12`)[]
#  THEN REWRITE_TAC[NOT_SUC;SUC_SUB1]);;
OK..
goal proved
|- !in out.
    PARITY_IMP(in,out) ==>
    (out 0 = T) /\ (!t. out(SUC t) = (in(SUC t) => ~out t | out t))

Previous subproof:
goal proved
```

Armed with **PARITY_LEMMA**, we can now move quickly towards the final theorem. We will do this proof in one step using the ML function **prove_thm** described in Section 7.2.

```
#let PARITY_CORRECT =
# prove_thm
#  (`PARITY_CORRECT`,
#   "!in out. PARITY_IMP(in,out) ==> (!t. out t = PARITY t in)",
#   REPEAT GEN_TAC
#    THEN DISCH_TAC
#    THEN IMP_RES_TAC PARITY_LEMMA
#    THEN IMP_RES_TAC UNIQUENESS_LEMMA);;
PARITY_CORRECT =
|- !in out. PARITY_IMP(in,out) ==> (!t. out t = PARITY t in)

#close_theory();;
() : void
```

This completes the proof of the parity checking device.

9 Conclusions

The example in the preceding section, though simple, is representative of larger scale proofs using the HOL system. For descriptions of such proofs see Avra Cohn's paper on the verification of the major state machine of the Viper microprocessor [3], Jeff Joyce's paper on the verification of a simple microcoded computer [11], John Herbert's proof of the ECL chip of the Cambridge Fast Ring [10] and Tom Melham's proof of a simple local area network [13]. All of these examples show that HOL can be used to prove non-trivial digital systems correct.

Current research at Cambridge is looking at new ways of modelling hardware in higher-order logic (*e.g.* the representation of low-level circuit behaviour); ways of expressing and relating behavioural abstractions; and methods of executing formal specifications (for animation and simulation).

In addition, we are trying to make the HOL technology available to the industrial community. This involves improving (and possibly reimplementing) the research prototype system to make it acceptably efficient and rugged. It also involves preparing documentation and tutorial material.

Acknowlegements

Members of the Hardware Verification Group at Cambridge have contributed to the development of HOL by using the system and suggesting improvements to it. Special thanks to Avra Cohn, Inder Dhingra, Roger Hale, John Herbert, Jeff Joyce and Tom Melham. Avra Cohn and Larry Paulson pointed out errors and infelicities in an earlier version of this paper.

References

[1] A. J. Camilleri, T. F. Melham and M. J. C. Gordon, *Hardware Verification Using Higher-Order Logic*, University of Cambridge Computer Laboratory, Technical Report No. 91, 1986.

[2] A. Church, *A Formulation of the Simple Theory of Types*, Journal of Symbolic Logic 5, 1940.

[3] A. J. Cohn, *A Proof of Correctness of the Viper Microprocessor: The First Level.* In: *VLSI Specification, Verification and Synthesis*, edited by G. Birtwistle and P.A. Subrahmanyam (this volume).

[4] G. Cousineau, G. Huet and L. Paulson, *The* ML *Handbook*, INRIA, 1986.

[5] M. Gordon, R. Milner L., Morris, M. Newey and C. Wadsworth, *A Metalanhuage for Interactive proof in LCF*, Fifth ACM SIGACT-SIGPLAN Conference on *Principles of Programming Languages*, Tucson, Arizona, 1978.

[6] M. Gordon, R. Milner and C. P. Wadsworth, *Edinburgh LCF: A Mechanised Logic of Computation*, Lecture Notes in Computer Science, Springer-Verlag, 1979.

[7] M. Gordon, *HOL: A Machine Oriented Formulation of Higher-Order Logic*, University of Cambridge Computer Laboratory, Technical Report No. 68, 1985.

[8] M. Gordon, *Why Higher-order Logic is a Good Formalism for Specifying and Verifying Hardware*. In: *Formal Aspects of VLSI Design*, edited by G. Milne and P. A. Subrahmanyam, North-Holland, 1986.

[9] F. K. Hanna and N. Daeche, *Specification and Verification Using Higher-Order Logic*. In: *Formal Aspects of VLSI Design*, edited by G. Milne and P. A. Subrahmanyam, North-Holland, 1986.

[10] J. Herbert, Ph.D. Thesis, University of Cambridge, to appear 1987.

[11] J. J. Joyce, *Verification and Implementation of a Microprocessor*, In: *VLSI Specification, Verification and Synthesis*, edited by G. Birtwistle and P.A. Subrahmanyam (this volume).

[12] A. Leisenring, *Mathematical Logic and Hilbert's ϵ-Symbol*, Macdonald & Co. Ltd., London, 1969.

[13] T. Melham, Ph.D. Thesis, University of Cambridge, to appear.

[14] R. Milner, *Implementation and Application of Scott's Logic for Computable Functions*, Proceedings of the ACM Conference on *Proving Assertions about Programs*, SIGPLAN notices 7,1, 1972.

[15] R. Milner, *A Theory of Type Polymorphism in Programming*, Journal of Computer and System Sciences, 17, 1978.

[16] R. Milner, *A Proposal for Standard ML*, Proceedings of the *1984 ACM Symposium on LISP and Functional Programming*, Austin, Texas, 1984.

[17] L. Paulson, *A Higher-Order Implementation of Rewriting*, Science of Computer Programming 3, 119-149, 1983.

[18] L. Paulson, *Logic and Computation*, Cambridge University Press, to Appear, 1987.

4

Formal Verification and Implementation of a Microprocessor

Jeffrey J. Joyce

University of Cambridge
Computer Laboratory
Corn Exchange Street
Cambridge, CB2 3QG
England.

Introduction

Mike Gordon has described the specification and verification of a microcoded computer using the LCF_LSM hardware verification system [8]. We have subsequently redone this example in higher-order logic using the HOL system [10].

In this paper we present the specification of Gordon's computer in higher-order logic and a brief explanation of its formal verification. A more detailed discussion of the formal verification may be found in [16]. We also describe several related examples of hardware verification based on Gordon's computer and other microprocessor designs. Finally, we report experience in using a formal specification to implement Gordon's computer as a 5,000 transistor CMOS microchip.

1 Specification-Driven Design

Previous discussions of this example in [8] and [16] have presented "finished" specifications of the computer. However, formal specification can also play an active role in the design process by precisely capturing informal descriptions and motivating cleanly derived views of abstract behaviour. In this paper, we describe the specification-driven design of Gordon's computer showing how an informal description can be refined into a formal specification written in the HOL language.

Basic Operation

The principal function of the computer will be to execute programs. However, we will also need a way to load programs into the memory of the computer as well as a means of starting and interrupting the execution of programs. This suggests a design with two modes of operation: an 'idle' mode for loading a new program into the memory and a 'run' mode for program execution.

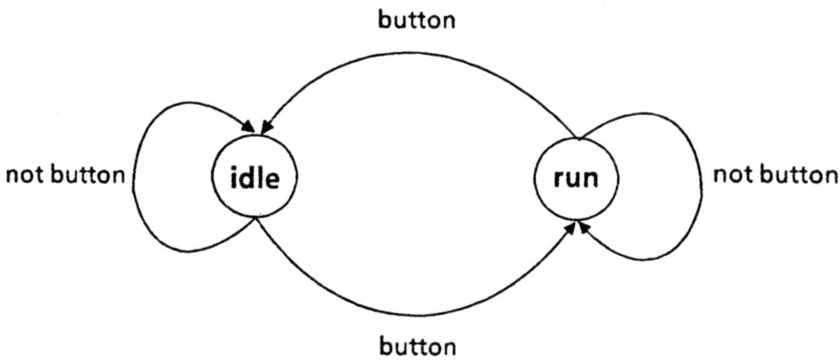

Figure 1: Finite State Machine Behaviour

To make the computer change from idle mode to run mode and *vice versa*, we will include a push button on the front panel of the computer. A transition from idle mode to run mode starts the execution of a program while a transition from run mode to idle mode corresponds to the interruption or termination of an executing program. The state graph in Figure 1 informally captures this first approximation to the basic operation of the computer. This behaviour can also be described in higher-order logic by the following HOL specification.

```
|- ABSTRACT_MACHINE (button_abs,idle_abs) =
   !t:time.
    idle_abs (t+1) =
     idle_abs t =>
      (button_abs t => F | T) | (button_abs t => T | F)
```

The above HOL specification defines a predicate `ABSTRACT_MACHINE` which describes the basic operation of the computer. As the design of the computer evolves we will we revise this definition until we obtain the final behavioural specification of the computer. As it appears above, the predicate `ABSTRACT_MACHINE` constrains the behaviour of `button_abs` and `idle_abs` which are boolean signals. These signals are modelled as functions from discrete time, isomorphic with the natural numbers, to boolean values, `T` and `F`. Hence, `idle_abs (t+1)` is the boolean value of the `idle_abs` signal at time `(t+1)`. The `idle_abs` signal indicates whether the computer is in idle mode or run mode at any time. In addition to booleans, a signal may be a mapping to bi-state and tri-state words or even to an entire memory state. '`!`', '`?`' and '`@`' are ASCII-printable characters which denote universal and existential quantifi-

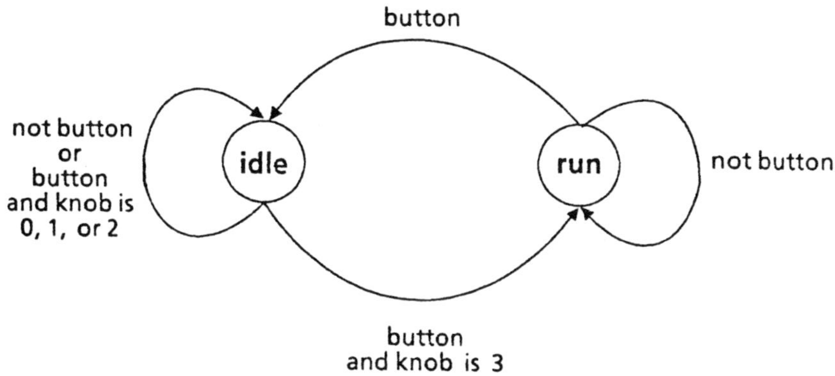

Figure 2: Revised Finite State Machine Behaviour

cation and Hilbert's ϵ-operator in the HOL system. A conditional expression has the form, "`A => B | C`". Given this explanation, it is easy to see that the above HOL specification formally captures the behaviour represented by the state diagram in Figure 1.

We must now consider how a program is to be loaded into memory while the computer is in idle mode. First, we need a set of sixteen switches so that the user can input memory addresses and data values. We will also need a pair of registers to buffer switch inputs. These are the 13-bit program counter PC, and 16-bit accumulator ACC. To store a 16-bit value into a memory location, a 13-bit address is read from the switches into the PC register and a 16-bit data value read from the switches into the ACC register. Following these two steps, the value of the ACC register can be stored in memory at the location given by the PC register. We refine our use of the button by adding a knob which selects one of four possible operations when the button is pushed. These four possible operations are: load PC, load ACC, store ACC at PC, and start program execution at PC.

The state graph is revised in Figure 2 to reflect these refinements. The execution of an instruction is represented by a single transition in the graph because we have not yet proposed an instruction set for the computer.

At this point our design consists of a memory with a 13-bit address space and a 16-bit word size, a 13-bit PC register, a 16-bit ACC register, 16 switches, a four position knob and a push button. Collectively, these signals along with the **`idle_abs`** signal completely describe the state of the abstract computer at any time. The revised HOL specification not only captures the new transitions introduced in the revised state graph but also provides precise details about the change of state for each transition. The HOL specification defines the next state of the computer as a function of its previous state and the current inputs.

```
|- ABSTRACT_MACHINE
    (button_abs,knob_abs,switches_abs,
     memory_abs,pc_abs,acc_abs,idle_abs) =
   !t:num.
    (memory_abs(t+1),pc_abs(t+1),acc_abs(t+1),idle_abs(t+1)) =
     (idle_abs t =>
      (button_abs t =>
       ((VAL2(knob_abs t) = 0) =>
         (memory_abs t, CUT16_13(switches_abs t), acc_abs t, T) |
        (VAL2(knob_abs t) = 1) =>
         (memory_abs t, pc_abs t, switches_abs t, T) |
        (VAL2(knob_abs t) = 2) =>
         (STORE13(pc_abs t)(acc_abs t)(memory_abs t),
          pc_abs t, acc_abs t, T) |
         (memory_abs t, pc_abs t, acc_abs t, F)) |
       (memory_abs t, pc_abs t, acc_abs t, T)) |
      (button_abs t =>
       (memory_abs t, pc_abs t, acc_abs t, T) |
       EXECUTE (memory_abs t, pc_abs t, acc_abs t)))
```

The above HOL specification describes the behaviour of the computer in terms of a relationship between signals of the abstract computer. For example, when the computer is in the idle state, the knob is in position 1 and the button is pushed, the next state of the computer is given by the following 4-tuple. The 4-tuple indicates that the accumulator will be loaded with the current switch inputs while the memory state and program counter will be unchange and the computer will remain in the idle state.

```
(memory_abs t, pc_abs t, switches_abs t, T)
```

Our use of the suffix "_abs" to name signals in the abstract computer emphasises the fact that we are describing these signals from an abstract point of view with respect to time. This convention will also be useful in distinguishing signals of the abstract computer from signals in the concrete implementation. The definition of ABSTRACT_MACHINE implies that every transition occurs in a single time step. However, when we consider a microcoded implementation of the computer, we shall see that state transitions actually require varying amounts of real time, or more precisely, varying numbers of clock cycles. However, it must be emphasized that this abstract view, where a state transition occurs in a single time step, is not inaccurate. Instead, much of our presentation is concerned with showing that this abstract view of sequential behaviour is logically consistent with the sequence of events in a concrete implementation of the computer.

Instruction Set Semantics

Having designed the basic idle/run operation of the computer, we can now proceed with the design of the instruction set. Using a 3-bit opcode and 13-bit operand address instruction format, the following instruction set is proposed.

000—address—	HALT	Stops execution
001—address—	JMP L	Jump to address
010—address—	JZR L	Jump to address if ACC = 0
011—address—	ADD L	Add contents of address to ACC
100—address—	SUB L	Subtract contents of address from ACC
101—address—	LD L	Load contents of address into ACC
110—address—	ST L	Store contents of ACC in address
111—address—	SKIP	Skip to next instruction

We formally specify the semantics of this instruction set in terms of the state change which result from the execution of an instruction. **EXECUTE** is a function which returns a 4-tuple consisting of the next values for the memory, program counter and accumulator and idle/run status given the current state of the computer.

```
|- EXECUTE (memory_val,pc_val,acc_val) =
    let opcode = VAL3(OPCODE(FETCH13 memory_val pc_val)) in
    let address = CUT16_13(FETCH13 memory_val pc_val) in
     ((opcode = HALT) => (memory_val, pc_val, acc_val, T) |
      (opcode = JMP) => (memory_val, address, acc_val, F) |
      (opcode = JZR) =>
       ((VAL16 acc_val = 0) =>
         (memory_val, address, acc_val, F) |
         (memory_val, INC13 pc_val, acc_val, F)) |
      (opcode = ADD) =>
       (memory_val, INC13 pc_val,
        ADD16 acc_val (FETCH13 memory_val address), F) |
      (opcode = SUB) =>
       (memory_val, INC13 pc_val,
        SUB16 acc_val (FETCH13 memory_val address), F) |
      (opcode = LD) =>
       (memory_val, INC13 pc_val, FETCH13 memory_val address, F) |
      (opcode = ST) =>
       (STORE13 address acc_val memory_val,
        INC13 pc_val, acc_val, F) |
       (memory_val, INC13 pc_val, acc_val, F))
```

The above definitions of **ABSTRACT_MACHINE** and **EXECUTE** form a complete target level specification of the proposed computer. A functional simulation of the abstract machine could easily be derived from this specification, however our interest lies in formally relating this target level specification to a detailed specification of a register-transfer level implementation of the computer.

2 Concrete Implementation

Many different implementations of the above computer are possible. For example, Daniel Weise has described a "RISC" implementation of the computer where every state transition occurs in exactly one clock cycle [18]. However, we follow Gordon's original presentation of this example by implementing the computer with a microcoded control unit.

The register-transfer level implementation of the computer is shown in Figure 3. In addition to the PC and ACC registers which are a part of the abstract computer, several other internal registers have been added. The registers, the memory, an arithmetic-logic unit and the external switches are connected to a 16-bit data bus through tri-state buffers. The control unit consists of a 32-word by 30-bit microcode ROM, decoding logic and a 5-bit microcode program counter.

The microcode is written to realize the abstract behaviour of the computer using this architecture. There are a total of fifteen possible transitions in the revised state graph of Figure 2 when the single transition for the execution of an instruction is expanded into eight different transitions. Each of the fifteen transitions are implemented by a sequence of microinstructions whose net effect is given by the definitions of `ABSTRACT_MACHINE` and `EXECUTE`. Furthermore, each sequence will start and finish at one of two locations in the microcode depending on the start and finish states, *i.e.* 'idle' or 'run', of the corresponding transition. In the final version of the microcode shown in Figure 4, the 'idle' and 'run' states correspond to locations 0 and 5.

The register-transfer level implementation of the computer can be formally described in the HOL language using hierarchical specification [4]. This method of hardware specification within higher-order logic is based on the composition of smaller circuits and hiding internal signals. Circuit composition and internal signal hiding are modelled by logical conjunction and existential quantification. The use of existential quantification to hide internal signals is explained in [17].

Behavioural specification in HOL involves the use of logical constants, *e.g.* ADD16, which are informally understood to name certain operations [1]. Unlike the specification of a simulation model, it is not always necessary to provide details about these operations. In other cases some properties of the operation may be stated or the operation may even require a full definition. Whereas a simulation model must be completely operational, it is possible to formally

[1]The distinction between a formal theory and its interpretation is discussed in [11].

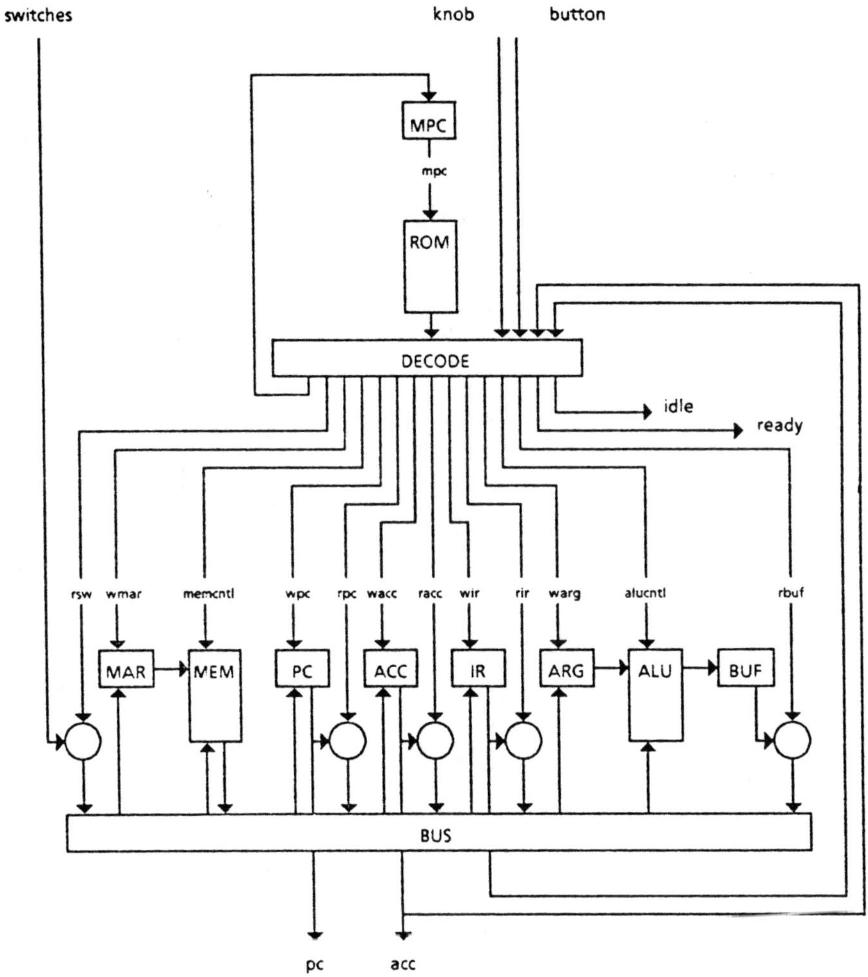

Figure 3: Register-Transfer Level Implementation of Gordon's Computer

Location	Control	Signals	Next Location	Explanation
0	ready	idle	test_button (1,0)	begin idling cycle
1			jump_knob 1	decode knob position
2	rsw	wpc	jump 0	switches → PC
3	rsw	wacc	jump 0	switches → ACC
4	rpc	wmar	jump 7	PC → MAR
5	ready		test_button (0,6)	begin instruction execution
6	rpc	wmar	jump 8	PC → MAR
7	racc	write	jump 0	ACC → MEM (MAR)
8	read	wir	jump 9	MEM (MAR) → IR
9			jump_opcode 10	decode instruction
10			jump 0	HALT
11	rir	wpc	jump 5	JMP, IR → PC
12			test_acc (11,17)	JZR
13	racc	warg	jump 19	ADD, ACC → ARG
14	racc	warg	jump 22	SUB, ACC → ARG
15	rir	wmar	jump 24	LD, IR → MAR
16	rir	wmar	jump 25	ST, IR → MAR
17	rpc	inc	jump 18	SKIP, PC + 1 → BUF
18	rbuf	wpc	jump 5	BUF → PC
19	rir	wmar	jump 20	IR → MAR
20	read	add	jump 21	ARG + MEM (MAR) → BUF
21	rbuf	wacc	jump 17	BUF → ACC
22	rir	wmar	jump 23	IR → MAR
23	read	sub	jump 21	ARG - MEM (MAR) → BUF
24	read	wacc	jump 17	MEM (MAR) → ACC
25	racc	write	jump 17	ACC → MEM (MAR)
26			jump 0	unused
27			jump 0	unused
28			jump 0	unused
29			jump 0	unused
30			jump 0	unused
31			jump 0	unused

Figure 4: Microcode Program Pseudo-Code

reason about hardware from an abstract point of view where certain operational details are suppressed.

At the leaf level, primitive components are specified in terms of their sequential behaviour. For example, the behavioural specification of a 16-bit register is shown below.

```
|- REG16 (input,load,output) =
   !t:time. output (t+1) = ((load t) => (input t) | (output t))
```

The intermediate level of specification consists of structural specifications for the data path and the control unit in terms of primitive components. The top level specification defines the structure of the entire computer as the composition of the data path and the control unit. Existential quantification is used to hide the the control signals connecting the control unit to the data path. For readability, `CONCRETE_MACHINE` is defined as a predicate on signals grouped into target level signals, internal state signals, and the `ready` signal. Following the terminology used by Avra Cohn in the formal specification of the VIPER microprocessor [5], the internal state signals, `mpc`, `mar`, `ir`, `arg` and `buf`, are collectively known as the 'transient'. As we shall see, the `ready` signal has a fundamental role in relating the behaviour imposed by the definition of `CONCRETE_MACHINE` to the abstract behaviour of the computer.

```
|- CONCRETE_MACHINE
     (button,knob,switches,memory,pc,acc,idle)
     (mpc,mar,ir,arg,buf) ready =
   ?rsw wmar memcntl wpc rpc wacc racc wir rir warg alucntl rbuf.
     CONTROL_UNIT
      MICROCODE
      (button,knob,acc,idle,mpc,ir,ready,rsw,wmar,
       memcntl,wpc,rpc,wacc,racc,wir,rir,warg,alucntl,rbuf) /\
     DATA_PATH
      (switches,memory,pc,acc,mar,ir,arg,buf,rsw,wmar,
       memcntl,wpc,rpc,wacc,racc,wir,rir,warg,alucntl,rbuf)
```

The complete specification of the implementation is given in an appendix to this paper and a detailed explanation may be found in [16]. This specification results in a predicate, `CONCRETE_MACHINE`, which defines a relationship on a set of signals from one clock instant to the next. This behaviour is imposed by the structure of the implementation and the behaviour of primitive components.

To summarize our progress so far, we have specified both the abstract and concrete behaviours for the computer. `ABSTRACT_MACHINE` defines the sequential behaviour of an abstract computer in terms of state transitions. On the other hand, CONCRETE_MACHINE defines the sequential behaviour of a concrete implementation in terms of clocked sequential components. A central feature of these formal specifications is that the abstract computer and its concrete implementation operate at different time scales. To prove that the computer has been implemented correctly, we need to show that the abstract behaviour of the computer is logically consistent with its concrete implementation. A fundamental step in the formal verification of the computer will be to establish a mathematical relationship between the abstract time scale and the concrete time scale.

The concrete behaviour of the computer consists of executing sequences of microinstructions whose net effect correspond to state transitions in the abstract computer. These sequences are represented in Figure 5 which shows the state graph for the concrete machine. Location 0 in the microcode corresponds to start of a sequence of microinstructions implementing a transition which begins in the 'idle' state and finishes in either the 'idle' or 'run' state. Similarly, location 5 corresponds to the start of a transition from the 'run' state. Therefore, when the microinstruction program counter reaches either location 0 or location 5, the state of the concrete implementation must be consistent with the state of the abstract computer. The ready signal was specifically included in the implementation to signal when the microinstruction program counter reaches location 0 or location 5. This explains the significance of the ready signal in relating the behaviour of the concrete implementation to the behaviour of the abstract computer: the ready signal of the concrete implementation is the implicit clock signal of the abstract computer.

To formally describe the relationship between the time scale of the abstract computer and the concrete time scale, we need to define a function which computes the concrete time of the *nth* point on the abstract time scale. More precisely, this function will compute the concrete time of *nth* occurrence of the ready signal.

In these proceedings, Tom Melham defines a function, TimeOf, which constructs a temporal abstraction function from a signal such as ready [17]. (TimeOf ready) is a function which maps abstract time to points on the concrete time scale. For completeness, the HOL definitions of TimeOf and the relation IsTimeOf are shown below, however the reader is referred to [17] for an explanation of these definitions and related theorems.

```
|- IsTimeOf p 0 t = p t /\ !t'. t' < t ==> ~(p t')
|- IsTimeOf p (n+1) t =
     p t /\
     ?t'. t' < t /\ IsTimeOf p n t' /\
          !t''. (t' < t'' /\ t'' < t) ==> ~(p t'')

|- TimeOf p n = @t. IsTimeOf p n t
```

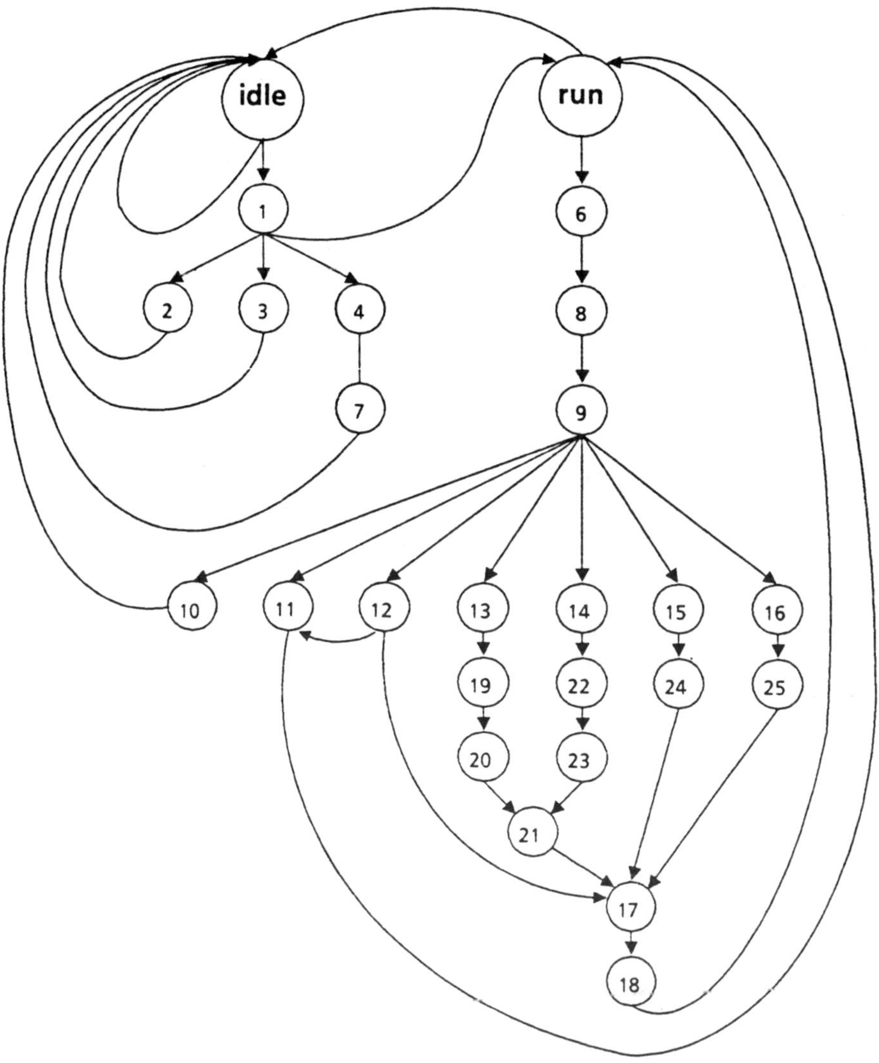

Figure 5: State Graph for the Concrete Implementation

Using `TimeOf` to construct a temporal abstraction function from the `ready` signal and function composition, signals of the abstract computer can be formally related to the corresponding signals in the concrete implementation. A formal relationship between abstract and concrete signals allows the correctness of the implementation with respect to the abstract computer to be stated by the following theorem. The correctness of the implementation depends on assumptions about the stability of external inputs, in particular, the signals `knob` and `switches`. The predicate STABLE is used to formally express these assumptions.

```
|- STABLE (t1,t2) s =
   !t. ((t1 < t) /\ (t < t2)) ==> ((s t) = (s t1))

|- CONCRETE_MACHINE
     (button,knob,switches,memory,pc,acc,idle)
     (mpc,mar,ir,arg,buf) ready /\
   (button_abs = button o (TimeOf ready)) /\
   (knob_abs = knob o (TimeOf ready)) /\
   (switches_abs = switches o (TimeOf ready)) /\
   (memory_abs = memory o (TimeOf ready)) /\
   (pc_abs = pc o (TimeOf ready)) /\
   (acc_abs = acc o (TimeOf ready)) /\
   (idle_abs = idle o (TimeOf ready)) /\
   (!t. STABLE(TimeOf ready t,TimeOf ready (t+1)) knob) /\
   (!t. STABLE(TimeOf ready t,TimeOf ready (t+1)) switches)
   ==>
   ABSTRACT_MACHINE
    (button_abs,knob_abs,switches_abs,
     memory_abs,pc_abs,acc_abs,idle_abs)
```

3 Formal Verification

A detailed description of the formal verification of Gordon's computer may be found in [16]. In this paper, we simply sketch an outline of the proof.

To prove the correctness statement given at the end of the previous section, we need to show that the values of the signals `button_abs`, `knob_abs`, `switches_abs`, `memory_abs`, `pc_abs`, `acc_abs` and `idle_abs` are related from one point in time to the next according to the definition of `ABSTRACT_MACHINE`.

The first step is to rewrite the hierarchical specification of the concrete implementation into a set of output equations for each of the externally visible state signals: **memory**, **pc**, **acc**, **idle**, **mpc**, **mar**, **ir**, **arg**, **buf** and **ready**. This step, called "expansion", is performed automatically by a derived rule of inference provided the by HOL system. For example, expansion results in the following equation for the output of the accumulator showing that ACC is controlled by bit 22 of the microinstruction word [2].

```
!t.
 acc(t + 1) =
  (CNTL_BIT 22(FETCH5 MICROCODE(mpc t)) => bus t | acc t))
```

This set of output equations can be used to derive theorems about future values of signals. In effect, the register-transfer implementation can be symbolically simulated through a sequence of states using case analysis on the initial state and current inputs to examine specific sequences of microinstructions. The crux of the proof is to show that for every possible state transition at the target level, the net effect of the the corresponding sequence of microinstructions preserves the effect of the target level state transition.

For example, if the computer is at the beginning of the run cycle and about to execute an ADD instruction, then assuming the interrupt button is not pushed, the following theorem can be proved about the state of the computer at the end of the cycle with respect to its initial state at time 't'. Note that the end of the cycle occurs in ten time units.

```
... |- (memory t+10, pc t+10, acc t+10, idle t+10)
     = (memory t, INC13(pc t), ADD16 (acc t) OP, F)

where OP =  FETCH13
             (memory t)
             (CUT16_13(FETCH13(memory t)(pc t))))
```

Another important set in the proof is to show that the **ready** signal is true infinitely often. This is a consequence of the fact that from any location in the microcode, control will eventually return to either location 0 or location 5 and the fact that the implicit clocking of register-transfer level primitives is assumed to be infinite. In practical terms, this implies that the computer will reset properly from a random location in the microcode on power-up. As a step in the formal proof, showing that the signal **ready** is infinite results in several useful theorems described in [17]. Using these theorems, results from symbolic simulation of the concrete implementation can be translated into statements about the behaviour of the implementation with respect to the abstract time scale. In this manner, we can obtain, for example, the following theorem about the abstract computer when an ADD instruction is executed.

---

[2] Complete expansion would actually expand the terms "CNTL_BIT" and "bus".

```
... |- (memory_abs t+1, pc_abs t+1, acc_abs t+1, idle_abs t+1)
     = (memory_abs t, INC13(pc_abs t), ADD16 (acc_abs t) OP, F)

    where OP =  FETCH13
                 (memory_abs t)
                 (CUT16_13(FETCH13(memory_abs t)(pc_abs t))))
```

This strategy of case analysis and symbolic simulation is straightforward in the case of Gordon's computer because there are exactly fifteen different possible sequences of microinstructions between adjacent points of time on the abstract time scale. A more realistic design would involve more than just fifteen unique sequences of microinstructions. For instance, Warren Hunt's FM8501 processor has a 'wait state' where control loops until an asynchronous memory process responds to a request. In this case, analysis of a microinstruction sequence involving a memory request would require a proof by induction on the length of the wait state.

The verification of the VIPER microprocessor [5] employs a similar strategy of case analysis and symbolic simulation to prove that event sequences at a finer grain of time correctly implement major state transitions in the target machine. In the next section we describe several other related examples of hardware verification beginning with Gordon's LCF_LSM version of the computer proof.

4 Related Examples of Hardware Verification

LCF_LSM

The LCF_LSM version of Gordon's computer example uses special terms to denote sequential behaviour. These terms consist of output and next state equations in terms of inputs and current state. A sequential behaviour for the concrete implementation is obtained from primitive sequential behaviours using the LCF_LSM analogues of circuit composition and internal signal hiding [7]. The target level behaviour is also specified as a sequential device using LCF_LSM device behaviour terms.

As in the case of the HOL specification, the behaviour of the target level device and the concrete implementation operate at different but related time scales. A special LCF_LSM term, '`until c do t`' is a temporal abstraction mechanism used to relate the sequential behaviour of the concrete implementation to the abstract behaviour of the target level computer. The semantics of this type of term are stated in [7]. Informally, an '`until c do t`' term sequences a behaviour denoted by 't' until the condition 'c' becomes true. In the computer example, this term is used to step the concrete implementation through a sequence of microinstructions implementing a target level operation. This allows correctness to be stated as follows in the LCF_LSM language.

```
|- ABSTRACT_MACHINE (memory_state,pc_state,acc_state,idle_state)
   == until ready do
        CONCRETE_MACHINE
         (WORD5 (idle_state->0|5),
          memory_state,pc_state,acc_state,
          mar_state,ir_state,arg_state,buf_state)
```

The LCF_LSM correctness statement for the computer proof restricts the initial state of the microinstruction program counter to the start of either the idle or run cycles depending on the boolean variable **idle_state**. In effect, this statement of correctness only shows that the major state transitions of the abstract computer are correctly implemented by the corresponding microinstruction sequences of the concrete machine. This falls short of a statement of 'total' correctness for the computer implementation. For example, the LCF_LSM correctness statement ignores the problem of resetting the computer. The original specification of the computer implementation in [8] left the unused part of the microcode ROM (locations 26 to 31) unspecified making it is easy to construct a case where the computer fails to power up correctly, *e.g.* it could loop forever in the unused part of the microcode. Another shortcoming is that the need for stable switch and knob inputs is not made apparent by the LCF_LSM proof. The UNTIL rule implicitly introduces an assumption that all inputs remain stable between major state transitions whether or not this matters while these assumptions must be stated explicitly in the HOL correctness statement.

VERIFY

Harry Barrow's VERIFY system closely resembles the LCF_LSM style of hardware specification and has been used to verify Gordon's computer [1]. This includes a temporal abstraction mechanism which is similar to the '**until c do t**' terms of LCF_LSM. The VERIFY system, implemented in Prolog, has been very successful in automating the verification of non-trivial designs using enumeration and symbolic manipulation strategies to prove the equivalence of an implementation and its desired behaviour. Automation is partially achieved by requiring the user to state behaviours (*i.e.* sub-goals) at most intermediate levels in a hierarchical design. Nevertheless, the verification of Gordon's computer required considerably less human intervention than the LCF_LSM version.

Silica Pithecus

Daniel Weise uses Lisp to describe the formal specification of a computer based on Gordon's example [18]. However, Weise has replaced the microcoded implementation in Gordon's example with an implementation that executes a target level instruction every clock cycle (*e.g.* the program counter has its own incrementation circuit). This modification simplifies the proof of correctness at the register-transfer level, however Weise's primary interest lies in establishing the correctness of the register-transfer implementation with respect to lower level circuit models. Weise does not discuss the target level specification of his computer or the formal verification of the register-transfer specification with respect to a target level specification.

Boyer-Moore

Using the Boyer-Moore theorem prover, Warren Hunt has formally verified the FM8501 microprocessor which is similar in complexity to a PDP-11 [14]. Hunt also specifies behaviour using an approach similar to LCF_LSM; clocked sequential devices are modelled as recursive functions where each recursive call corresponds to a clock cycle.

The significant difference between the FM8501 and Gordon's computer is that memory is modelled more realistically as an independent asynchronous process. In higher-order logic, an asynchronous memory process could be straightforwardly modelled using existential quantification. In a quantifier-free logic such as Boyer-Moore logic, free variables correspond to universally quantified variables while functions are used in place of existential quantification. The absence of existential quantification in Boyer-Moore logic motivates the use of an oracle to predict asynchronous responses from the memory process. Here, an oracle is a function from time to asynchronous events. In a higher-order logic version of the FM8501 example, induction would be used to prove that microcode 'wait states' terminate upon response from the memory process.

In addition to verifying the register-transfer level behaviour with respect to the target level specification, Hunt establishes the correctness of gate level implementations for register-transfer level devices such as the arithmetic-logic unit.

5 Specification-Driven Implementation

An 8-bit version of Gordon's computer was implemented as a 5,000 transistor CMOS microchip in December 1985 while the author was a visiting student at Xerox PARC. The purpose of this exercise was to study the role of formal specification in the implementation of a design. The design was small enough to be completed in two months but was sufficiently varied to be representative of many aspects of VLSI design. The fabricated chip was tested in April 1986 at the University of Calgary and found to be fully functional.

Besides halving the size of the data path down to an 8-bit version, the most significant difference between the formally specified design and the actual implementation was the use of a two-phase non-overlapping clock. We are currently studying how to prove that the clocked sequential behaviour of the register-transfer level specification is an abstract view of the two-phase non-overlapping clocking scheme. In addition to the formally specified control unit and data path, a test structure and a small static RAM memory were included in the implementation.

The layout of the "TAMARACK" microprocessor, the name given to the implementation, is shown in Figure 6. Four separate subsystems are easily recognized in the layout. From top to bottom, these are: control unit, test structure, data path and 32-word by 8-bit RAM memory. Several other features are also easily identified such as the control signal wires and the microcode ROM at the top right-hand corner of the layout.

An important decision in the design of the TAMARACK microprocessor was to use ratioless, complementary, restored, static CMOS logic wherever possible. A very simple layout style was also used even where this resulted in less than optimal space and time utilization [3]. Our desire for correct functionality over marginally better performance justified our use of simple circuitry and simple layout. Another motivation for simple circuitry was to see how much of a varied system could be designed using a simplified view of circuit behaviour. We found that only a very few parts of the design warranted more complex circuitry. The microcode ROM was implemented as a pseudo-NMOS NOR array and a 6-transistor XOR gate was used to implement the full-adder in the arithmetic-logic unit. Both of these departures resulted in considerable savings of circuit area using well-understood variations of CMOS logic [19]. However, a simple model of circuit behaviour could not be used to design the RAM memory. In this case, a detailed electrical model of the RAM memory was developed. The electrical model was not a structural representation of the RAM memory but rather a model based on analysis of worst-case behaviour.

In many respects the usefulness of our formal specification of Gordon's computer was less than expected. The register-transfer level specification did not significantly influence circuit design and layout at the transistor level. Furthermore, there was no way to machine-check the consistency of the layout with the formal specification. A slight mis-ordering of control signals in the layout was only discovered by visual inspection shortly before the design was submitted for fabrication. This indorses the need for a tighter integration of hardware verification systems with conventional VLSI CAD tools.

One of the most interesting discoveries of this implementation exercise was that the formal verification of the design missed a design error: there is no reset button to initialize the microinstruction program counter (this is different from the function of the interrupt button). The HOL version of the computer proof improves upon the original LCF_LSM version by ensuring that for any

[3]The same principles were used in the implementation of the much larger SECD chip, University of Calgary, 1986. The SECD chip is a 20,000 transistor processor which executes a purely functional language called "Lispkit" [12].

Figure 6: Microphotograph of the TAMARACK microprocessor

initial state of the microinstruction program counter, even an address into the 'unused' part of the microcode, the computer will eventually reach the start of either the 'idle' or 'run' cycles, *i.e.* locations 0 or 5. However, our formal specification implied that the microinstruction program counter would assume a proper state on power-up and not some undefined value.

After the design was submitted for fabrication and long after its formal verification, we decided to simulate the design using a switch-level simulator involving a more accurate model of a signal value [3]. The design failed to simulate properly because the initial state of the mpc signal was undefined and there was no way to force the signal to a defined state. Fortunately, when the actual chips were returned and tested we found that the mpc signal tended to address 0 due to electrical factors not modelled at the switch-level and the chip worked correctly.

Why did formal verification fail to discover this design error ? Clearly, the actual verification process, *i.e.* formal deduction, was correct. The source of the problem was the incorrect assumption that the mpc signal was bi-stable. This assumption was introduced by our behavioural specification of the MPC as a primitive component.

Bi-stability is an abstract view of more complex forms of data in digital circuits, *e.g.* tri-state data, voltage values, etc. This abstraction is valid only under certain conditions. For instance, we could prove that the output of a register such as the MPC is bi-stable provided that the input to the register is bi-stable. In the formal specification of the computer we simply assumed that the output of the MPC was bi-stable. If instead we had been required to establish that the output of the MPC was bi-stable by showing that the input to the MPC was bi-stable (following a reset), then the need for a reset button would have inevitability been discovered.

More generally, verification failed here because our register-transfer level primitives were not adequate models for reliable verification. Behaviours at the register-transfer level should be derived from some lower level model such as a switch-level model. [15] describes the verification of a CMOS full-adder using a data abstraction from a 4-value logic (low, high, float and error) up to bi-stable values. Behavioural models of register-transfer level components derived from a lower level model would clearly state the conditions under which the behaviour is valid. A library of register-transfer component behaviours and their respective validity conditions would then serve as a basis for a hardware verification methodology for VLSI design.

6 Summary and Future Research

The verification procedure requires approximately six hours of running time on a lightly loaded, 14 megabyte VAX 11/780. The proof required about two man-months of effort which included learning how to specify and verify hardware in higher-order logic and how to use the HOL proof assistant system.

The main overhead of producing the initial specification and writing the ML programs to do the necessary inferences need only be done once. Thereafter, the finished proof could be easily modified to accommodate small changes in the hardware. The correctness of incremental changes to a design can be verified by editing the specification and verification procedure, usually in a very minor way, and re-running the proof as a batch job. This contrasts with the simulation approach which may require less initial effort but every modification requires the complete re-run and debugging of the entire simulation test suite. An experienced designer at Xerox PARC wisely observed that most 'bugs' in a VLSI design are due to last-minute modifications when time constraints do not permit complete re-simulation. Therefore, the initial effort to produce a formal verification may be rewarded in the long term.

Fundamental Circuit Models

Future research in this area will likely focus on better circuit models [20]. Our implementation of Gordon's computer suggests that it is possible to design non-trivial hardware with a simple model of circuit behaviour especially when functionality is more important than performance, *e.g.* safety-critical systems. Therefore, a fundamental circuit model for the purposes of formally reasoning about circuit behaviour is a reasonable prospect.

Computer-Assisted Theorem Proving

The formal verification of non-trivial designs clearly requires computer assistance. Some of the theorems in the HOL version of Gordon's example were several pages in length but could be manipulated with ease using the HOL system. Nevertheless, the HOL system required considerable human intervention to guide the verification process. Other hardware verification systems such as VERIFY have largely automated the verification process but there will always be designs for which automatic verification fails. Although the automation of tedious low-level inference steps is universally agreed upon, the complete automation of the verification process is not necessarily desirable. We believe that the mere generation of a theorem in a formal system is unconvincing when taken as proof that a circuit has been properly designed. Behavioural specification of primitive components will always be approximations of physical reality. More importantly, the implication of a correctness statement may not be fully understood. Rather, it is the verification process itself and not the final result which is the more plausible guarantee of correct design. Future research in this area should focus on providing powerful tools which allow designers to reason accurately and efficiently about hardware designs.

Specification-Driven Design

Most examples of hardware verification are based on "finished" designs. Little is known about the role of formal specification and verification in the actual design process. For instance, it is unclear how the derivation of a behaviour from a structure would significantly influence the design of the structure (other than to reject an incorrect design). However, we suggest that hardware verification has a potentially useful role in clarifying the informal abstractions used by designers.

The design of complex hardware is often viewed as a hierarchy of informal abstractions. For instance, Neil Weste and Kamran Eshraghian present a relatively simple working set of abstractions about the behaviour of integrated circuits, *e.g.* CMOS logic structures, timing strategies [19]. At a much higher level, the specification-driven design of Gordon's computer used abstraction to describe the target-level behaviour of the computer suppressing details about its concrete implementation. Once the abstract behaviour was formally specified in HOL, a microcode program was developed according to a mathematically defined relationship between the abstract and concrete time scales.

More complicated hardware designs would involve a hierarchy of abstractions about data, time and behaviour [17]. For example, various memory management strategies could be formally verified resulting in simpler and more abstract models of memory which could then be used to consider the correctness of the system at a higher level. An abstract view of the record allocation process in a Lisp machine would be designed to suppress details about garbage collection. Establishing the validity of this abstraction would involve showing that an abstract view of the processor state is left untouched by garbage collection.

Finally, we suggest that formal abstraction may serve as a metric for "good" hardware design. Abstractions are usually only valid under certain conditions. For instance, the target-level behaviour of Gordon's computer is an abstract view of its register-transfer level behaviour only under assumptions about the stability of the knob and switch inputs. The formal verification of Gordon's computer results in a correctness statement in which these assumptions look untidy. It could be argued that a better design for the computer would have latched the knob and switch inputs at the start of the idle cycle eliminating the need for assumptions about the stability of these inputs. Hence, a correctness statement with fewer validity conditions may be an indication of a "better" design.

Acknowledgements

I am grateful to Mike Gordon and Tom Melham for their helpful advice on this example and on using the HOL system. Xerox PARC generously fabricated the CMOS implementation of the TAMARACK microprocessor; thanks especially to Brian Preas, Rick Barth and Louis Monier. Rick Schediwy and Glenn Stone worked on testing the fabricated chip and Graham Birtwistle has helped to clarify my presentation of this work.

This research is funded by the Cambridge Commonwealth Trust, Government of Alberta Heritage Fund and the Natural Sciences and Engineering Research Council of Canada.

References

[1] Barrow, H., "VERIFY: A Program for Proving Correctness of Digital Hardware Designs", Artificial Intelligence, Vol. 24, No. 1-3, December 1984.

[2] Birtwistle, G., Joyce, J., Liblong, B., Melham, T., and Schediwy, R., "Specification and VLSI", Formal Aspects of VLSI Design: Proceedings of the 1985 Edinburgh Conference on VLSI, G.J. Milne and P.A. Subrahmanyam. eds., North-Holland, Amsterdam, 1986.

[3] Bryant, R., "An Algorithm for MOS Logic Simulation", Lambda Magazine, Fourth Quarter, 1980.

[4] Camilleri, A., Gordon, M., and Melham, T., "Hardware Verification using Higher Order Logic", Technical Report No. 91, Computer Laboratory, The University of Cambridge, June 1986.

[5] Cohn, A. "A Proof of Correctness of the Viper Microprocessor: The First Level", Specification, Verification and Synthesis, January 1987, (this volume).

[6] Gordon, M., "Representing a Logic in the LCF Metalanguage", Tools and Notions for Program Construction, edited by D. Neel, Cambridge University Press, 1982.

[7] Gordon, M., "LCF_LSM, A System for Specifying and Verifying Hardware", Technical Report No. 41, Computer Laboratory, The University of Cambridge, September 1983.

[8] Gordon, M., "Proving a Computer Correct using the LCF_LSM Hardware Verification System", Technical Report No. 42, Computer Laboratory, The University of Cambridge, September 1983.

[9] Gordon, M., and Herbert, J., "A Formal Hardware Verification Methodology and its Application to a Network Interface Chip", Technical Report No. 66, Computer Laboratory, The University of Cambridge, 1985.

[10] Gordon, M., "HOL: A Proof Generating System for Higher Order Logic", Specification, Verification and Synthesis, January 1987, (this volume).

[11] Hanna, F., and Daeche, N., "Specification and Verification of Digital Systems using Higher Order Predicate Logic", IEE Proceedings, Vol. 133, Part E, No. 5, September 1986.

[12] Henderson, P., "Functional Programming: Application and Implementation", Prentice-Hall, 1980.

[13] Herbert, J., "Application of Formal Methods to Digital System Design", Ph.D Thesis, Computer Laboratory, University of Cambridge, (forthcoming 1987).

[14] Hunt, W., "FM8501: A Verified Microprocessor", Ph.D. Thesis and Technical Report No. 47, Institute for Computer Science, The University of Texas at Austin, February 1986.

[15] Joyce, J., "Formal Verification of a CMOS Full-adder in the HOL System", Internal Report, University of Cambridge. April 1987.

[16] Joyce, J., Birtwistle, G., and Gordon, M., "Proving a Computer Correct in Higher Order Logic", Technical Report No. 100, Computer Laboratory, The University of Cambridge, December 1986.

[17] Melham, T., "Abstraction Mechanisms for Hardware Verification", Specification, Verification and Synthesis, January 1987, (this volume).

[18] Weise, D., "Formal Multilevel Hierarchical Verification of Synchronous MOS VLSI Circuits", Ph.D Thesis, Massachusetts Institute of Technology, August 1986.

[19] Weste, N., and Eshraghian, K., "Principles of CMOS VLSI Design", Addison-Wesley Publishing Company, 1985.

[20] Winskel, G., "Models and Logic of MOS Circuits", Specification, Verification and Synthesis, January 1987, (this volume).

Appendix

The formal specification of the register-transfer level implementation of Gordon's computer is a higher-order logic 'theory' consisting of constants, axioms, and definitions.

The following constants are used to specify the behaviour and implementation of Gordon's computer.

EL	:num→* list→*	nth element of a list
SEG	:num#num→* list→* list	sublist of a list
V	:bool list→num	number denoted by a bit list
VAL<n>	:word<n>→num	number denoted by a word
WORD<n>	:num→word<n>	word representing a number
BITS<n>	:word<n>→bool list	list of bits in a n-bit word
FLOAT16	:tri_word16	high impedance value
DEST_TRI16	:tri_word16→word16	tri-state into bi-state
MK_TRI16	:word16→tri_word16	bi-state into tri-state
U16	:tri_word16→tri_word16	joining two tri-state signals
FETCH5_30	:mem5_30→word5→word30	ROM memory fetch
FETCH13_16	:mem13_16→word13→word16	RAM memory fetch
STORE13_16	:word13→word16 →mem13_16→mem13_16	RAM memory store
PAD13_16	:word13→word16	pad 3 most significant bits
CUT16_13	:word16→word13	cut 3 most significant bits
INC16	:word16→word16	16-bit incrementation
ADD16	:word16→word16→word16	16-bit addition
SUB16	:word16→word16→word16	16-bit subtraction
OPCODE	:word16→word13	extract opcode field
MICROCODE	:mem5_30	microcode ROM

The constants EL, SEG, V, VAL<n>, WORD<n> and BITS<n> are built into the HOL system. These constants name operations on lists and conversions between natural numbers and their binary representations. Currently, complete theories about lists and n-bit words have not been developed. Instead, each of these constants has an improvised evaluation rule which evaluates applications of the constant to primitive data objects.

The constants FETCH5_30, FETCH13_16, STORE13_16, INC16, ADD16, SUB16 and OPCODE are 'understood' to name certain operations but have no properties in the formal theory. Properties for the constants PAD13_16, CUT16_13, FLOAT16, DEST_TRI16, MK_TRI16, U16 and MICROCODE are described by the following axioms.

```
|- !w:word13. CUT16_13 (PAD13_16 w) = w

|- !w:word16. DEST_TRI16 (MK_TRI16 w) = w

|- !w:tri_word16. (FLOAT16 U16 w = w) /\ (w U16 FLOAT16 = w)

|- FETCH5 MICROCODE #00000 = #000000000000000110000000001001
|- FETCH5 MICROCODE #00001 = #000000000000000000001000000011
|- FETCH5 MICROCODE #00010 = #010001000000000000000000000000
|- FETCH5 MICROCODE #00011 = #010000010000000000000000000000
|- FETCH5 MICROCODE #00100 = #001000100000000000011100000000
|- FETCH5 MICROCODE #00101 = #000000000000000100011000000001
|- FETCH5 MICROCODE #00110 = #001000100000000000100000000000
|- FETCH5 MICROCODE #00111 = #000100001000000000000000000000
|- FETCH5 MICROCODE #01000 = #000010000100000000100100000000
|- FETCH5 MICROCODE #01001 = #000000000000000000101000000100
|- FETCH5 MICROCODE #01010 = #000000000000000000000000000000
|- FETCH5 MICROCODE #01011 = #000001000010000000010100000000
|- FETCH5 MICROCODE #01100 = #000000000000000001000101011010
|- FETCH5 MICROCODE #01101 = #000000001001000001001100000000
|- FETCH5 MICROCODE #01110 = #000000001001000001011000000000
|- FETCH5 MICROCODE #01111 = #001000000010000001100000000000
|- FETCH5 MICROCODE #10000 = #001000000010000001100100000000
|- FETCH5 MICROCODE #10001 = #000000100000010001001000000000
|- FETCH5 MICROCODE #10010 = #000001000000000100001010000000
|- FETCH5 MICROCODE #10011 = #001000000010000001010000000000
|- FETCH5 MICROCODE #10100 = #000010000000100001010100000000
|- FETCH5 MICROCODE #10101 = #000000010000001001000100000000
|- FETCH5 MICROCODE #10110 = #001000000010000001011100000000
|- FETCH5 MICROCODE #10111 = #000010000000110001010100000000
|- FETCH5 MICROCODE #11000 = #000010010000000001000100000000
|- FETCH5 MICROCODE #11001 = #000100001000000001000100000000
|- FETCH5 MICROCODE #11010 = #000000000000000000000000000000
|- FETCH5 MICROCODE #11011 = #000000000000000000000000000000
|- FETCH5 MICROCODE #11100 = #000000000000000000000000000000
|- FETCH5 MICROCODE #11101 = #000000000000000000000000000000
|- FETCH5 MICROCODE #11110 = #000000000000000000000000000000
|- FETCH5 MICROCODE #11111 = #000000000000000000000000000000
```

The rest of the formal theory consists of definitions. INC13, CNTL_BIT, CNTL_FIELD, B_ADDR, A_ADDR and TEST are defined to improve the readability of the formal specification. The remaining definitions describe structure and behaviour in the design.

```
% ------------------------------------------------------------ %
% -- Auxiliary Definitions ----------------------------------- %

|- INC13 w = CUT16_13(INC16(PAD13_16 w))

|- CNTL_BIT n w = EL n (BITS30 w)
|- CNTL_FIELD (m,n) w = WORD2 (V (SEG (m,n) (BITS30 w)))
|- B_ADDR w = WORD5 (V (SEG(3,7)  (BITS30 w)))
|- A_ADDR w = WORD5 (V (SEG (8,12) (BITS30 w)))
|- TEST w = V (SEG(0,2) (BITS30 w))


% ------------------------------------------------------------ %
% -- Primitive Behaviours ------------------------------------ %

|- REG16 (input,load,output) =
    !t:time. output (t+1) = ((load t) => (input t) | (output t))
|- ACC = REG16
|- IR = REG16
|- ARG = REG16

|- REG13 (input,load,output) =
    !t:time.
     output (t+1) = ((load t) => (CUT16_13 (input t)) | (output t))
|- MAR = REG13
|- PC = REG13

|- BUF (input,output) = !t:time. output (t+1) = input t

|- GATE16 (input,control,output) =
    !t:time.
     output t = (control t => MK_TRI16 (input t) | FLOAT16)
|- GO = GATE16
|- G2 = GATE16
|- G3 = GATE16
|- G4 = GATE16
```

```
|- GATE13 (input,control,output) =
    !t:time.
     output t =
      (control t => MK_TRI16 (PAD13_16 (input t)) | FLOAT16)
|- G1 = GATE13

|- MEM (memory,mar,bus,memcntl,memout) =
    !t:time.
     (memout t =
      ((VAL2 (memcntl t) = 1) =>
       MK_TRI16 (FETCH13 (memory t) (mar t)) | FLOAT16)) /\
     (memory(t+1) =
      ((VAL2 (memcntl t) = 2) =>
       STORE13 (mar t) (bus t) (memory t) | memory t))

|- ALU (arg,bus,alucntl,alu) =
    !t:time.
     (alu t =
      ((VAL2 (alucntl t) = 0) => bus t |
       (VAL2 (alucntl t) = 1) => INC16 (bus t) |
       (VAL2 (alucntl t) = 2) => ADD16 (arg t) (bus t) |
       SUB16 (arg t) (bus t)))

|- BUS(memout,g0,g1,g2,g3,g4,bus) =
    !t:time.
     bus t = DEST_TRI16
      (memout t U16 g0 t U16 g1 t U16 g2 t U16 g3 t U16 g4 t)

|- ROM microcode (mpc,rom) =
    !t:time. rom t = FETCH5 microcode (mpc t)

|- MPC (nextaddress,mpc) = !t:time. mpc (t+1) = nextaddress t

|- DECODE
    (rom,knob,button,acc,ir,nextaddress,rsw,wmar,memcntl,
     wpc,rpc,wacc,racc,wir,rir,warg,alucntl,rbuf,ready,idle) =
    !t:time.
     (nextaddress t =
      (((TEST(rom t) = 1) /\ (button t)) =>
        B_ADDR(rom t) |
       ((TEST(rom t) = 2) /\ (VAL16(acc t) = 0)) =>
        B_ADDR(rom t) |
       (TEST(rom t) = 3) =>
```

```
        WORD5(VAL2(knob t) + VAL5(A_ADDR(rom t))) |
       (TEST(rom t) = 4) =>
        WORD5(VAL3(OPCODE(ir t)) + VAL5(A_ADDR(rom t))) |
       A_ADDR(rom t))) /\
     (rsw t = CNTL_BIT 28 (rom t)) /\
     (wmar t = CNTL_BIT 27 (rom t)) /\
     (memcntl t = CNTL_FIELD (25,26) (rom t)) /\
     (wpc t = CNTL_BIT 24 (rom t)) /\
     (rpc t = CNTL_BIT 23 (rom t)) /\
     (wacc t = CNTL_BIT 22 (rom t)) /\
     (racc t = CNTL_BIT 21 (rom t)) /\
     (wir t = CNTL_BIT 20 (rom t)) /\
     (rir t = CNTL_BIT 19 (rom t)) /\
     (warg t = CNTL_BIT 18 (rom t)) /\
     (alucntl t = CNTL_FIELD (16,17) (rom t)) /\
     (rbuf t = CNTL_BIT 15 (rom t)) /\
     (ready t = CNTL_BIT 14 (rom t)) /\
     (idle t = CNTL_BIT 13 (rom t))


% --------------------------------------------------------- %
% -- Intermediate Level Specifications ------------------- %

|- DATA_PATH
    (switches,memory,pc,acc,mar,ir,arg,buf,rsw,wmar,
     memcntl,wpc,rpc,wacc,racc,wir,rir,warg,alucntl,rbuf) =
    ?g0 g1 g2 g3 g4 memout alu bus.
     MEM(memory,mar,bus,memcntl,memout) /\
     MAR(bus,wmar,mar) /\
     PC(bus,wpc,pc) /\
     ACC(bus,wacc,acc) /\
     IR(bus,wir,ir) /\
     ARG(bus,warg,arg) /\
     BUF(alu,buf) /\
     G0(switches,rsw,g0) /\
     G1(pc,rpc,g1) /\
     G2(acc,racc,g2) /\
     G3(ir,rir,g3) /\
     G4(buf,rbuf,g4) /\
     ALU(arg,bus,alucntl,alu) /\
     BUS(memout,g0,g1,g2,g3,g4,bus)
```

```
|- CONTROL_UNIT
     microcode
     (button,knob,acc,idle,mpc,ir,ready,rsw,wmar,
      memcntl,wpc,rpc,wacc,racc,wir,rir,warg,alucntl,rbuf) =
    ?rom nextaddress.
     ROM microcode (mpc,rom) /\
     MPC (nextaddress,mpc) /\
     DECODE
      (rom,knob,button,acc,ir,nextaddress,rsw,wmar,memcntl,
       wpc,rpc,wacc,racc,wir,rir,warg,alucntl,rbuf,ready,idle)


% ---------------------------------------------------------- %
% -- Top Level Specification ----------------------------- %


|- CONCRETE_MACHINE
     (button,knob,switches,memory,pc,acc,idle)
     (mpc,mar,ir,arg,buf) ready =
   ?rsw wmar memcntl wpc rpc wacc racc wir rir warg alucntl rbuf.
     CONTROL_UNIT
      MICROCODE
      (button,knob,acc,idle,mpc,ir,ready,rsw,wmar,
       memcntl,wpc,rpc,wacc,racc,wir,rir,warg,alucntl,rbuf) /\
     DATA_PATH
      (switches,memory,pc,acc,mar,ir,arg,buf,rsw,wmar,
       memcntl,wpc,rpc,wacc,racc,wir,rir,warg,alucntl,rbuf)
```

5

Towards a framework for dealing with system timing in Very High Level Silicon Compilers[1]

P.A.Subrahmanyam
AT&T Bell Laboratories

Abstract. The design of a VLSI system typically involves many decisions relating to several levels of timing detail. Examples include choices relating to timing disciplines, architectural and circuit design paradigms, the number of clock phases, the number of amplification stages, the sizes of specific transistors, and the temporal behavior demanded of specific input signals (such as stability criteria they must obey). This paper discusses a framework for dealing with system timing issues in the context of very high level silicon compilers intended to aid in mapping abstract behavioral specifications into VLSI circuits. The paradigm considered allows the details of system timing to be *gradually* introduced as a design progresses. During the early stages of a design, no timing details need be explicit, and the only temporal constraints arise due to functionality, causality and stability criteria. As a design evolves, and decisions relating to resource allocation, computation schedules, timing and clocking disciplines are made, the temporal constraints on signals evolve correspondingly. If the external environment in which a system resides imposes any *a priori* temporal constraints, such constraints can be explicitly included in the initial specification, and accounted for in the synthesis process. The framework supports synchronous and asynchronous (self-timed) disciplines, multiphase-clocks, as well as techniques that bear on optimizations of performance metrics such as power and area.

1. Introduction: System Timing Issues in VLSI Design

With an increase in the sophistication of tools dealing with lower level VLSI design tasks (such as layout editors and circuit simulators), there is growing interest in design tools that map higher level behavioral specifications into mask descriptions for circuit fabrication. An important issue in this context is that of *system timing*, which deals with the relative sequencing of subcomputations, the detailed behavior of signal voltages over time, and overall system performance.

---

1. May 1986, Revised

The VLSI design process spans the behavioral, structural and physical dimensions as well as several abstraction levels in each of these dimensions. The design of a non-trivial system typically involves many decisions relating to several levels of timing detail. Examples include choices relating to timing disciplines, such as synchronous versus asynchronous (or self-timed[21]) disciplines; architectural and circuit design paradigms, such as static versus dynamic circuits in CMOS; the number of clock phases; the number of amplification stages; the sizes of specific transistors; and the temporal behavior demanded of specific input signals, such as intervals of time over which they must be stable. Design decisions that concern these and other timing issues, particularly when taken in combination, clearly have a major impact on the performance attributes, if not the correct functioning of a system. It is therefore important to be cognizant of the impact of each decision on the overall system timing in order to avoid making conflicting decisions that result in an erroneous functioning of the system. This requires a way to correlate timing decisions of various kinds from some uniform perspective.

Unfortunately, however, no well-defined framework exists for dealing with the various aspects of system timing. In particular, no coherent formal framework exists that can serve as the basis for an agent dealing with system timing issues in the context of high level synthesis. In practise, therefore, considerations relating to overall system timing tend to be somewhat intangible, and are often introduced into a design in *ad hoc* ways and at ill-defined points. As a consequence, most work on high level design synthesis tools, such as very high level silicon compilers, has hitherto not addressed the issue of global system timing, let alone consider the issue of synthesizing circuits optimized for performance. While several existing design tools deal with physical and geometric constraints, such as chip area and floorplanning constraints, few tools deal with the spectrum of temporal constraints relating to the overall system behavior. Even at the lower levels, designers of digital VLSI circuits have few tools available for the analysis of system timing and the optimization of circuit performance. Instead, a designer relies extensively on design intuition and low level circuit analysis tools such as circuit simulation and/or critical delay path analysis. Such analysis is often very labor intensive and somewhat error prone. Further, the results obtained tend to be suboptimal in terms of area/delay or power/delay tradeoff. Only recently have some tools begun to evolve in this area that use mathematical programming techniques (such as linear programming, nonlinear programming, and convex programming) to address some limited aspects of these problems[18, 5].

This paper describes a rudimentary framework in which various aspects of system timing can be dealt with in a cohesive manner. The issues addressed here include the following: What are the different classes of external timing specifications, and how are such timing criteria specified? What are the temporal abstractions that typically arise in a digital system design, and how are they specified and modeled? When, and how, are various timing details introduced into a design? Are there, or should there be, any criteria that influence these decisions? Is there a conceptual framework and an associated design paradigm that allows for an appropriate consideration of the temporal levels of abstraction involved? and, What are the formal underpinnings of such a framework?

More specifically, the objectives of the framework developed here are:

- to identify some of the important levels of temporal abstraction involved in system design, and to indicate how they may be formally modeled;
- to associate each level of temporal abstraction with a class of design decisions that can potentially influence system timing details at this level;
- to provide a formal means to infer the constraints induced on system timing by some important classes of design decisions; and conversely, to suggest how existing timing constraints (either externally decreed or generated as a result of earlier design decisions) may influence subsequent design decisions;
- to provide a way to correlate the details relating to system timing at various levels of abstraction via their influence on a *common* representation of timing information based on partial specifications of waveforms;
- to relate the timing parameters introduced by the more "abstract" models to the more "concrete" timing parameters that are associated with lower level circuit models, and ultimately semiconductor models.

In addition, we identify some categories of timing information that can be generated during the synthesis process, and that can serve as useful input to lower level timing analyzers.

1.1 Related Research

Some aspects pertaining to system timing have been studied in great depth; examples include simulation (at the functional, switch, electrical and device levels [3, 16, 10]), static timing analysis[25], and the impact of transistor sizing on circuit timing characteristics[5, 18]. Similarly, descriptive mechanisms for dealing with timing, such as temporal logic, have been examined in depth in various contexts, e.g.,[14, 15]. Unfortunately, however, not much attention has yet been given to examining when and how different timing decisions are made in the context of the overall VLSI synthesis process, and how the cumulative effects of such design decisions may be reasoned about. The lack of research in this area may partly owe to the fact that design aids for high level behavioral synthesis have not yet matured to a point where they demand a sophisticated approach to system timing.

Before discussing the details of the proposed framework, we first make some relevant observations about the interrelation between circuit design and system timing. These observations lead to a list of characteristics that are useful in such frameworks.

1.2 Some Observations

It is desirable for design aids such as very high-level silicon compilers to avoid the premature introduction of explicit timing information during specification and/or synthesis. This is because typically, during the early stages of a system design/description, the major concern is one of functionality, and inessential details of timing are best suppressed. When dealing with system descriptions at a lower level, however, the temporal aspects assume greater importance and visibility. This suggests the need for a framework that supports *gradual* introduction of the relevant timing information as a design evolves. Further, it is important that any framework for dealing with system timing not impose any *a priori* restrictions on the timing discipline adopted. More specifically, it should allow for synchronous and asynchronous (self-timed) circuits, local and global clocks, multiple clocks, and multiphase-clocks (such as those used in dynamic CMOS circuits).

Unfortunately, a majority of existing design aids typically (i) include explicit (and detailed) information about eventual circuit timing as part of their input specifications, or (ii) provide only for certain default mechanisms, or (iii) ignore timing issues altogether. For the reasons alluded to above, such approaches are somewhat unsatisfactory.

1.3 Characteristics of the proposed framework

The design paradigm and associated framework advocated here attempt to meet the criteria just enumerated, and allows the details of system timing to be gradually introduced as a design progresses. At the highest level, no timing details need be explicit, and the only temporal constraints are causal. As a design evolves, and decisions relating to the implementation and timing disciplines are made, the temporal constraints on signals evolve correspondingly. If, on the other hand, the external environment in which a system resides imposes any *a priori* temporal constraints, the framework allows for such constraints to be explicitly included in the initial specification, and accounted for in the synthesis process.

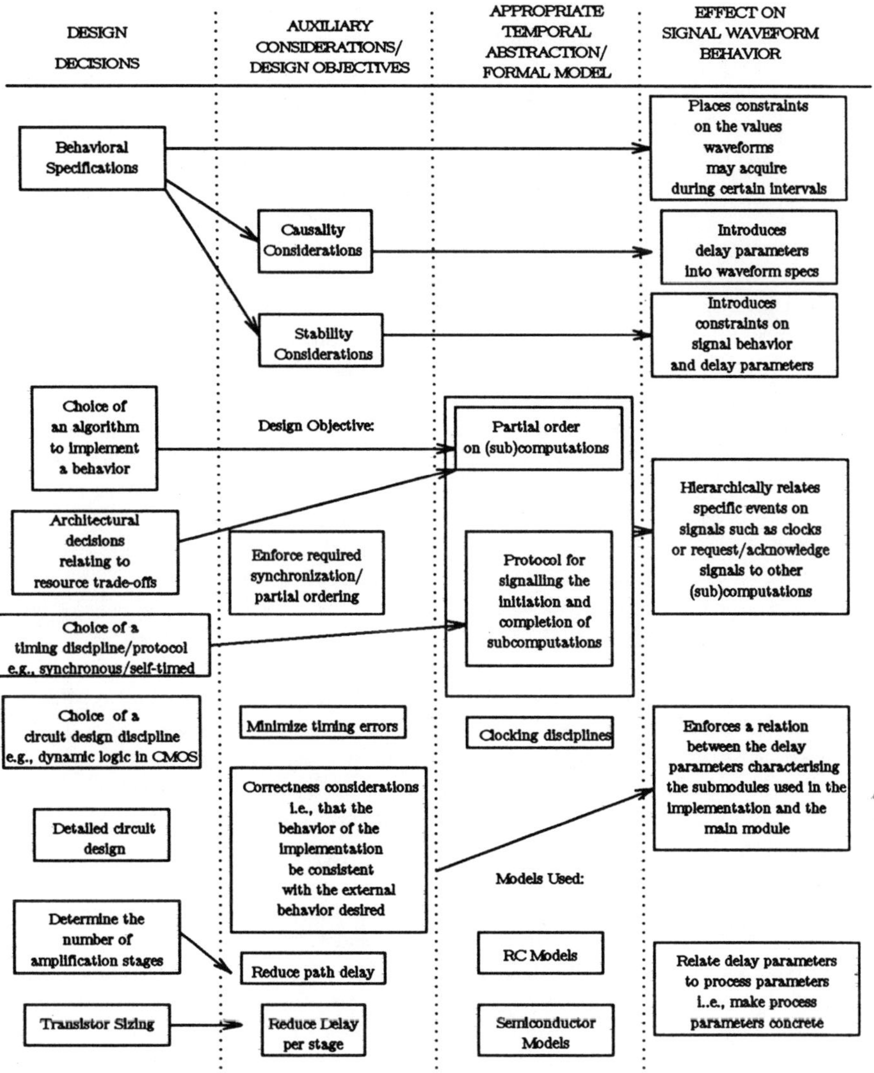

Figure 1. The evolution of system timing details

2. Design evolution and the introduction of timing details

Before elaborating on the technical details of the proposed framework, we will first motivate (and summarize) the approach adopted. We begin by indicating how increasing degrees of timing detail are introduced into the design description as a design evolves (cf. Figure 1). In order to do this, we need to make certain assumptions about how a design evolves, and more specifically, we need to made make assumptions about those aspects of the design process that have a major impact on system timing.[2] Some aspects of such a design evolution are listed in the first column of Figure 1; the associated effects on system timing details are listed in the rightmost column. The first two entries in the second column are examples of auxiliary considerations that are used in generating some of the temporal constraints; the remainder of the second column lists some of the local design objectives that influence the design decisions (in horizontal proximity) in the first column. The first two entries in the third column are examples of temporal abstractions that are associated with the design decisions in the first column. The last two entries in the third column are examples of models that are used in circuit analysis; such analysis influences the design decisions and procedures listed in the first column.

Initially, a design commences with a behavioral specification that details the functionality desired of the circuit to be synthesized.[3] The temporal constraints induced on signals at this stage stem from causality considerations implied by intrinsic data dependencies, as well as functionality and stability criteria. Inference and approximation techniques for computing such constraints are discussed in Section 4. It should be noted that the parameters characterising the constraints generated at this stage are not (yet) related to any underlying circuit models.

The choice of a particular algorithm to implement a specified behavior induces a temporal partial order on the subcomputations involved. As architectural design decisions relating to resource trade-offs are made, additional temporal constraints are imposed by the need to share resources in the design. Examples of such constraints and their ramifications are discussed in Section 5.

The choice of timing disciplines (synchronous or asynchronous) introduces the next level of timing detail. A timing discipline provides a protocol for indicating the initiation and/or completion of a set of events that comprise a (sub)computation. A specific protocol establishes a relation between the duration of subcomputations and the states of some distinguished signals, *e.g.*, clocks (for synchronous systems) and request/acknowledge signals (for

2. Although experienced human designers appear to have subliminal (and not always explicit) ways of accounting for such information, researchers in design automation do not appear to have hitherto addressed this problem in a rigorous manner (to our knowledge).

3. The initial specifications may additionally include the desired performance criteria (possibly along several dimensions such as area, response time, pin count, and throughput), and interfacing specifications such as interface protocols.

self-timed systems). As a consequence, such protocols provide a mechanism for synchronizing the subcomputations in a system, and enable the temporal constraints to be developed hierarchically. Some forms of circuit design styles are related to a specific choice of a timing discipline, and may in turn impose constraints on clocking disciplines. For instance, the decision to use clocked dynamic circuits in CMOS typically implies the need for a precharge phase and a compute phase, and consequently a certain minimal number of clock phases. Additionally, the structural details of a circuit that is used to implement a function may impose further temporal constraints on signals, such as stability constraints relating to setup and hold times for signals. These issues are dealt with in Section 6.

Within the confines of a fixed architecture, the need to optimize certain performance metrics such as area, time, (or some function thereof), and an analysis of the critical paths in an architectural design, may further influence design decisions relating to clocking disciplines (such as multi-phase clocks, multiple clocks, local clocks, and global clocks). Lower level timing optimizations based on RC circuit models of increasing sophistication are performed once the overall architecture has been outlined. Such optimizations influence factors such as the number of amplification stages used by a logic design, and the sizes of the individual transistors in each stage. Further, they have a direct effect on controlling many of the parameters that influence the details of how various signals evolve over time. These issues are elaborated upon in Sections 7 and 8. Additionally, considerations at this level relate the more "abstract" timing parameters introduced during the earlier design phases to more "concrete" process parameters that are associated with semiconductor models.

The above scenario highlights some of the design decisions that can potentially have a major impact on system timing. To summarize, these include: the formulation of a behavioral specification for a system; design decisions pertaining to the system architecture (i.e., the choice of the set of modules used to implement a desired behavior); the choice of timing disciplines; the choice of clocking disciplines (including the number of clock phases); and further optimizations that influence the number of amplification stages and the sizes of the individual transistors in each stage. Further, a partial order on the evolution of these design decisions is suggested.

Given this setting, we will, in subsequent sections, identify different levels of temporal abstraction in system design and indicate how they can be formally modeled in an algebraic/logical setting. We will also indicate how it is possible to mechanically *infer* some of the temporal constraints arising out of a design decision, as opposed to such constraints being explicitly specified. There are several noteworthy consequences of the existence of such a framework (consisting of the set of models and inferencing techniques). Firstly, it provides a way of correlating the timing decisions at various levels via their (cumulative) effect on signal waveform behaviors over time. Secondly, the information inferred may be used in *incrementally* checking for the consistency of the timing constraints induced on a signal waveform from different sources. Should an inconsistency exist, it enables an early detection of this fact, and serves to identify the set of design decisions that contribute to this inconsistency. The framework also assists in computing relations between the delay parameters of the modules used in an implementation and the delays characterising the overall system; and in conveying information relating to circuit partitioning and timing protocols to lower level tools such as timing analyzers. Thirdly, the existence of such a formalism potentially implies that both the generation of the temporal constraints and the consistency checking process can benefit from mechanical assistance, provided appropriately constrained primitives are used; however, we do not elaborate on this issue here. The details of how the various design decisions are actually made is an important related concern that is beyond the scope of this paper; we occasionally comment, however, on some of the criteria influencing such decisions. An example illustrating some of these issues is contained in Section 9.

It is important to bear in mind that the overall VLSI (system) design process is often iterative in nature. Consequently, design details that evolve at a later stage may potentially affect earlier decisions, for instance layout considerations may influence some aspects of the architectural design. The conceptual model presented here for the introduction of timing details accommodates this viewpoint, and nothing in our subsequent development precludes such iterations in design. Indeed, each level of timing elaboration can impact design decisions at earlier stages; in optimistic scenarios, such redesign is avoided (or at least localized) by appropriate decisions during the early stages of design. The application of the proposed framework in the context of analysis and synthesis is discussed in Section 10.

We next elaborate on some facets of the proposed framework.

3. Representing Timing Characteristics

We first introduce a low level representation of the behavior of signals over time, and then discuss the correlation between such low level representations and the system behavior. Our subsequent discussions will relate to how the temporal aspects of design decisions at different levels of abstraction may be modeled, and how they (ultimately) influence this signal representation.

3.1 Signal and System Behavior

3.1.1 Signals A signal may be viewed as a function from a time domain (denoted `Time`) that describes appropriate instants of time, to a set of signal voltage values (denoted `Voltage`). Thus,

$$\texttt{Signal: Time} \rightarrow \texttt{Voltage}$$

Signal voltage values usually form a continuous domain. However, in the context of digital design, we will assume here that such values can be abstracted to a discrete domain of values, e.g., {0, 1, X} or {0, 1}. As a simple example, consider a function Discretize:Voltage→ {0,1,X} of a function that maps the continuos range of voltages [0.0–5.0] to {0, 1, X}; voltages less than 0.5 are mapped to the value 0, voltages between 0.5 and 4.5 to the value X, and voltages above 4.5 are mapped to the value 1.

As the details of a system evolve, constraints evolve on the detailed nature of various signals. A signal waveform is constrained when an explicit statement is made about any of its characteristics (such as its value or shape) during some interval of time. Such statements may be formally made by using appropriately defined predicates or functions, as we will elaborate upon later. Very often, these constraints are only partial, in that the behavior of the signal waveforms outside of these intervals is unrestricted.

Further abstractions of such discrete representations are usually required to relate them to elements of other problem specific domains, such as integers, stacks of integers, and arrays. We will not be directly concerned with such data abstractions in this paper, although they are naturally handled within the current framework.

3.1.2 System Behavior The behavior of a system may be described at many different levels of abstraction. At one extreme, system behavior may be described in terms of the values in some abstract input domains that are presented at the various input ports, and the functions computed on these values and made available at (some or all of) the output ports. At the other extreme, system behavior may be described in the terms of the relations (that must hold) between the signal voltage waveforms at its input and output pins.

An important difference between these two perspectives is that in the first, there is no explicit mention of timing, whereas in the second there is a very detailed concern with how signals behave over time (since signals waveforms represent signal voltages as they evolve over time). One of the objectives of this paper is to explore a framework for timing in the context of synthesis that enables us to bridge this gap. As a first step, we examine the temporal implications of the functional/behavioral specification of a system.

4. Temporal Implications of Causality Considerations

A major assumption we make relates to causality: we assume that an output of a function (or module) can be computed (*i.e.*, become available) only after the inputs it depends upon are available. While we will typically assume that a delay is associated with such a computation, we will make no *a priori* assumptions about the detailed nature of this delay.

We now outline some of the ramifications of such an assumption. More specifically, we give examples of ***constraints*** on the input and output waveforms that are implied by such an assumption. We will not discuss in detail an algorithm that enables the computation of such causality constraints in general, but rather convey the intuition underlying it.

Our subsequent discussion uses the notion of signals being stable over intervals of time having specified durations. We next review the definition of these notions before proceeding further.

4.1 Notation

The terminology used here draws upon work in Temporal logic[6,7,15]. Two instants t_1 and t_2 of `Time` serve to define an ***interval*** i of ***duration*** t_2-t_1. The interval is defined to be semi-open, starting at t_1, and ending "just before" t_2, and is sometimes denoted $[t_1,t_2)$.

A signal is said to be ***stable*** during an interval of time if its value remains unchanged during this interval. The stability of a signal can be captured by means of the predicate

$$\texttt{Stable}\colon (\texttt{Time} \times \texttt{Time}) \times \texttt{Signal} \rightarrow \texttt{Boolean}$$

defined by

$$\text{Stable}(t_1,t_2)(\text{s}) \equiv_{def} \text{for all t } (t_1 \leq \text{t} < t_2)\ (\text{s(t)} = \text{s}(t_1))$$

which embodies the fact s maintains a constant value from time t_1 to just before t_2.

We will denote the fact that a signal s has the constant value c in domain D during an interval $[t_1,t_2)$, by $\texttt{constant}(t_1,t_2)(s)(c)$. Formally, the predicate

$$\texttt{constant}\colon (\texttt{Time} \times \texttt{Time}) \times (\texttt{Signal} \rightarrow \texttt{D} \rightarrow \texttt{Boolean})$$

is defined by

$$\texttt{constant}(t_1,t_2)(\text{s})(\text{c}) \equiv_{def} \text{for all t } (t_1 \leq \text{t} < t_2)\ (\text{s(t)} = \text{s}(t_1) = \text{c})$$

which may also be written as

$$\texttt{constant}(t_1,t_2)(\mathrm{s})(\mathrm{c}) \equiv_{def} \mathrm{stable}(t_1,t_2)(\mathrm{s}) \ \& \ \mathrm{s}(t_1){=}\mathrm{c}$$

4.2 Causality constraints and system timing

We now elaborate, via an example, the implications of causality assumptions on the timing behavior of the input/output signals of a module.

We will denote by d_1 the time taken for an output value (*i.e.*, the signal at an output port) to stabilize once the stable input values are available. If there is a change in one or more inputs, the output may change; however, this change takes some time, which duration we will denote by d_2. Intuitively, d_1 and d_2 can be thought of being the times taken for the output to "react" to a change in the input values. This reaction time may of course be different for each output (and, as discussed below, for each combination of inputs/outputs); we will here restrict ourselves to considering only one output. It is relatively straightforward to generalize the arguments to the multiple output case.

If the cardinality of each of the input domains is D, and there are N inputs, the total number of input combinations (or input states) is D^N. The total number of input state transitions possible is therefore $D^N*(D^N-1)$. In general, the reaction times for an output may be different for each of these transitions. In such a case, we will denote by d_1^i and d_2^i the reaction times associated with the i-th transition. A transition *into* a given state may be denoted by the pair <previous state, current state>, whereas a transition *out of* the current state may be denoted by the pair <current state, next state>. For each state transition, the corresponding output signal may (1) remain unchanged; (2) change to a different value; or (3) exhibit a "glitch" i.e., change to a different value for some period and then return to the original value. Obviously, it is *expected* that the output(s) of a properly functioning circuit will not exhibit any glitches, and exhibit a change to a different value only when this is consistent with the specified behavior.

Consider a NAND module which implements a 2-input NAND function. The NAND function may be defined, for instance, as shown below.

```
nand(0,0)=1
nand(0,1)=1
nand(1,0)=1
nand(1,1)=0
```

As an example, consider a time interval when the inputs of a two input

NAND gate have the value 00. If p denotes the previous state and n the next state, the basic causality assumption associated with this interval yields the following constraint on the signals s_x, s_y, and s_z

$$\exists\, t_1,t_2 \text{ such that } \mathtt{constant}(t_1,t_2)(s_x)(0)\ \&\ \mathtt{constant}(t_1,t_2)(s_y)(0) \supset \\ \mathtt{constant}(t_1+d_1^{p,00},\ t_2+d_2^{00,n})(s_z)(1)$$

where $\supset$ denotes logical implication.

Intuitively, this constraint says that if the inputs s_x and s_y of the NAND gate are stable at the value 00 during the interval from t_1 to t_2, then the output s_z must acquire the value 1 at some time $t_1+d_1^{p,00}$ after this interval, where the actual delay in responding to the state 00 may depend upon the previous state p. Further, this value of s_z is maintained until a time $t_2+d_2^{00,n}$ where the constant $d_2^{00,n}$ may depend upon the next state n.

It is important to note that there are two distinct factors that lead to this constraint. The first has to do with causality: if the value of s_z is specified to depend on the values of s_x and s_y, then s_z must, at some time, react (correctly or incorrectly) to the values of s_x and s_y in some (and all) intervals. The second has to do with functionality, *i.e.*, s_z reacting properly: this mandates that if s_x and s_y both have the value 0 then s_z needs to attain the value 1, consistent with the definition of the NAND function.

In essence, the causality assumption serves to "parameterize" a design by the variables d_1^i and d_2^i (for each possible transition index i), and introduces constraints on the signal waveform. The form of the causality constraints that arise from considering a *single* output is given by the proposition below, which follows from the arguments we have gone through above. We denote the vector of signal waveforms corresponding to the inputs by that comprise the input domain D by $\vec{s}_D$, and the signal waveform corresponding to the output on which a function F is computed by s_F. The variables p,c,n that range over the domain D are intended to connote the previous, current, and next state of the inputs of the module. We use $\mathtt{constant}(t_1,t_2)(\vec{s}_D)(\vec{c})$ to denote the fact that all components of $\vec{s}_D$ are stable during the interval $[t_1,\ t_2)$ at the value in the corresponding component of $\vec{c}$.

Proposition 1. Given a function $F{:}D \rightarrow R$, the temporal constraints on the signal waveforms $\vec{s}_D$ and s_F that arise from causality considerations are given by

$\forall\ c, p, n \in D\ \exists\ t_1, t_2$
$\texttt{constant}(t_1, t_2)(\vec{s}_D)(\vec{c}) \supset \texttt{constant}(t_1 + d_1^{p,c},\ t_2 + d_2^{c,n})(s_F)(F(\vec{c}))$

where the $d_1^{p,c}$ and $d_2^{c,n}$ are parameters that characterise the module implementing F. Further, $\forall$ c $\in$ D $d_1^{c,c} = 0$, $d_2^{c,c} = 0$. ∎.

To simplify notation, we will henceforth drop explicit use of the vector symbol.

4.2.1 Reducing the number of independent parameters For a specific function, the actual number of independent variables may be reduced to symmetries in the function implemented and/or the circuit structure used to implement it. Even for small values of N (where N denotes the number of inputs), this set can be quite large. For example, if N=2, and the domain of each input is {0,1}, the number of transitions is $2^2 * (2^2 - 1) = 4 * 3 = 12$, and the number of parameters is 24 (i.e., independent values of d_1 and d_2 for each of the 12 transitions). To account for finite signal transition, it is necessary to have at least a 3-valued domain, such as {0,1,X}. As a consequence, it seems desirable to reduce the ensuing complexity, by approximating the behavior of the signals in some appropriate manner.

Given a function $F: D1 \times D2 \times \cdots \times Dn \rightarrow R$, where R may in some cases decomposed as $R1 \times \cdots \times Rm$, the delay parameters characterizing an implementation of F may depend, in the most general case, on all of the $m+n$ values in the domain and range. A simplification may be achieved by reducing the number of values in the domain and range the delay parameters are assumed to be dependent upon. Three obvious simplifications are to consider dependency only on (i) the domain values $D1 \times D2 \times \cdots \times Dn$; (ii) the range values R; (iii) an explicitly specified subset of the range/domain values (particular to the function being considered). Additionally, it is possible to make the assumption that the delay parameters are completely independent of the actual values in the domain and range.

The example discussed earlier in this section in fact corresponds to the first approximation. That is, the assumption was that the delay parameters for a two-input NAND gate depended upon the state of both the input signals. In the rest of this section, we elaborate on the ramifications of the other approximations indicated above.

4.2.1.1 Approximations based on output values Consider now the implication of the assumption that the delay parameters depend only upon the range R, *i.e.*, upon the value computed by the function F. This approximation has the effect of partitioning all of the state transitions into equivalence classes that correspond to transitions into and out of *different* stable output values. Thus, in the case of Boolean valued functions, we can partition the set of transitions into two classes, corresponding to the stable output values 0 and 1. Formally, this may be stated as follows.

Approximation R. The set of parameters $d_1^{p,c}$ and $d_2^{p,n}$ where $p,c,n \in D$ are approximated by the set of parameters d_1^r and d_2^r where $r \in R$. ■

It should be noted that this approximation will reduce the number of independent parameters only if the cardinality of the domain $D = D1 \times \cdots \times Dn$ is less than the cardinality of the range R. This condition is typically met for a nontrivial boolean valued function with more than one input.

Let the domain D be partitioned into subsets D_k such that d_i and d_j are in the same partition only if they are mapped into the same value by the function F. That is, $d_i, d_j \in D_k$ if and only if $F(d_i) = F(d_j)$. We will denote the set of such partitions D_k by D_F. Note that this partition depends on the semantics of the function F.

Proposition 2. If approximation R is used, the causality constraints on the signals of a module implementing a function $F: D \rightarrow R$ have the form

$$VD_k \in D_F \; [\, (\mathbf{V}_{d_i \in D_k} \exists\, t_1, t_2 \; \texttt{constant}(t_1, t_2)(s_D)(d_i) \supset$$
$$\texttt{constant}(t_1 + d_1^{F(d_i)},\; t_2 + d_2^{F(d_i)})(s_F)(F(d_i))] \; ■$$

As an illustration, consider again the 2-input NAND gate. There are only two stable values (0 and 1) that the output signal s_z can assume. The output value 1 can result from any of the input combinations 00, 01, and 10. The output value 0, on the other hand, can result only from the input combination 11. In other words, we may interpret this as stating that

- If a NAND gate is functioning properly, then
  - i. if either of the inputs is 0 during some interval $[t_1, t_2)$, then the output must attain the value 1 during a subsequent interval defined by $[t_1 + d_1^{NAND,1}, t_2 + d_2^{NAND,1})$, where d_1^0 and $d_2^{NAND,1}$ are parameters characterising the NAND gate;
  - ii. if both of the input are 1 during some interval, then the output must attain the value 0 during a subsequent interval defined by $[t_1 + d_1^1, t_2 + d_2^{NAND,0})$, where $d_1^{NAND,0}$ and d_2^1 are parameters characterising the NAND gate.

This may be stated formally by the two formulas below:

$\exists\, t_1,t_2$ such that $\mathtt{constant}(t_1,t_2)(s_x)(0)$ or $\mathtt{constant}(t_1,t_2)(s_y)(0) \supset$
$\mathtt{constant}(t_1+d_1^{NAND,1},\ t_2+d_2^{NAND,1})(s_z)(1)$

$\exists\, t_1,t_2$ such that $\mathtt{constant}(t_1,t_2)(s_x)(1)$ & $\mathtt{constant}(t_1,t_2)(s_y)(1) \supset$
$\mathtt{constant}(t_1+d_1^{NAND,0},\ t_2+d_2^{NAND,0})(s_z)(0)$

In summary, using approximation R as discussed above, causality considerations for a 2-input NAND gate introduce the parameters $d_1^{NAND,0}$, $d_1^{NAND,1}$, $d_2^{NAND,1}$, $d_2^{NAND,0}$ into its signal behavior, and yield the following temporal constraints on the signals waveforms s_x, s_y, and s_z:

NAND_SIGNAL_BEH($d_1^{NAND,1}$, $d_1^{NAND,0}$, $d_2^{NAND,1}$, $d_2^{NAND,0}$)(s_x,s_y,s_z) $\equiv$
$\exists\, t_1,t_2$ such that $\mathtt{constant}(t_1,t_2)(s_x)(0)$ or $\mathtt{constant}(t_1,t_2)(s_y)(0) \supset$
$\mathtt{constant}(t_1+d_1^{NAND,1},\ t_2+d_2^{NAND,1})(s_z)(1)$ &
$\exists\, t_1,t_2$ such that $\mathtt{constant}(t_1,t_2)(s_x)(1)$ & $\mathtt{constant}(t_1,t_2)(s_y)(1) \supset$
$\mathtt{constant}(t_1+d_1^{NAND,0},\ t_2+d_2^{NAND,0})(s_z)(0)$ ■

For future reference, and to reduce clutter, we will rename the parameters $d_1^{NAND,1}$, $d_2^{NAND,1}$, $d_1^{NAND,0}$, $d_2^{NAND,0}$ that characterise a NAND gate by n_1,n_2,n_3,n_4, (where the letter n is intended to be a mnemonic for NAND), and abbreviate the entire set as $\vec{n}$. With this notational change, the causal constraints on the NAND signal behaviors may be expressed by

NAND_SIGNAL_BEH(n_1, n_3, n_2, n_4)(s_x,s_y,s_z) $\equiv$
$\exists\, t_1,t_2$ such that $\mathtt{constant}(t_1,t_2)(s_x)(0)$ or $\mathtt{constant}(t_1,t_2)(s_y)(0) \supset$
$\mathtt{constant}(t_1+n_1,\ t_2+n_2)(s_z)(1)$ &
$\exists\, t_1,t_2$ such that $\mathtt{constant}(t_1,t_2)(s_x)(1)$ & $\mathtt{constant}(t_1,t_2)(s_y)(1) \supset$
$\mathtt{constant}(t_1+n_3,\ t_2+n_4)(s_z)(0)$ ■

Note that the approximations made above have reduced the number of independent parameters for a 2-input NAND gate from 24 to 4. Similarly, for a 3-input NAND gate, the number of parameters is reduced from $2*2^3(2^3-1) = 112$ to 4. It is important to stress that causality assumptions merely serve to *introduce* these parameters into the signal behavior of a NAND gate, without imposing any constraints on their values.

4.2.1.1.1 Some Remarks It is interesting to note that Hanna and Daeche, in their analysis discussed in[6] *commence* with a signal behavior for a NAND module that is very similar to the derived behavior NAND_SIGNAL_BEH above. They observe that this is quite a complicated definition (although, with familiarity, readily comprehensible); however, they argue that this complexity is intrinsic to a digital designers notion of the behavior of a NAND gate. They also state, in the context of such examples: "We note that, in formulating this definition, several choices had to (be) exercised, as anyone starting off to frame such a definition *ab initio* will very soon discover! Thus, while we believe that the proposed definition is a reasonable one, it is not unique: many subtle variations are possible."

Our discussion in this section indicates how signal behaviors such as those assumed in[6] can be ***inferred*** from a description of the function to be computed, given certain assumptions about causality and stability criteria. Further, the notion of "approximations" introduced above (for reducing the number of independent delay parameters that characterising a module) rationalises the use of the seemingly arbitrary parameters assumed in[6].

4.2.1.2 Approximations depending on identifying control signals Consider a "conditional" function defined as follows:

$$\text{cond(P,X,Y)} = \text{if P then X else Y}$$

Such a function may be implemented by a multiplexer with ports p,x,y,z corresponding to the inputs P,X,Y and the multiplexer output, denoted Z. It is possible to view the output of this function as being "controlled" by the value of the predicate P, and to assume that the delay parameters depend on the value of s_p but not on the values of the signals s_x and s_y. Such an approximation leads to the causality constraint

$$\texttt{constant}(t_1,t_2)(s_p)(0)\ \&\ \texttt{stable}(t_1,t_2)(s_y) \supset$$
$$\texttt{constant}(t_1+d_1^{p=0},\ t_2+d_2^{p=0})(s_z)(s_y(t_1))$$

$$\texttt{constant}(t_1,t_2)(s_p)(1)\ \&\ \texttt{stable}(t_1,t_2)(s_x) \supset$$
$$\texttt{constant}(t_1+d_1^{p=1},\ t_2+d_2^{p=1})(s_z)(s_x(t_1)) \ \blacksquare$$

This may be generalized to approximations where the set {*D1*, ..., *Dn*, *R1*, ..., *Rm*} is partitioned into two sets, say {*C1*,....,*Cp*} and {*V1*,...,*Vq*}, where the delay parameters are modeled as being dependent only on the values of the signals in the set {*C1*,...,*Cp*} and not on the values of the signals in the set {*V1*,...,*Vq*}. Note, however, that the signals in {*V1*,...,*Vq*} may be required to obey other (independent) constraints, such as stability constraints. The resulting causality constraints may then be formulated in a manner similar to the example above; we omit the formal statement for brevity.

4.2.1.3 Approximations independent of values in the Domain and Range In order to further reduce the complexity of delay model, the delay parameters may be assumed to be independent of the actual values at the input and output.

Proposition 3. (Independent Delay Approximation). If it is assumed that the delay parameters characterising a module are independent of the values in the domain and range, then the causality constraints on the signals of a module implementing a function $F{:}D \rightarrow R$ have the form:

$$\texttt{stable}(t_1,t_2)(s_D) \supset \texttt{constant}(t_1+d_1^F,t_2+d_2^F)(s_F)(\mathrm{F}(s_D(t_1))) \blacksquare$$

4.2.1.3.1 Approximating by a single delay In the extreme case, each output may be characterised by a single delay parameter under the assumption that $d_1^F=d_2^F=d^F$. Such a model may be acceptable in situations where a reduction in the complexity of the model is the primary consideration, and where a finer degree of timing detail is not important.

Proposition 4. (Single Delay Approximation). The causality constraints on the signals of a module implementing a function $F{:}D \rightarrow R$, when a single parameter d_F is used to approximate the delay induced by the module, have the form:

$$\text{for all } t \;\; s_F(t+d^F) = \mathrm{F}(s_D(t)) \blacksquare$$

4.2.2 Summary In this section, we considered four categories of approximations aimed at reducing the number of delay parameters used to characterise a module. Such approximations enable a reduction in the complexity of the resulting delay model that needs to be analyzed during subsequent stages of a design. The disadvantage stems from the fact that the actual values of the delay parameters need to be conservative, in they must reflect the worst case behavior of the entire set of signal transitions that would otherwise have been treated individually. In essence, a reduction in the complexity of the delay model is obtained at the expense of having to use worst case approximations for the actual values of the delays.

It is important to note that the nature of delay parameters introduced into the module description at this stage has no *a priori* correlation with any underlying technology or implementation mechanism. However, when a design is fleshed out, the actual values that are assumed by the delay parameters depend on the details of an implementation, and ultimately on the process parameters of the underlying technology. This relation will be discussed in subsequent sections.

We next examine some further temporal constraints on the signals that arise from stability considerations.

4.3 Stability Criteria

It is important that a properly functioning circuit be devoid of any spurious and perhaps transient outputs (sometimes called "glitches"). If the durations d_1^i and d_2^i are non–zero, then it becomes necessary to impose restrictions on signal waveforms to prevent such glitches.

Consider, for example, a transition from input state p to c to n. We will assume that each state transition goes through the state $\perp$.[4] Consequently, let the input states be p, $\perp$, c, $\perp$, n during the intervals $[t_1,t_2)$, $[t_2,t_3)$, and $[t_3,t_4)$, $[t_4,t_5)$, $[t_5, t_6)$ respectively. Consider an output signal s_F that is intended to carry the value of a computed function F. Causality constraints imply that the value $F(p)$ be maintained *until* time $t_2+d_2^{p,c}$ and that the value $F(c)$ be maintained *from* time $t_3+d_1^{p,c}$ until $t_4+d_2^{c,n}$. If $F(p) \neq F(c)$ then these two intervals must not overlap in order prevent a conflict in the values the output signal is expected to assume. That is, we must have $t_3 + d_1^{p,c} \geq t_2 + d_2^{p,c}$, *i.e.*

$$t_3\text{-}t_2 \geq d_2^{p,c} - d_1^{p,c}$$

where t_3-t_2 defines the transition time between the states p and c.

Further, for the output $F(c)$ to be perceived at the output, we must have $t_3+d_1^{p,c} \leq t_4+d_2^{c,n}$ *i.e.*,

$$t_4\text{-}t_3 \geq d_1^{p,c} - d_2^{c,n}$$

where t_4-t_3 defines the time for which the input is stable at the value c.

Such a stability criterion may be formally stated as follows.

Proposition 5. Given a function $F{:}D \rightarrow R$, the temporal stability constraints arising out of a consideration of stability criteria are:

$\forall\, p,c,n \in D\ \exists\, t_1 < t_2 < t_3 < t_4$
$\texttt{constant}(t_1,t_2)(\vec{s}_D)(\text{p})$ & $\texttt{constant}(t_2,t_3)(\vec{s}_D)(\perp)$ & $\texttt{constant}(t_3,t_4)(\vec{s}_D)(\text{c})$
& $\text{F(p)} \neq \text{F(c)} \supset t_3\text{-}t_2 \geq d_2^{p,c} - d_1^{p,c}$ & $t_4\text{-}t_3 \geq d_1^{p,c} - d_2^{c,n}$ ■

Of course, other implications, quite unrelated to stability criteria, may also follow from this precondition, but this does not concern us here.

It should be noted, in reference to this example, that we have presumed that the outputs are expected to respond to ***steady state inputs,*** and that this imposes constraints on the time taken for ***transitions*** between steady states. If, on the other hand, the outputs are intended (or designed) to respond to ***transitions*** in the input states, such as in an edge-triggered device, then analogous reasoning yields constraints on the ***durations*** that inputs are required to maintain a steady state. Further, the approximations

---

4. Intuitively, this is justified by the fact that transitions between voltages that represent a logical 0 to voltages represent a logical 1 must involve intermediate voltage values that may be viewed as corresponding to an "undefined" state denoted X (or $\perp$).

discussed in the preceding subsection in connection with causality criteria may be applied to the discussions related to stability in many cases; we refrain from elaborating on these issues any further.

4.3.1 An Example: Causality and Stability criteria for an inverter The function `not` computed by an inverter (Figure 2) is defined by:

```
not(0)=1
not(1)=0
```

We will denote the signals at the ports corresponding to the input and output by s_x, s_y. The causality considerations, in conjunction with a "domain" approximation (cf. Proposition 1) lead to the following temporal constraints:

$$\texttt{constant}(t_1,t_2)(s_x)(0) \supset \texttt{constant}(t_1+d_1^{0,1},\ t_2+d_2^{0,1})(s_y)(1)$$
$$\texttt{constant}(t_1,t_2)(s_x)(1) \supset \texttt{constant}(t_1+d_1^{1,0},\ t_2+d_2^{1,0})(s_y)(0)$$

For concreteness, consider the following values for the delay parameters (obtained by running SPICE with some default parameters): $d_1^{0,1} = 3.05$ ns; $d_2^{0,1} = 0.65$ ns; $d_1^{1,0} = 3.3$ ns; $d_2^{1,0}) = 0.35$ ns. For an input transition 0, the value $d_1^{0,1}$ is defined to be the time for the output to stabilize at not(1) *i.e.* 0 (*i.e.*, reach 10 per cent of VDD = 0.5 volts), once the input is stable at 1 (*i.e.*, has reached 90 per cent of VDD = 4.5 volts). The value $d_2^{0,1}$ is defined to be the time at which the output is no longer equal to not(0) *i.e.*, no longer equal to 1 (*i.e.*, is less than 4.5 volts) after the time when the input is no longer 0 (*i.e.*, increases beyond 0.5 volts).

Stability criteria then imply that

1. the input transition time from 0 to 1 be $\geq d_2^{0,1} - d_1^{0,1} = 0.65\text{-}3.05 =$ -2.4ns, and that the input be stable at 1 for a period exceeding $d_1^{0,1}$ - $d_2^{1,0} = 3.05\text{-}0.35 = 2.6$ns;

2. the input transition time between 1 and 0 $\geq d_2^{1,0} - d_1^{1,0} = 0.35\text{-}3.3 =$ -2.95ns & that the input be stable at 0 for a period exceeding $d_1^{1,0}$ - $d_2^{0,1} = 3.3\text{-}0.65 = 2.55$ns.

Thus, the input transition times are in fact unconstrained, since the inequalities are always satisfied by virtue of the right hand sides being negative. On the other hand, the stability criteria imply that the input is constrained to be stable for a period of at least (approximately) 2.6ns for the output to drive another gate. It is interesting to note that this agrees well with observed empirical simulations that indicate that the output has glitches when the input pulse widths are less than about 3 ns.

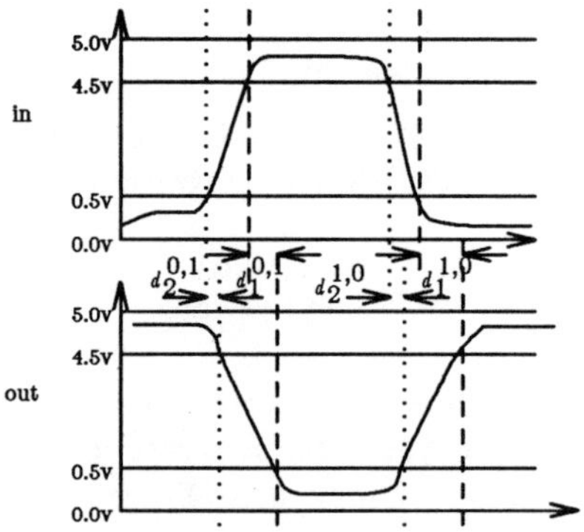

Figure 2. Inverter Behavior

4.4 Summary

In summary, we have indicated in this section the nature of the constraints induced on signal waveforms by causality, functionality and stability considerations.

- Causality constraints arise from the assumption of finite delays between cause (the availability of inputs) and effect (the availability of computed output). Causality considerations serve to *introduce* various *delay parameters* that characterize signals waveforms; they do not constrain the actual values of these parameters.

  Initially, these delay parameters have no direct correlation with any implementation details. As a design evolves, however, such a correlation also evolves; ultimately, the delay parameters depend on the details of the system architecture and the process parameters of the underlying technology. A spectrum of approximation techniques was introduced to control the number of delay parameters explicitly modeled.

- The need to exhibit a specified functionality imposes constraints on the *values* output waveforms can acquire during certain intervals, given that the input waveforms have certain values during related intervals.

- The need to avoid conflicts between the constraints arising from causality and functionality considerations give rise to stability constraints on signal waveforms that govern the *durations* of some intervals (as a function of the delay parameters).

5. Further Levels of Abstraction in System Timing

5.1 Partial temporal orders induced by resource allocation

Up to this stage, we have outlined the effect on system timing behavior due to causality, functionality and stability criteria by indicating how they impose constraints on various signal waveforms. Much of this level of timing detail depends only on the specified functionality of the system, and not on any implementation decisions. As a design progresses, the details of the system architecture gradually evolve. We will next comment on the impact of architectural design decisions on system timing.

The fully unraveled data flow graph corresponding to a computation embodies much of the available parallelism. The only partial ordering (i.e., sequencing) amongst the subcomputations arises from essential data dependencies. Resource allocation constraints may, in turn, impose further partial temporal orders amongst the subcomputations involved. As a very simple example, the functional description of the computation (a+b)+(c+d) permits both the additions a+b and c+d operations to be performed in parallel; however, if only one adder module is available, the two additions have to be done in some (in this case, arbitrary) sequence. Analogously, a shared data bus imposes constraints on the concurrent bus accesses.

The temporal partial orders amongst subcomputations may be reasonably described using primitives that are characteristic of temporal logic. Alternatively, they may be explicitly described by means of constraints that must hold on the times when specific events occur. As an example, we next consider next the computations corresponding to two classes of commonly occurring expressions in a generic functional language (or even in an imperative language).

Notation. Given an expression e, let t^{e}_{init} denote the time at which the computation corresponding to the expression e is initiated, let d_e denote the time taken for this computation, and let t^{e}_{valid} denote denote the time till when the result of e will be valid (*i.e.*, beyond which the result of e will be invalid, for example, due to the discharging of a node). Further, if e has the form $f(e_1,..., e_n)$, and f is a "strict" function, *i.e.*, a function whose evaluation requires all of its arguments to be evaluated, we will denote by t^{f}_{init} the time at which the computation of f is commenced, and by d_f the duration of this computation.

Consider first the case when all of the subcomputations $\{e_i$, i=1..n$\}$ involved in the computation of $e = f(e_1,..., e_n)$ are independent and may therefore be evaluated in parallel. If f is indeed a strict function, its computation cannot be initiated until *after* all of the results of the computations corresponding to e_i are available; further, its computation must be initiated *before* they become invalid, *i.e.*, during the such results are valid. Consequently, the temporal partial orders induced on the subcomputations involved may be expressed as follows:

$$t^{f}_{init} \geq t^{e_i}_{init} + d_{e_i}, \text{ for all i in } \{1,..,n\}$$

$$t_{init}^{e_i} \leq t^{e_{i\,valid}}, \text{ for all i in } \{1,..,n\}$$

Consider now the case when architectural design decisions lead to resource constraints such that two functions, say f_i and f_j, both need to share a resource to which they must have mutually exclusive access. Then in the computation of an expression having the form $e = f(...,f_i(e_i),...,f_j(e_j),...)$, it follows that f_i and f_j cannot be executed in parallel. Stated differently, f_j cannot be initiated until the execution of f_i has completed, or the other way around. Formally, this implies that the following constraint embodying this temporal ordering must be satisfied:

$$t^{f_{i\,init}} \geq t^{f_{j\,init}} + d_{f_j(e_j)} \text{ or } t^{f_{j\,init}} \geq t^{f_{i\,init}} + d_{f_i(e_i)}$$

The intent of these examples was to convey some intuition relating to the nature of partial temporal orders on the subcomputations in a system that are induced by resource trade-offs decisions during architectural design. These temporal ordering are in addition to the temporal dependencies implied by the data flow graph corresponding to the underlying computation.

5.1.1 Timing and Allocation Functions Given a data flow graph of a computation, a ***resource allocation function*** details how the operators corresponding to the nodes in such a graph are allocated to hardware modules that implement these operators. A ***timing*** or ***scheduling function***, details when (*i.e.*, in what order) the various operations are actually performed in these modules. Such a schedule is typically implemented in hardware by state machines that may either be centralised (in the "control" part of an architecture) or distributed (as in systolic architectures). One of the tasks in synthesis involves the choice of allocation and timing functions that minimize an appropriate cost function. For certain classes of data flow graphs, *e.g.*, graphs corresponding to uniform recurrence relations[19], timing and allocation functions may be computed algorithmically. In general, however, heuristic techniques or global optimization techniques are needed. Heuristic or rule-based techniques often use local optimization techniques. Global knowledge can, however, be brought to bear on the solution process by using analyses of partially fleshed out designs. Since the computation graph is hierarchical, this hierarchy can be exploited in the corresponding analysis. Clustering techniques are useful in obtaining a measure of proximity between the various modules.

Consider a set of modules M_i that implement the functions F_i corresponding to the nodes in the computation graph. Let d_i be the delay associated with an execution of the module M_i (using a single delay model), and let d be the delay associated with the overall computation. One aspect of the resource allocation problem consists of choosing the number N_i of modules M_i to be used in the actual implementation. In addition, the variables corresponding to edges need to be allocated to signals, buses, registers, or memory. The cost C of an implementation may be defined in terms of a multi-dimensional vector that includes criteria such as area (depending upon the N_i, M_i and the area for interconnections), the delays d_i and d. The timing schedule governs the overall delay d, while the resource allocation affects the flexibility available in the choosing a timing schedule.

The resource allocation problem maybe viewed as a multi-dimensional microcode placement problem. In the simplest case, a subcomputation is allocated to grid location (i,j) if it is executed during "time step" t_i and uses module M_j. The interconnection density, and the number of buses required may be approximated by appropriate functions on a specific allocation on such a grid. Such a placement problem may be solved by using optimization techniques developed in this context, *e.g.*,[13]. Global optimization methods such as simulated annealing have also been found to yield reasonable solutions in this context. The dimensionality of the grid can be increased to allow for other independent factors of consideration, such as conditional sharing of resources.

In a similar vein, the scheduling problem is related to that of microcode optimization[4]. If explicit values for the delays d_i are known, it is possible to use solution techniques (for example, integer programming techniques,) to solve for the variables t^e_{init} and obtain an admissible schedule of computations that obey the specified constraints[27]. We will not elaborate any further on this issue here.

We next discuss how design decisions relating to timing disciplines influence a further level of timing detail. This process establishes a relation between temporal partial orders on computations and the evolution of signal waveforms.

6. Timing Disciplines

6.1 Events and Computations

We define an *event* to be a predicate "of interest" defined on (one or more) signal waveforms. An example of an "interesting" event is the transition in the value of a signal, say from representing a boolean value of 0 to a boolean value of 1, or vice versa.

The overall computation performed by a system may typically be hierarchically decomposed into "smaller" (sub)computations. An operation or subcomputation that is "meaningful" in the application domain may then be defined in terms of a sequence of relevant "lower level" events. Further, and more important, the initiation and the completion of such a (sub)computation needs to be signaled by specific events that are explicitly intended for this purpose. Examples of such distinguished events are transitions in clocks (in synchronous system designs), and transitions in request/acknowledge signals (in self-timed, or asynchronous systems).

A timing discipline is essentially a protocol that governs the manner in which the initiation and completion of (sub)computations is indicated. A timing discipline therefore relates a computation an identifiable (set of) event(s) on some distinguished signal(s), such as clocks or protocol lines.

6.2 Elaboration of Time Intervals; timing disciplines.

An additional level of timing detail in a design is introduced when a specific timing discipline is chosen. It establishes a relation between temporal orderings amongst subcomputations to event predicates defined on signal waveforms. Formally, the "next time instance" is interpreted (defined) as being the next "real" time instance when some specific function (predicate) p:Time→ Boolean becomes true. While we provide some simple (but typical) examples below, such a predicate may in general be any arbitrary predicate defined on waveforms. In this context, it is useful to define the function NextTime as follows:

$$\text{NextTime: Time} \times \text{Time} \times (\text{Time} \rightarrow \text{Boolean}) \rightarrow \text{Boolean}$$

where

$$\text{NextTime}(t_1,t_2)(\text{p}) \equiv_{def} t_1 < t_2 \ \& \ \text{p}(t_2) \ \& \text{ for all t } t_1 < \text{t} < t_2 \neg \text{p(t)}.$$

Intuitively, NextTime(t_1,t_2)(p) says that t_2 is the first instance at which the predicate p is true *after* time t_1; note that it does not say anything about the value of p *at* t_1.

We now indicate how this notion supports both synchronous (clocked) and self timed timing disciplines.

Consider, for example, a clocked latch that is used to implement a register. The predicate p may be interpreted as being true at the times "when the clock rises". If the function defining the clock behavior is *clk*:Time→ Boolean, then the first instance after t_1 when the clock rises next is described by

$$\text{NextTime}(t_1,t_2)(\text{Rising}(clk))$$

where Rising(*clk*) is defined to be true at time *t* if the signal *clk* is true at time *t* and false just before.

In general, given a function *s*: Time→ LogicalSignalValues,[5] Rising(*s*) is defined to be true at time t if s is true at time t and false just before. The functionality of Rising is

$$\text{Rising: (Time} \rightarrow \text{Boolean)} \rightarrow \text{(Time} \rightarrow \text{Boolean).}$$

Consider now an asynchronous circuit implementation, using say, a request-acknowledge protocol as in[8]. The predicate p may in this case be defined to be true when the value of the *request* signal makes the desired transition.[6]

6.3 The clocked register

Consider a latch or a register. At a reasonably abstract level, the functionality of the latch that is most relevant is that it "stores" a value presented at its input port, and makes this value available at its output port at the (abstract) next time instance. In other words, the register may be characterized by

$$\text{Reg}(in,out) \equiv \texttt{always} \text{ for all t } (out(\text{t}+1) = in(\text{t})).$$

Once a decision about timing disciplines has been made, it is possible to amalgamate this into the register definition as follows:

$$\text{Reg}(in,out)(clk) \equiv$$
$$\texttt{always} \text{ for all } t_1,\ t_2 \text{ (NextTime}(t_1,t_2)(\text{Rising}(clk)) \supset out(t_2) = in(t_1))$$

Note here that the parameter *clk* is now explicitly manifest in the definition of Reg, and the dependence on time of the signals in and *out* has now become more explicit; all of the signals *in*, *out*, and *clk* have the functionality Time→ Boolean.

6.4 On synchronous and self-timed disciplines

A timing discipline is often used in an attempt to minimize (ideally, to eliminate) timing errors that can arise due to races and hazards in circuits. Some of these errors cannot always be caught by simulation, particularly in tightly designed circuits, where there is a low tolerance for deviations in timing margins. Such deviations may arise due to fabrication process variations, thermal noise, etc. Two commonly used classes of timing disciplines are self-timed and synchronous (or clocked) timing disciplines.

---

5. The basic formalism we are using allows for multiple valued logic; however, we do not consider this any further in this paper.

6. An algebraic characterization of Seitz's "weak conditions" for self timed circuits may be used to prove the correctness of such protocols[20].

Self timed timing disciplines have the advantage of yielding designs that are delay independent, and consequently modules that use self-timed protocols are *composable*. This delay independence is typically achieved by adding auxiliary logic elements, such as Muller C elements, that aid in detecting when the preconditions needed for initiating a computation (event) are met. Special signals for indicating the initiation and completion of events, called request and acknowledge signals, are often used in self timed protocols[8]. The drawback with using self timed mechanisms is the added area and delay that is introduced by the auxiliary logic elements.

Synchronous timing disciplines, on the other hand, use special global signals called *clocks* in order to sequence the (sub)computations in a system. The use of a synchronous timing discipline tends to localize timing problems that may be encountered to the segments between clocked elements.

In technologies such as CMOS, however, there are typically fairly stringent constraints relating to overlaps of clock lines that need to be met. For example, in very tight designs the overlap between clocks and the rise and fall times of clock edges may need to be less than a gate delay. Such requirements are difficult to ensure as geometries are scaled down. Indeed, they are further exacerbated by the increasingly dominant role of capacitive and resistive effects that tend to contribute to clock skew and slower transition times. As a consequence, one is either forced to adopt a conservative circuit design discipline that provides larger margins for error, or invest considerable effort in fine tuning clocks, using large clock drivers (to improve rise and fall times) and providing clock buffers (to minimise skewing effects). The drawback associated with the latter technique is the increased area taken up by clocking structures; the drawback associated for the former is increase in delay that is introduced by conservative design styles. This drawback is partially compensated for in "dynamic" design styles in CMOS by exploiting the presence of a continuously running clock in various ways in order to avoid the necessity of fully complementary logic (thereby reducing area). However, a deleterious secondary effect of this strategy is to increase the power consumed by such circuits.

Some common examples of dynamic circuit design paradigms in CMOS include: DOMINO logic[11], which uses only n-transistors for computation and provides non-inverting logic; NORA logic[12], which allows the use of both n- and p-type transistors and provides both inverting and non-inverting logic; and the CLIC design style[17] that addresses a problem associated with the NORA technique, namely extreme sensitivity to clock transition and overlap times.

6.4.1 Trade-offs Rough analysis[23] indicates that synchronous disciplines may yield comparatively faster response times when the size of the circuits is relatively small, and that self timed disciplines may be better for larger circuits. In small circuits with total delays of approximately 10-15 "normalized" time units,[7] relatively tight design margins can be used; as a consequence, the overhead associated with self timed design is not justifiable. However, in larger circuits with larger total delays, the timing margins required in a synchronous design style may tend to exceed the delay introduced by auxiliary elements needed for a self timed protocol[23]; moreover, the area overhead of such auxiliary elements tends to be relatively low in this context.

An improvement in technology may easily invalidate a very tight synchronous design, whereas this is less likely to be the case for a self timed circuit. Indeed, the clocking structures in a tightly designed synchronous circuit may have to be substantially modified when a technology is scaled down, whereas a completely self timed circuit will, as a consequence of its delay independence, yield somewhat more robust designs across improvements in technology.

6.5 Considerations in Choosing Timing Disciplines

The actual decision relating to the timing discipline used depends on the environment a system will be embedded in, constraints on interfacing protocols, envisioned system speed, possible clock skews, etc. Quite often, it may be desirable to design subsystems of a system in a "locally synchronous" fashion, while the interfaces between such systems may be self-timed[21]. Additionally, in some high performance systems, designers choose to adopt self timed protocols in view of the fact that the specifications of commercial parts (such as memories) are usually conservative: adopting a self timed interface, if possible, enables the design to exploit the actual working speed of the system, which may be better than its specified speed[9].

In the context of automated synthesis systems, such considerations may conceivably be accounted for in an agent dealing with timing disciplines. We refrain from delving into any further details on this issue.

6.6 On the parameters characterising an implementation

Once the appropriate timing disciplines have been decided upon, the circuits used to implement the functional modules can be detailed. The interdependencies between the signals in such a circuit may impose further constraints on the behavior of signals over time, e.g., a signal might be required to be stable over some period of time. If s_i, x_i denote

7. Where the unit of time corresponds to the delay through an inverter that is driven by equally sized inverter.

corresponding objects in an implementation domain (*e.g.*, signals) and an abstracted domain (*e.g.*, {0,1,X}), and Abs is an abstraction relation (function) that relates (maps) objects in the implementation domain to corresponding objects in the abstract domain, then a "consistent" implementation R' of an n-ary relation R in the abstract domain is typically homomorphic, *i.e.*,

$$\text{Abs}(R',R) \ \& \ \text{Abs}(s_i,\ x_i) \supset \text{Abs}(R'(s_1,...,s_n),R(x_1,\ ....\ ,\ x_n))$$

In general, however, in order for this condition to hold, certain additional constraints relating s_i, x_i, R and R' may have to be satisfied, *i.e.*, $\text{Abs}(R',R)\ \&\ \text{Abs}(s_i,\ x_i)\ \&\ \text{constraints}(s_i,\ x_i,\ R',\ R) \supset \text{Abs}(R'(s_1,...,s_n),R(x_1,\ ....\ ,\ x_n))$ Thus, certain constraints may be imposed on the values of the parameters characterising R and R'. Here, we will be primarily concerned with the relations that must hold between the timing parameters characterising the modules used in the implementation and the timing parameters characterising the desired behavior.

We next illustrate the nature of such constraints and parameter dependencies.

6.6.1 An Example Consider, as an example, the specification of an edge-triggered D-type flip-flop, shown in Figure 3 (cf.[6]). This device has 3 ports: a data input port (D), a data output port (Q), and a strobe or clock port (C). It is typically characterised in manufacturer's specification sheets by a set of (idealized) waveforms, and associated with certain parameters. Examples of such parameters are maximum clock rates, and minimum data setup and hold times, as well as maximum propagation times.

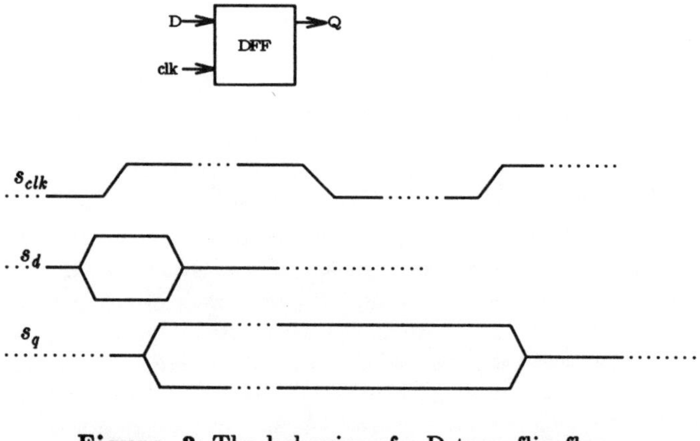

Figure 3: The behavior of a D-type flip-flop

Let s_{clk}, s_d, s_q denote the signals at the ports corresponding to the clock, the input and the output. A detailed, but rigorous, specification of the D-flip flop may have the following form, where the annotations within braces provide an informal explanation of the associated expressions.[8]

DFF_SIGNAL_BEH($a_1,a_2,a_3,a_4,a_5,a_6,a_7,a_8$)($s_{clk}$, s_d, s_q) =
$\forall\ t_0,t_1,t_2,t_3,t_4,t_5$
    constant(t_0,t_1)(s_{clk})(0) {*if the clock is low from* t_0 to t_1}
  & t_1-$t_0 \geq a_1$ {*and the minimum clock low time is* a_1}
  & NextTime(t_1,t_2)(s_{clk})(1) {*and the clock rises at* t_2}
  & t_2-$t_1 \leq a_2$ {*and maximum clock rise time is* a_2}
  & constant(t_2,t_3)(s_{clk})(1) {*and the clock remains high from* t_2 to t_3}
  & t_3-$t_2 \geq a_3$ {*and the minimum high time is* a_3}
  & NextTime(t_3,t_4)(s_{clk})(0) {*and the clock falls at* t_4}
  & constant(t_4,t_5)(s_{clk})(0) {*and the clock remains low from* t_4 to t_5}
  & NextTime(t_5,t_6)(s_{clk})(1) {*and rises again at* t_6}
  & stable(t_1-a_5,t_1+a_6)(s_d)
        {*and the data is stable for a period overlapping the clock rise time*}

$\supset$ {*then*}

stable(t_1+a_7, t_5+a_8)(s_q) {*the output remains stable during this period*}
& $s_q(t_1+a_7) = s_d(t_1)$ {*and has the same value as the data input at* t_1}

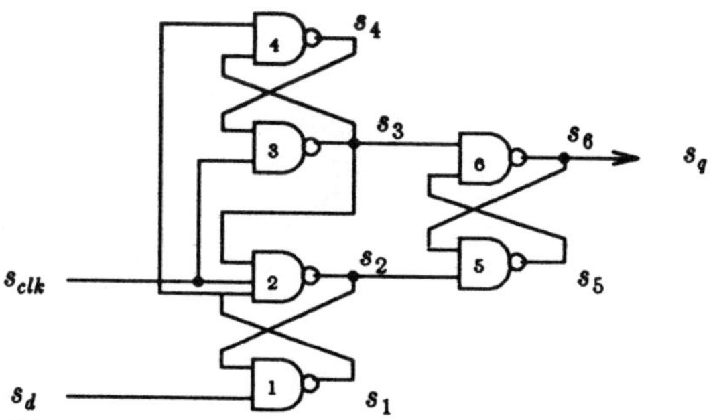

Figure 4: An implementation of a D-flip-flop

A standard implementation of such a D-type flip flop is indicated in Figure 4. Formally, this may be described by the predicate IMPL_SIGNAL_BEH below. Recall that we denote the 4 parameters characterising a NAND gate by $\vec{n}$; since the implementation uses only NAND gates, the implementation is also parameterised by $\vec{n}$.

8. This is a variant of the definition of a flip-flop used in [6].

$$\text{IMPL\_SIGNAL\_BEH}(\vec{n})(s_{clk},s_d,s_q,s_1,s_2,s_3,s_4,s_5) =_{def}$$
$$\text{NAND\_SIGNAL\_BEH}(\vec{n})(s_2,s_d,s_1)$$
$$\& \text{ NAND3\_SIGNAL\_BEH}(\vec{n})(s_3,s_{clk},s_1,s_2)$$
$$\& \text{ NAND\_SIGNAL\_BEH}(\vec{n})(s_4,s_{clk},s_3)$$
$$\& \text{ NAND\_SIGNAL\_BEH}(\vec{n})(s_1,s_3,s_4)$$
$$\& \text{ NAND\_SIGNAL\_BEH}(\vec{n})(s_q,s_2,s_5)$$
$$\& \text{ NAND\_SIGNAL\_BEH}(\vec{n})(s_3,s_5,s_q)$$

Of course, it needs to be shown that this implementation indeed exhibits the behavior of a D-type flip flop. It can be shown that proving this involves proving the following implication:[9]

$$\text{IMPL\_SIGNAL\_BEH}(\vec{n})(s_{clk},s_d,s_q,s_1,s_2,s_3,s_4,s_5) \supset$$
$$\text{DFF\_SIGNAL\_BEH}(a_1,a_2,a_3,a_4,a_5,a_6,a_7,a_8)(s_{clk}, s_d, s_q)$$

It should of course be immediately obvious that the parameters $a_1,..,a_8$ characterising the flip flop must be related in some manner to the parameters $\vec{n}$ (*i.e.* $n_1,...,n_4$) characterising the NAND gates used in the implementation. It can indeed be shown that this configuration of NAND gates (*i.e.*, the implementation) satisfies the desired behavior *only if* the parameters $\vec{n}$ and $\vec{a}$ are appropriately constrained. The normal manner in which such constraints may be computed is by *accumulating the weakest conditions that need to be made in order for a proof of correctness to be carried out.* A set of the constraints relating the parameters of the flip-flop and those of the NAND gate is given by the relation DFF_CONSTRAINTS described below, and is obtained in course of the proof of this implementation described in[6].

DFF_CONSTRAINTS($\vec{a}$,$\vec{n}$) ≡

$$a_1 > 2n_1 + 2n_3$$
$$a_1 \geq a_5 + n_1$$
$$a_3 > 2n_1 + 2n_3$$
$$n_4 + n_2 \geq a_4 + n_1$$
$$a_5 > 2\,n_1 + n_3$$
$$a_6 > n_1 + n_3$$
$$a_7 \geq n_1 + 2\,n_3$$
$$n_4 + n_2 \geq a_4 + a_8$$
$$n_2 > 0$$

---

9. This amounts to approximating the homomorphism function by an implication. We refrain from commenting in detail here on the general nature of proofs that need to be carried out. A skeleton of a detailed proof that this implementation satisfies the behavior of the flip flip (as embodied in the predicate DFF_SIGNAL_BEH) may be found in[6].

It is possible to express these constraints as a function that defines the parameters $\vec{a}$ in terms of the parameters of the implementation primitives, *i.e.*, $\vec{n}$. We illustrate this below.

DFF_CHARACTERISTICS($\vec{n}$) ≡

$$
\begin{aligned}
&a_1 > 2n_1 + 2n_3\\
&a_1 - a_5 \geq n_1\\
&a_3 > 2n_1 + 2n_3\\
&a_4 \leq n_4 + n_2 - n_1\\
&a_5 > 2\ n_1 + n_3\\
&a_6 > n_1 + n_3\\
&a_7 \geq n_1 + 2\ n_3\\
&a_4 + a_8 \leq n_4 + n_2\\
&n_2 > 0
\end{aligned}
$$

Given a set of constraints that need to be satisfied, two possibilities arise. First, the required set of constraints may be unsatisfiable *i.e.*, there may exist no combination of values for the timing parameters ($\vec{n}$ and $\vec{a}$ in this example) that are consistent with the given set of constraints. In such a case the proposed implementation is an incorrect one. Alternatively, a set of values satisfying the constraints and consistent with available devices may in fact exist, in which case the implementation may provide the specified behavior, subject to using devices that have the appropriate parameters.

7. What should the clock width be? Clocking disciplines

Given the semantics of an architectural description, its combinational and sequential components can be identified. The timing delays introduced by the elements in the combinational component of circuit may then be computed. Such a delay may either be already available as an attribute of the circuit element (obtained, say, from a standard library), or computed from the layout details, given appropriate circuit timing models and the process fabrication parameters. The delays along various paths through the combinational circuit component can consequently be computed, and the worst case delay obtained. Existing critical path analysis tools, such as those found in[5, 25] can be used in this context. In synchronous systems, this worst case delay typically tends to limit the speed of the overall system, by setting a limit on the minimum clock width that can be safely used.

At this stage of development of the design, it may either be the case that the delays involved are acceptable (*i.e.*, consistent with the desired timing requirements) or that they are not. In the latter case, two of the alternatives that may be considered are (i) complete architectural redesign at a higher level, or (ii) "local" architectural reconfigurations in the portions of the circuit that contain critical paths. In a synchronous design, it might be the case that (i) much of the circuit may have acceptable delays, except for a few "critical" paths; or that (ii) a certain operation must be performed in a stipulated interval (say, a single virtual clock period), but that the design style adopted to implement it (such as pre-charged/dynamic logic), or a pre-existing subsystem being used for this operation (e.g., an off-the-shelf FIFO chip), consists of a sequence of sub-operations that need to be initiated by clock-pulses. Such situations lead to the introduction of additional clocks (or clock phases). As to whether these clock phases are distributed globally or regenerated locally is governed by trade-offs such as the clock skew that may arise versus the area needed to regenerate the phases, etc. Again, such decisions need to be made by an agent that is concerned with clocking disciplines. If the actual performance of an implementation is deemed unsatisfactory, the architectural design has to be suitably modified.[10]

8. Improving performance

Given a gate-level architecture for a module, it is typically possible to improve, within bounds, the response time of a circuit corresponding to this architecture. This involves reasoning about system timing details at even finer levels. The combinational parts of a circuit are of primary concern in the context of synchronous designs, since they determine the minimum clock width possible, and thereby govern the overall performance of a circuit.

Performance improvements are usually guided by an appropriate circuit model. The correlation of the actual improvements obtained to the projected improvement derived from a model are obviously dependent on the accuracy of the model employed. We next indicate the flavor of some RC models that may be used to fine tune the performance of circuits; our intent here is neither to be exhaustive, nor to prescribe any specific set of models, but merely to indicate some of the possibilities.

8.1 Estimating the number of stages and the gain per stage

The delay incurred in each stage of a typical MOS circuit is directly proportional to the capacitive load driven, and inversely proportional to the current produced by the driver. While the current delivered by any stage can be increased by increasing the size of the transistors used by that stage (*i.e.*, by widening the transistors used), this also has the effect of increasing the effective capacitance observed at the inputs of this stage, thereby slowing down the preceding stage.

10. Throughout this paper, the word "performance" refers only to the timing characteristics of a module; other attributes such as area and power consumption, are explicitly used when needed.

A technique is needed, therefore, in order that enables a rapid computation of the optimal number of stages in a circuit, and further, the sizes of the individual transistors in each stage of such a circuit.

Roughly speaking, if G is the total amplification required, d is the total delay incurred, and there are n stages in the circuit, it is can be easily shown that the overall circuit delay is minimized by dividing the overall delay equally between the stages. Each stage would provide $G^{1/n}$ of the amplification, and contribute d/n to the delay, where d is roughly proportional to $G^{1/n}$. Thus, the total delay through the circuit is proportional to $n^*G^{1/n}$, and an estimate for the optimal value for the number of stages, for a specific value of G, is given by that value of n for which $n^*G^{1/n} = (n+1)^*G^{1/(n+1)}$.

It is therefore necessary to have a technique for computing the overall gain G that is required, and to provide appropriate guidelines for restructuring a circuit should an increased number of stages be required. When additional stages are added to a circuit, it is simplest to add them in the form of inverters, while making sure that proper functionality is maintained[24]. Such inverters provide the necessary electrical amplification while minimally affecting the logical function. It is of course possible to restructure the circuit while increasing the number of stages; while the restructured circuit may appear quite different, it is still performing basically the same function.

We next provide examples of two such techniques. The first example indicates how performance improvements may be obtained by using conventional logic optimizations may be used in conjunction with analysis techniques based on RC models of MOS circuits. Such an analysis technique accommodates the potential alteration of circuit structures, but does not directly assist in indicating when and how such a modification may be performed. The second example indicates how multiple stages of amplification may be used in order to provide an improvement in the performance of the circuit.

8.2 Examples of logic optimization and cascaded gain stages

As indicated above, performance improvements may be obtained by logic optmizations and cascading multiple gain stages. Detailed RC models may be used to compute the gain required per stage, depending on the attributes one desires to optimize, such as power dissipation or size. For concreteness, we first summarize computations relating to two examples, drawn from[26].

8.2.1 Restructuring a circuit for improved gain Consider the implementation of an 8-input AND gate. Three alternative circuits for this purpose are shown in Figure 5. These consist of an 8-input NAND gate with an inverter; two 4-input NANDs and a 2-input NOR; and four 2-input

NANDs, two 2-input NORs, and an inverter. Using a straightforward RC model, it can easily be shown that the approximate average of the rise and fall times of an m-input NAND/NOR gate, normalised against an inverter's average rise and fall time, is given by

$$T_{AVE(NOR)} = T_{AVE(NAND)} = [m^2 C_D + 2(m+1)C_L + mC_D]/[2(C_D + C_L)]$$

where C_D = capacitance of a unit drain area, and C_L = load capacitance on gate owing to routing and fanout. This expression indicates that the best circuit for implementing an 8-input AND gate is circuit 3 for small to medium loads (e.g., $C_D = .5C_L$), while for dominant values of C_L, circuit 2 performs better.

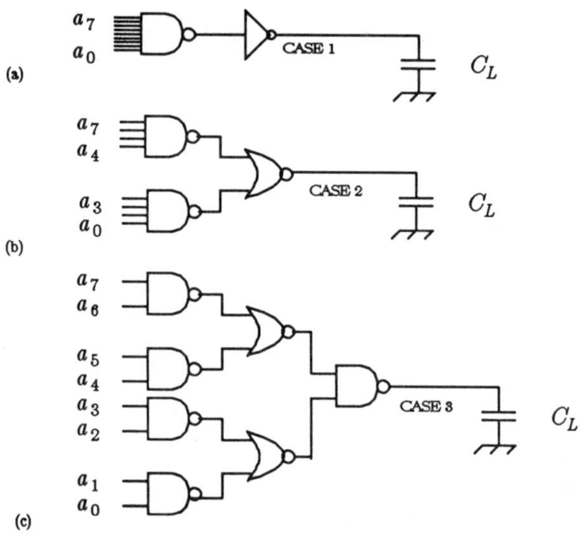

Figure 5. 3 alternative circuits for and 8-input AND gate

8.2.2 Cascading gain stages As a second example, consider the implementation of an XOR gate using a 2-input NAND gate, an OR-AND gate, and an inverter; this circuit is shown in Figure 6. If the average stage ratio (i.e., the ratio by which successive transistor widths are multiplied) is denoted G, the average delay through the series of gates can be shown to be

$$T_{AVERAGE} = 0.5[2GR_n C_L + GR_p C_L + 2GR_n C_L + 2GR_p C_L + (R_n C_{LOAD})/G^2 + (R_p C_{LOAD})/G^2]$$

where C_L is the load capacitance in each individual stage; C_{LOAD} is the final load capacitance; R_n and R_p are the individual stage n- and p-transistor resistances. Depending upon the value of C_{LOAD} / C_L, the average stage ratio that is optimal for this circuit may be determined. In this particular

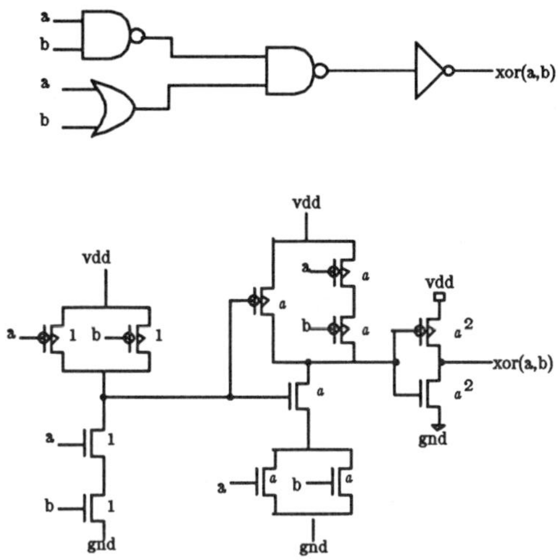

Figure 6. Cascading Gains

example, G $= [C_{LOAD}/(R_n+R_p)(2R_n C_L+1.5R_p C_L)]^{1/3}$. For example, if $C_{LOAD}=100C_L$, G=5 yields a circuit that is almost twice as fast as one in which G=1.

Once a transistor level design has been finalized, performance may be improved locally by transistor resizing. A viable technique for transistor resizing is the one adopted in TILOS[5]; another alternative is the technique discussed in [18]. If A is the sum of transistor sizes and T the longest delay through the circuit and it is desired to minimize metrics such as A subject to T<K, T subject to A<K or AT^K, then it can be shown that the resulting optimization problem has efficient solution techniques, based on the fact that the functions describing A and T are a special class of convex functions called posynomials.

8.3 Approximate Estimations Using RC Models

By using an approximate RC timing model, it is possible to estimate how many stages of logic are required to optimize the delay in a given logic function. It is also possible to estimate a sizing for each transistor that reduces the overall delay. One such model, developed in[24] considers only response time (and no other performance metrics); it enables an approximate computation of the gain required by a circuit, given the external capacitive load to be driven. This information can then be used to compute the optimal number of stages. Any restructuring of the circuit is restricted to the introduction of inverters and associated reorganizations in order to preserve functionality.

In essence, a parameter called the "logical effort" is associated with every logic function; intuitively, it serves to characterise the "normalised virtual load" contributed by a logic function (when it is implemented in a specific manner, in a given technology). The logical effort of an inverter is assumed to be 1, and the logical effort of any other function measures how much "worse" it is than an inverter in delivering output current, for an equivalent amount of input capacitance. Equivalently, the logical effort of an implementation is the effective input capacitance when it produces the same amount of current as an inverter of unit size. The logical effort of a function depends upon a specific circuit chosen to implement a logical function, the circuit topology, and the fabrication process. It does not therefore directly accommodate alternative "architectures" for a circuit such as the possibilities for the 8-input AND gate indicated in Figure 5.

Given the logical effort of individual gates, the logical effort of a compound circuit may be computed by combining the logical effort of the component stages. If $\sum_i C_i$ and $\prod_i C_i$ denote circuits with the elements C_i in parallel and series respectively, then the logical effort function G_L for the overall circuit is defined as follows:

$$G_L(\sum_i C_i) = \sum_i G_L(C_i)$$
$$G_L(\prod_i C_i) = \prod_i G_L(C_i)$$

Compound circuits with less overall logical effort can be optimized to yield less total delay than logically equivalent circuits with a larger logical effort. The total gain required is obtained by multiplying the electrical gain required by the logical effort (gain) for a circuit. The number of stages for least delay is roughly proportional to the overall logical effort required. Once the overall logical effort is known, the actual number of stages can be computed by table lookup.

In summary, therefore, given a logical network, the above model provides a simple way of estimating the least delay. Given the delay per stage, the load to be driven, and the logical effort required, individual transistor sizes can be computed.

8.4 Summary

It is possible to improve performance within the confines of a relatively rigid architecture. This may be achieved by local logic optimizations, techniques that optimize the number of amplification stages (including buffer introduction), and techniques that resize transistors (either globally or within each stage). The performance trade-offs introduced by some of these techniques may be formulated as mathematical optimization problems (involving linear programming, convex programming, or nonlinear programming, based upon the specifics of the formulation). Examples of transistor resizing programs based upon such formulations include[5, 18].

9. An Example

We next consider a simple example that illustrates the interplay of some (but not all) of the different facets of the framework discussed in this paper.

9.1 A segment of a microprocessor-memory interface

Consider the block diagram of a simple microprocessor shown in Figure 7. The three components shown are the data path, the control path and the memory subsystem. The data path consists of a register file and an Arithmetic-Logic Unit (ALU) (connected to a multiplexer). The control path, as indicated, consists of a PLA driven finite state machine, although several alternatives are possible. The memory subsystem is capable of processing read and write operations; it is shown as an independent subsystem that interfaces with the microprocessor system bus via address and data latches.

In this section, we will examine the temporal aspects of the implementation of an operation to "fetch the next word from memory" (henceforth denoted GetNextWord). Before proceeding with an analysis of the GetNextWord operation, however, we first summarise the behavior that is *assumed* of the memory subsystem, and the protocol established for its use.

9.1.1 Behavior and protocol (assumed) for the memory subsystem Given a memory address, the memory subsystem performs either a read or a write operation (whichever is requested). The protocol assumed here for reading from memory is as follows:

1. the address is loaded into the memory address register MAR (connected to the system bus) using an enable signal denoted MARE; if the operation is a write operation, the memory data register must also be loaded using an enable signal MDRE.
2. the operation desired (Read or Write) is indicated by a read/write signal being set high (for read) and low (for write).
3. If the operation requested is a Memory Read, then after a time D^{Memory}_{access} following the initial time at which the address latch MAR was enabled, the memory system provides the specified word in the register IOR. If the operation requested is a Memory Write, the contents of the data register MDR are written onto the address MAR after a time D^{Memory}_{write}.

If we assume that the address register MAR is an edge triggered flip-flop, then the setup and hold time constraints associated with this latch must be satisfied. The first step in the above protocol therefore implies that the valid address must be stable on the system bus for a time at least D^{MAR}_{setup} before the enable signal MARE can be raised. Further, the address must remain valid (*i.e.* stable) on the bus until time D^{MAR}_{hold} after the enable signal

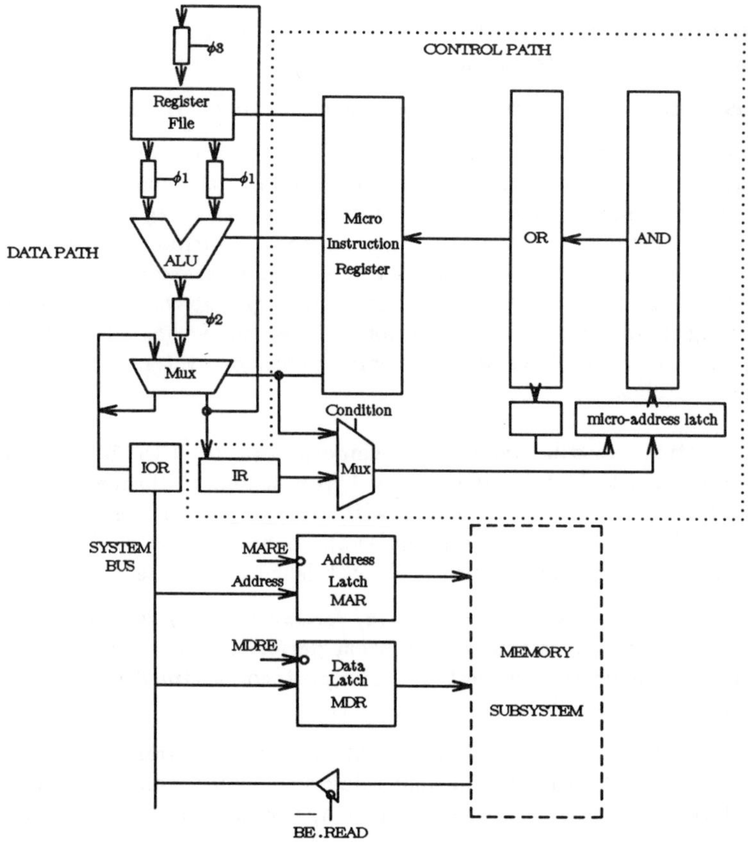

Figure 7. The Block Diagram of a Simple Microprocessor.

MARE has reached a high value.

This behavior of the memory subsystem, and the required protocol for a Read operation, may be described as follows:

[Memory Read]

$$\texttt{rising}(t, t+D^{MARE}_{rise})\,\texttt{(MARE)}$$
$$\texttt{\& stable}(t-D^{MAR}_{setup}, t+D^{MARE}_{rise}+D^{MAR}_{hold})\,\texttt{(SystemBus)}$$
$$\texttt{\& constant}(t, t+D^{MARE}_{rise}+D^{MAR}_{hold})\,\texttt{(ReadWrite)(1)}$$
$$\supset \texttt{IOR}(t+D^{MARE}_{rise}+D^{Memory}_{access}) = \texttt{Memory(SystemBus(}t\texttt{))}$$

In words, this says:

IF

- the enable signal MARE is raised at time t (and takes time D^{MARE}_{rise} to rise)
- and the system bus maintains a valid address from $t\text{-}D^{MAR}_{setup}$ to $t+D^{MARE}_{rise}+D^{MAR}_{hold}$
- and a memory READ is requested by holding the ReadWrite signal at 1 during this interval

THEN

- the word at the address specified on the system bus will be made available in the registed IOR after a delay D^{Memory}_{access} following the time when the enable signal MARE becomes high.

Note that a somewhat more complicated memory system accommodating the notion of "fast" and "slow" memory fetches can also be specified; however, we do not consider this system here.

Given the above assumptions relating to the behavior of the memory subsystem, we next describe the desired behavior of the GetNextWord operation, and analyze the evolution of a plausible implementation.

9.1.2 Desired Behavior of GetNextWord Informally, the objective of the GetNextWord operation is to fetch the word in memory at a computed (effective) address, and store this word either in the register IOR or in the register file. For concreteness, we assume that the word is stored in a register called IOB (for the I/O Buffer) that is in the register file. We also assume that the effective address is calculated by adding the contents of the register denoted PC (for the program counter) and the register denoted OA (for offset address); further, both these registers are assumed to be in the register file.

The behavior desired of the GetNextWord function may be described as follows:

```
GetNextWord(RegFile,Memory) ≡
      Next IOB = Memory[EffectiveAddress]
```

which basically reflects the fact that following the execution of the GetNextWord operation, the content of the register IOB is the same as the contents of memory at the computed effective address.

9.1.3 A computation method A straightforward algorithm for achieving the desired behavior involves the following steps:

- Compute the effective address by adding the contents of the register PC and OA in the register file;

- Issue a read operation to the memory subsystem;
- Write the result of the memory read operation into the IOB register.

Functionally, this may be expressed by

Write(Regfile, IOB, Read(Memory,EffectiveAddress(RegFile)))

where

EffectiveAddress(RegFile)=Add(Read(RegFile,PC),Read(RegFile,OA))

Alternatively, using more conventional imperative notation, the same computation may be expressed as:

RegFile[IOB] := MemoryRead[Regfile[PC] + Regfile[OA]]

9.1.4 Temporal Consequences of a computation method The choice of this algorithm implies that the effective address must be computed prior to the memory read operation.[11] However, the two reads from the register file may potentially be concurrent. Finally, the write operation into IOB follows the memory read. This partial order is embodied by the graph shown in Figure 8.

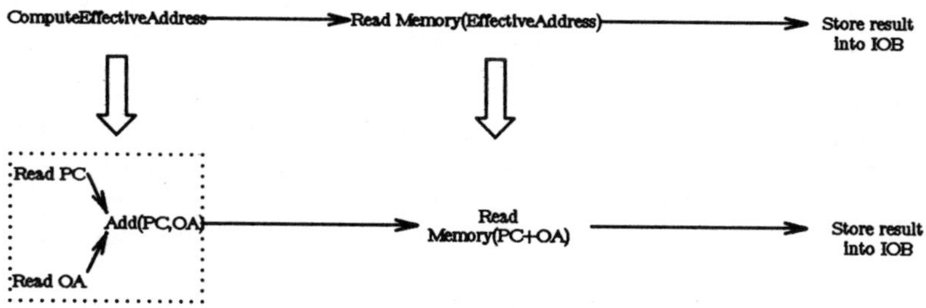

Figure 8. Temporal ordering induced by the algorithm

9.1.5 Resource Allocation Decisions Hardware resources are needed for implementing (i) the read operation on the register file, (ii) the Add operation, (iii) the processor-memory interface, and (iv) the memory subsystem. We assume the following design decisions pertaining to the resource allocation:

11. The actual temporal orders induced are of course independent of the notation in which the algorithm is expressed. However, a functional notation has the advantage of providing a more explicit representation of the potential parallelism.

- The logic associated with the Register file allows two simultaneous reads onto the two input ports of the ALU;
- The ALU is capable of performing the addition operation;
- A single system bus is shared by the microprocesor and the memory subsystem;
- The data and address are multiplexed on the system bus.

The first two allocation decisions do not impose any additional temporal constraints to the ones shown in Figure 8, since they allow the full parallelism permitted by the algorithm choice. However, because a single bus is used for carrying both the data and the address (and because the system bus may be shared by other subsystems) a protocol is needed for obtaining access to the system bus. We assume a simple protocol here that uses a distinguished signal denoted BusRequest for staking claim to the bus. This protocol stipulates that if the microprocessor wants to gain control of the bus, it should raise the BusRequest signal high.

The set of temporal partial orders arising as a result of the resource allocation constraints is depicted in Figure 9.

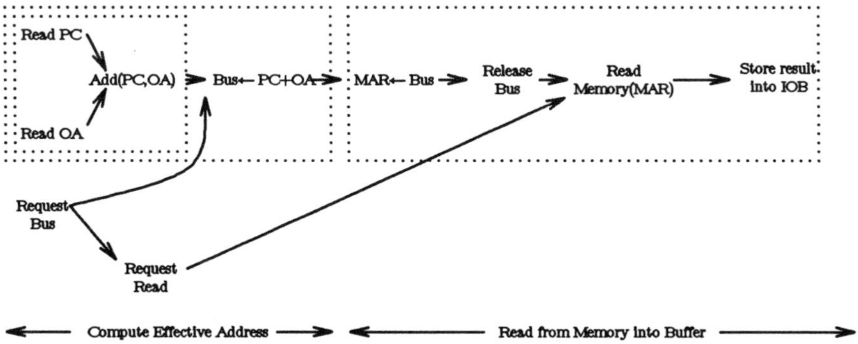

Figure 9. Temporal orders due to resource constraints (multiplexed bus)

9.1.6 Timing Requirements for the operation GetNextWord In the computation shown in Figure 8, there are 3 steps involved.

We assume that the mandated requirement is that the computation of the effective address be completed during the first major cycle, and that the memory read and store into the register IOB be completed during the next major cycle. This is indicated in Figure 11.

The requirement may be formally stated as follows, where

`Instruction(t)` denotes the instruction in the instruction register at time t

```
[GetNextWord Timing Requirement]
Rises(t)(φ1) & Instruction(t)=GetNextWord ⊃
  Bus(t+TP)=Add(RegFile[PC](t),RegFile[OA](t))
  & IOB(t+2*TP)=Memory[RegFile[PC](t)+RegFile[OA](t)](t)
```

The constraints implied by this protocol are that the delay incurred by the computation of Add(RegFile[PC](t),RegFile[OA](t)) is less than T_P, and that the memory access (and store into IOB) from the time the address is available on the system bus takes less time than T_P.

9.1.7 Timing/Clocking Discipline The clocking scheme adopted in a system typically corresponds to some of its "naturally occurring" logical operations. In the case of a microprocessor, the basic logical interval of concern is the microprocessor's microinstruction cycle. Depending upon the details of the processor, however, a microcycle may be further divided into two, three, or more phases. We assume here, for the sake of illustration, a three phase cycle.[12] In the data path, we assume that the register file is read during phase one, that the ALU performs its operation during phase two, and that the result written onto the register file (or multiplexed onto the system bus) during phase three. The control path may, for instance, update the microaddress latch during the first phase, access the AND plane during the second phase, access the OR plane and update the micro instruction register during phase 3. We next describe the behavior assumed of a 3-phase clock, and relevant aspects of the control path behavior.

9.1.7.1 Three Phase Overlapping Clock Behavior The behavior of a three phase overlapping clock with overlapping and single phase periods of T_O and T_S each is shown in Figure 10, and may be described as follows:

12. A 2 phase nonoverlapping clock is sometimes difficult to guarantee in tightly designed high performance systems, while a 2 phase overlapping clock leads to overly stringent timing constraints that restrict both the minimum and the maximum path delays. On the other hand, a three phase overlapping clock is relatively easy to generate, and does not lead to unreasonable constraints provided that the three phases do not simultaneously overlap during any one time interval.

Rises(t1)(ϕ_1)

for all t

Rises(t)(ϕ_1) $\supset$ `constant`(t+T_S+2*T_O)(ϕ_1)(1)
  & Falls(t+T_S+2*T_O)(ϕ_1) & Rises(t+T_S+T_O)(ϕ_2)
Falls(t)(ϕ_1) $\supset$ `constant`(t,t+2*T_S+T_O)(ϕ_1)(0) & Rises(t+2*T_S+T_O)(ϕ_1)

Rises(t)(ϕ_2) $\supset$ `constant`(t+T_S+2*T_O)(ϕ_2)(1)
  & Falls(t+T_S+2*T_O)(ϕ_2) & Rises(t+T_S+T_O)(ϕ_3)
Falls(t)(ϕ_2) $\supset$ `constant`(t,t+2*T_S+T_O)(ϕ_2)(0) & Rises(t+2*T_S+T_O)(ϕ_2)

Rises(t)(ϕ_3) $\supset$ `constant`(t+T_S+2*T_O)(ϕ_3)(1)
  & Falls(t+T_S+2*T_O)(ϕ_3) & Rises(t+T_S+T_O)(ϕ_1)
Falls(t)(ϕ_3) $\supset$ `constant`(t,t+2*T_S+T_O)(ϕ_3)(0) & Rises(t+2*T_S+T_O)(ϕ_3)

The above specification ignores the rise and fall times of the individual clocks; this may be explicitly modeled if necessary, but we refrain from doing this here. The "major" cycle has a period T_P = 3*(T_O+T_S). Each phase has a high time of T_S+2*T_O, and a low time of 2*T_S+T_O.

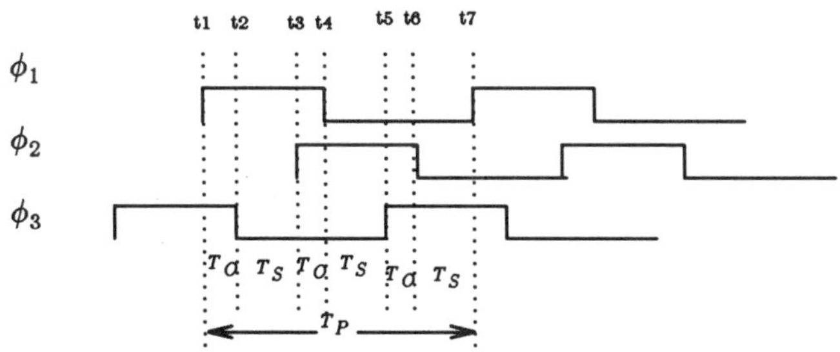

Figure 10. A 3 phase overlapping clock

9.1.8 Assumptions relating to the Control Unit Behavior We assume that the output of the microinstruction register MIR is stable for a period T_P, beginning with the time when ϕ_1 rises. The register MIR contains the subfields [R1,R2,W,OP,CNTRL]. These fields denote the registers R1,R2 to be read from the register file (during phase 1); the operations OP to be performed by the ALU (during phase 2); the gating control CNTRL for the multiplexer indicating whether IOR or ALU is to be gated onto the multiplexer output; and the register W in the register file that is written with the value on MuxOut (during phase 3).

If the functions NextMIR and NextState define the next microinstruction and next state (as a function of the current state and the value on MuxOut), then the behavior of the control unit may be described as follows:

```
Control(State) ≡_def
Rises(t)(φ1) & stable(t,t+TS+TO)(MuxOut) ⊃
    constant(t+TP,t+2*TP)(MIR)(NextMIR(State,MuxOut))
    constant(t+TP,t+2*TP)(State)(NextState(State,MuxOut))
```

The actual details of the functions NextMIR and NextState are not of immediate concern to us. We further assume the existence of an "Instruction Register" somewhere in the control that contains the next "main" instruction to be executed. The content of this register at time t is denoted Instruction(t).

9.1.9 Hierarchical temporal decomposition We restrict our discussion in the remainder of this section to a consideration of the operation GetNextWord. As discussed earlier, the execution of the GetNextWord instruction involves subcomputations whose details may be hierarchically considered. The protocol for each of the subcomputations shown in Figure 9 can be enumerated and their temporal implications analyzed using the techniques introduced earlier. The timing diagram in Figure 11 indicates some of the constraints generated in a graphical fashion. The temporal constraints arising out of a hierarchical analysis of the subcomputations involved in an execution of the operation GetNextWord above are summarized below. We have assumed an independent delay approximation for the ReadRegFile, ALU and MUX operations, with the associated delay constants being $d_1^{ReadReg}$, $d_2^{ReadReg}$; d_1^{ALU}, d_2^{ALU}; d_1^{MUX}, d_2^{MUX} respectively. d_1^{BusOut} denotes the time for the result from MuxOut to be driven onto the system bus. t_{init}^{BR} is the time at which a bus request is initiated.

Assume $\mathrm{Rises}(t_1)(\phi_1)$

$d_2^{ReadReg} > 0$

$d_1^{ReadReg} < T_S + 2*T_O$.

$d_2^{ALU} > 0$

$d_1^{ALU} < T_S + 2*T_O$.

$d_1^{MUX} < T_S + 2*T_O$.

$d_1^{MUX} + d_1^{BusOut} < T_S + 2*T_O$.

$t_1 < t_{init}^{BR} < t_4$.

Rises(t)(MemoryOperation) $\supset \; t_4 < t < t_6$

$$2*T_O + 2*T_S > d_1^{ALU} + d_1^{MUX} + d_1^{BusOut} + D_{setup}^{MAR} + D_{hold}^{MAR} + D_{rise}^{MARE} - (d_2^{MUX} + d_2^{BusOut})$$

$$d_1^{ALU} + d_1^{MUX} + d_1^{BusOut} + D_{setup}^{MAR} + D_{rise}^{MARE} + D_{access}^{Memory} + d_1^{MUX} + d_1^{WriteRegFile} < 5*(T_O + T_S)$$

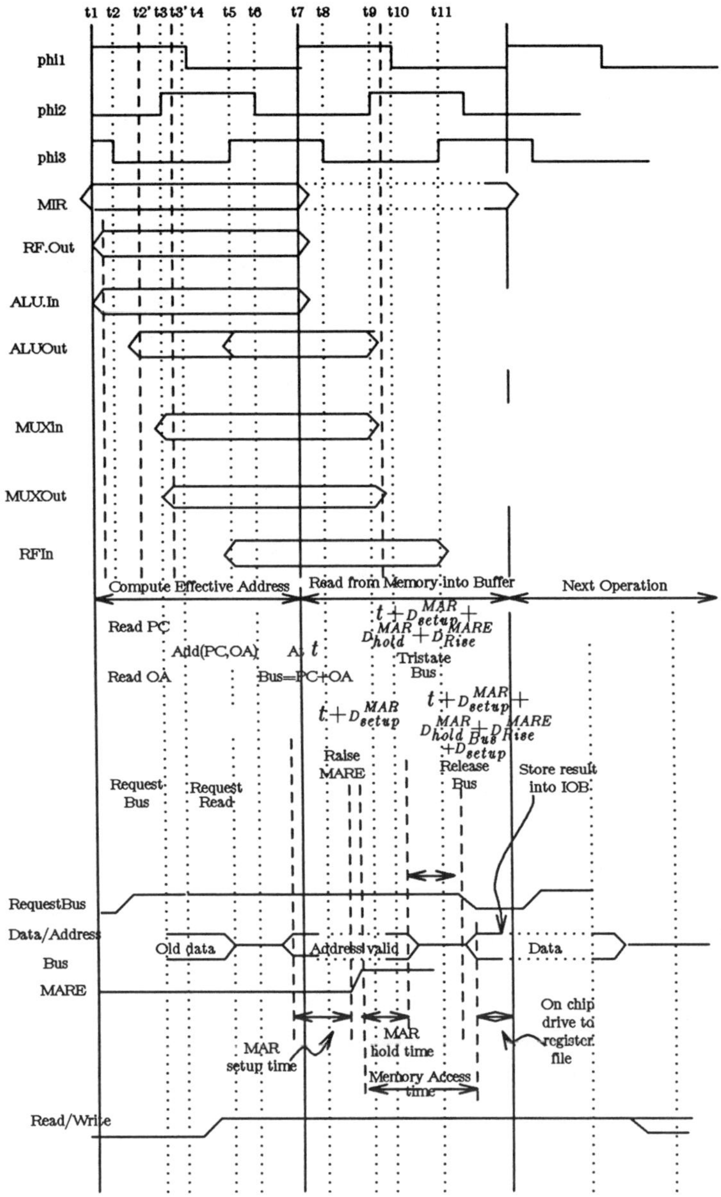

Figure 11. A timing diagram for the GetNextWord operation

These constraints essentially relate the clock periods T_O and T_S to the delay parameters associated with the components in the data path and the memory subsystem. These delay parameters in turn depend upon the details of the circuitry that implements the various subcomponents of the system. At much lower levels, the techniques for lower level circuit optimizations may be used to locally adjust the actual values of these delays; we will not elaborate upon this aspect here.

The example discussed in this section illustrates a design evolution analogous to that assumed in the development of the proposed framework.

- The desired behavior (GetNextWord semantics) and the performance requirements (operation completion within two cycles) were initially specified;
- An independent delay approximation (cf. Proposition 3) was repeatedly used to model the ramifications of causality assumptions relating to the subcomputations (and associated submodules) involved.
- An algorithm for implementing the behavior was chosen, and some decisions relating to resource allocation were assumed. The temporal ramifications of these decisions were then inferred.
- It was shown how the various subcomputations could be characterised (*i.e.*, specified and reasoned about) in isolation; it was indicated that there are two contexts in which such specifications of operations are relevant (for the user, and for the implementer).
- Examples of timing specifications at several levels of the design hierarchy were provided.
- A hierarchical representation of timing details was introduced, *e.g.*, temporal partial orders among subcomputations, major instruction cycles, microcyles and multiple clock phases.
- Both synchronous and asynchronous protocols were considered.

10. Discussion

In this section, we recapitulate some of the main points developed so far and comment on the role of the framework in the context of analysis and synthesis.

10.1 Considerations in Synthesis and Analysis

The primary objective in synthesis is to obtain an implementation that meets the specified behavior and performance criteria. Our main concern in this paper has been to examine the issues related to system timing that are pertinent in such a context. Since the overall design strategy typically involves some combination of top-down decomposition and bottom-up synthesis, we next individually consider the questions that arise in both of these design phases.

During the initial top-down decomposition phase, it is necessary to specify the external timing criteria; appropriately model design decisions that are explicitly related to system timing, and infer the temporal constraints that are implied by other design decisions; and be cognizant of the externally specified timing constraints in making various design decisions.

The corresponding questions of interest therefore include the following: What are the different classes of external timing specifications? and, How are such timing criteria specified? What are the different levels of temporal abstraction, and how are they specified and modeled? What timing constraints are implied by various design decisions? and conversely, How do timing constraints influence design decisions (if at all)?

During bottom-up synthesis, it becomes necessary to analyze various aspects of a partially or completely fleshed out design, in order to obtain information that provides useful feedback for appropriate design modifications. The questions of interest in such a context include: Given the characteristics of the submodules used in an implementation, what are the characteristics of the overall module? What local adjustments should be made that enable an improvement in performance? How exactly are such improvements to be made? What effect do the temporal consequences of design decisions have on the subsequent stages of a design?

In the remainder of the section, we summarize our discussion relating to these issues.

10.2 External Timing Specifications

The temporal and functional aspects of low-level behavioral descriptions are often inextricably intertwined. Nevertheless, the external specifications that relate to system timing may be roughly categorized as follows:

- A category dealing with the *logical* interface between a system and the external world. This category includes specifications of interfacing protocols that must be obeyed, *e.g.*, a dual-rail request-acknowledge protocol (cf.[20]).
- A category dealing with the *timing characteristics* of the *externally provided* input signals, such as periods during which they are valid, their transition times, etc.
- A category dealing with the *timing constraints* that an *implementation* is required to satisfy, such as response times and throughputs desired. Such requirements specify the desired timing behavior of various output signals with respect to the available input signal behavior.

The timing constraints (or performance criteria) that an implementation may be required to satisfy can be of three varieties:

- *concrete* timing constraints (synonymously, complexity criteria), such as absolute timing requirements that must be met by an operation e.g., a worst case upper bound of 50 milliseconds for some operation;
- *abstract* timing criteria, where the expression denoting the timing requirements is parameterized by variables appropriate to higher levels of abstraction e.g., a sorter of complexity O(NlogN), where N is the size of the input *i.e.*, the number of elements to be sorted;
- *relative* timing constraints, where the expression denoting the timing requirements uses intervals of time defined by other events or (sub)computations in the system, such as clock cycles of a microprocessor.

This classification is relevant when exploring heuristics that use performance criteria in guiding architectural decompositions. While all of these varieties of timing specifications are accommodated by the framework discussed here, only abstract and relative timing constraints (*i.e.*, the last two categories of timing constraints) are useful in the *initial* top-down design decomposition phase. More specifically, absolute timing constraints (such as 50 milliseconds) are not effective in guiding the synthesis strategy until a design is elaborated upto the point where the primitives used have known timing characteristics;[13] the timing constraints may then be correlated with the actual timing characteristics obtained by analysis, and the results used in guiding subsequent design strategy in an attempt to satisfy these constraints. Abstract timing criteria can effectively be used to prune the exploration of design alternatives, because it is possible to estimate the abstract complexity of an architectural choice without elaborating its constituents in detail. A detailed discussion of this issue is beyond the scope of the current paper; a brief discussion may be found in[2].

10.3 Temporal Abstractions: Specification, Modeling and Applications

In addressing the question "What are the different levels of temporal abstraction, and how are they specified, modeled, and used?", some of the important levels of temporal abstraction in the context of system design were identified. Examples include: signal voltage waveforms; abstracted

13. Such timing characteristics may be obtained from a library of available primitives, an analysis of a circuit layout, or by some alternative means.

signal waveforms; partial orders on the subcomputations in a system; and observable functionality.

Syntax for expressing the appropriate temporal concepts at various levels of abstraction was illustrated.[14] This syntax is relatively succinct, and is based on primitives that can be rigorously defined in algebra and logic. As a consequence, a clean semantic basis is also obtained, which assists in reasoning about various temporal constraints that arise during different stages in the design process. Such reasoning can be performed either entirely by humans, or benefit from machine assistance via the use of programs that aid in manipulating the logical/algebraic expressions involved[7, 22, 1, 6].

Further, we have provided ways to *infer* the temporal constraints introduced by various design decisions. As additional temporal constraints are introduced, it must be confirmed that they are not in conflict with the existing set of constraints; if they are, then some of the design decisions that have led to such a conflict must be revoked. An obvious strategy is to have an "incremental" consistency checker to ensure that the temporal consequences of each new design decision are consistent with the existing timing constraints. Nevertheless, an explicit statement of such constraints is quite useful even for humans, particularly when translated into a graphical form, e.g., as in timing diagrams. A designer can often gain considerable insight by inspecting such constraints. The designer may then either be convinced of the consistency of such specifications (although in a complex system this is not a very reasonable expectation), or modify relevant aspects of the design.

In general, the design decision that needs to be revoked or modified may not always be the last (*i.e.*, the "latest") decision that was made. In fact, isolating the set of conflicting constraints helps focus on the set of design decisions that lead to these constraints, and therefore assists in identifying the set of decisions that needs to be modified. It is relevant to note that nothing in the paradigm enforces a strict order on the various design decisions.

We next highlight some of the important correlations between design decisions and system timing. This also establishes a correlation between some of the levels of design abstraction and the levels of temporal abstraction, and has the cumulative effect of yielding a hierarchical temporal specification of a system.

---

14. The specific primitives used here are not unique, and several variants are possible.

10.4 Implications of Behavioral Specifications: Functionality, Causality, and Stability criteria

Given a desired behavior, the discussion in Section 4 indicated how the framework enables temporal constraints arising from functionality, causality and stability criteria to be inferred. In brief, recall that

- The need to exhibit a specified functionality imposes constraints on the *values* output waveforms can acquire during certain intervals, given that the input waveforms have certain values during related intervals.
- Causality constraints arise from the assumption of finite delays between cause (the availability of inputs) and effect (the availability of computed output). Causality considerations serve to *introduce* various *delay parameters* that characterize signals waveforms; they do not constrain the actual values of these parameters. Four categories of approximations were introduced in Section 4 that allow the number of delay parameters to be appropriately controlled, mainly by assuming decreased dependence of the delays on the actual transitions involved. Reducing the number of delay parameters enables a reduction in the complexity of the resulting delay model, at the expense of needing more conservative approximations in the actual values of the delays.[15]
- The need to avoid conflicts between the constraints arising from causality and functionality considerations give rise to stability constraints on signal waveforms that govern the *durations* of some intervals (as a function of the delay parameters).

It is important to note that the delay parameters introduced at this stage are quite independent of any lower level circuits models. However, as a design is refined, these parameters are related to those characterizing the submodules used in an implementation, and ultimately, to process parameters that arise from the semiconductor models used.

10.5 Temporal Constraints related to Architectural Design Decisions

An answer to the question "What are the timing constraints arising out of a design decision?" obviously depends on the specific nature of the design decision. The discussion in Section 5 indicated the nature of temporal constraints arising out of the selection of an algorithm, architectural decisions relating to resource allocations for implementing

15. The approximations are conservative because they must accommodate the worst case behavior of the entire set of signal transitions that would otherwise have been treated individually.

operations, and various flavors of synchronization protocols.

10.5.1 Partial temporal orderings induced by architectural decisions The fully unraveled data flow graph corresponding to a computation embodies much of the available parallelism; the only partial ordering amongst the subcomputations arises from data dependencies arising out the selection of an algorithm. The choice of a particular architecture, however, imposes constraints on the number of various resources available for a computation. Such resource constraints, in turn, impose further partial temporal orders amongst the subcomputations involved. Conventional flow analysis techniques, along with those indicated in this paper, may be used to infer the necessary partial orderings.

The implication of these temporal partial orderings in the context of synthesis is that an appropriate schedule of computations must be found and then enforced. "Mathematical programming" techniques, such as linear, integer and nonlinear programming techniques, can assist in computing such schedules. The enforcement of such schedules is typically done by using a "controller" (state machine) in conventional architectures, although other distributed mechanisms may be used, such as those commonly found in systolic architectures[20].

10.5.1.1 Partitioning timing allocations The temporal partial orders discussed above lead to subcomputations that must performed either in sequence or in parallel. Given that we have some timing constraint to be met by the entire computation, this constraint needs to be "partitioned" amongst the sequential and parallel subcomputations. This is obviously relevant if hierarchical design techniques are employed, since a subproblem can be attacked in isolation only if the relevant constraints are available.

If a computation e involves the subcomputations $e_1, \cdots, e_n$ in parallel, then the delay constraint for each of the subcomputations e_i is identical to delay constraint for e. A difficulty arises, however, in allocating constraints to subcomputations that are sequential, when nothing more than the semantics of the individual operations is known, *i.e.*, when the implementation of the subcomputaoins e_i is completely unexplored; in such a case, there is no knowledge about how much time each of the individual operations consume. One possible approach in such a context is to use estimates of lower bounds on the complexity of the computation involved. There are two problem with this approach. Firstly, lower bounds on complexities are difficult to obtain (even for humans, with or without machine assistance) and often seem to involve significant insight into the semantic nature of a computation; mechanising this process therefore appears a remote possibility at the present time. Secondly, lower bounds are often asymptotic in nature and quite dependent on the computational models assumed; as a consequence, the value of such results may be rather limited, particularly in the context of designing "real" circuits. The partitioning of timing constraints in such a case is therefore very much an open problem.

The situation is quite different (and fortunately somewhat more tangible) during subsequent stages of a design, when an implementation has at least partially been evolved, or when a library of alternative implementations for the operations concerned is available. In such a case, estimates are typically available for the set of implementations that have been explored thus far. Various heuristics or optimization procedures may then be used in the context of synthesis.

In general, it is required to optimize some function of a multi-dimensional performance vector (involving dimensions such as area, response time, power, pincount, and throughput). A relevant question in such a context is: How much better can we do along other dimensions while still meeting the stipulated timing constraints? Using the techniques discussed here, it is possible to obtain a set of constraints on the delay parameters characterising the different submodules used in an implementation. In many cases, the sets of such constraints are often linear, and the resulting optimization problem one of linear programming. Solution techniques for optimization problems can therefore be used in order to allocate concrete values to the delay parameters desired of submodules, and the process repeated in a hierarchical fashion. During the later stages of design evolution, a range of plausible values is typically available for some of the delay parameters, by virtue of existing implementations (or libraries thereof).

An "iterative" approximation to this problem is one that starts out with the implementation option that is consistent with the timing constraints, and then attempts to improve the objective function by making appropriate design modifications, ensuring that the timing constraints are not violated. From this perspective, the subcomputations e_i can be ranked in an order that indicates the relative (incremental) improvement in the cost function for a unit increase (i.e., degradation) in response time. The overall time T available for the sequence $e_1; \cdots ;e_n$ can be partitioned amongst the e_i in proportion to their best available computation times. Given an initial ranking of the subcomputations e_i, and starting with an allocation of times in proportion to the least times for each subcomputation, it is possible to attempt incremental degradations in time for each of the expressions, in the order of decreasing payoff in the optimization function. Several related variations of such a heuristic are possible, and experimental exploration will provide additional insight.

10.6 Timing disciplines

The relation between temporal orderings amongst subcomputations and signal waveform behavior is established to a finer degree of detail by the choice of a timing discipline. A timing discipline is essentially a protocol that governs the manner in which the initiation and completion of (sub)computations is indicated. A timing discipline therefore relates a computation to an identifiable (set of) event(s) on some distinguished signal(s), such as clocks or protocol lines. The exact correlation is specified in the framework by the predicates that associate computations with events on signal waveforms.

Considerations involved in choosing timing disciplines were discussed in Section 6, as well as issues that influence decisions relating to clocking disciplines in the case of synchronous designs. An explicit documentation of the timing protocols during the synthesis process is valuable in subsequent design in two contexts: (i) in assisting tools that are used for formal verification and/or timing analysis tools; (ii) viewed as additional temporal constraints on the signals.

10.7 Relations between timing parameters in an implementation

Once the appropriate timing disciplines and clocking strategies have been decided upon, the circuits used to implement the functional modules can be detailed. The interdependencies between the signals in such a circuit may impose further constraints on the behavior of signals over time, e.g., a signal might be required to be stable over some period of time. Moreover, in order for a specific implementation to exhibit a desired behavior, some relations must typically hold between the timing parameters characterising the modules used in the implementation and the timing parameters characterising the desired behavior.

During the initial top-down decomposition of a design, this relation may be viewed as a function that yields the timing parameters that must be used (*i.e.*, that need to be met) by the submodules used in the implementation. Conversely, once the details of the modules used by an implementation are actually available, the timing parameters characterising the composite module may be computed. Such a computation is therefore useful in "design iteration" phase, rather than in the initial top-down decomposition phase.

10.7.1 Improving performance Given a gate-level architecture for a module, it is typically possible to improve, within bounds, the response time of a circuit corresponding to this architecture. This involves reasoning about system timing using circuit models of increasing sophistication. The combinational parts of a circuit are of primary concern in the context of synchronous designs, since they determine the minimum clock width possible, and thereby govern the overall performance of a circuit. We summarized two techniques that assist in computing the optimal number of amplification stages for a design, and indicated the role of transistor sizing and static timing analyzers in this context.

10.7.2 Summary In summary, given the behavior and performance criteria, design decisions are made during the synthesis process that gradually flesh out an implementation. If a design decision explicitly concerns timing, then an explicit record is made of the timing protocols that are associated with the decision. If the decision results in a partitioning of the circuit into sequential and combinational parts, then the associated details are recorder for use by lower level timing analysis tools. When a design decision does not explicitly concern timing, the temporal constraints induced by the decision are inferred using the techniques outlined earlier. Architectural and other design decisions are influenced by the desired performance criteria.

The above procedure yields

- a hierarchical description of the temporal details associated with the design development;
- timing protocols/disciplines/constraints at several levels of abstraction.

This information may be used in

- incrementally checking for consistency of the timing constraints induced on a signal waveform from different sources;
- inferring relations between the delay parameters of the modules used in an implementation and the delays characterising the overall system;
- conveying information relating to circuit partitioning and timing protocols to lower level tools such as timing analyzers.

Once the timing details of a design are computed (say, to a first approximation using static timing analysis tools), two possibilities may arise. One is that the timing characteristics are consistent with the desired criteria, or "almost" consistent to within a factor (determined empirically) such that lower level optimizations such as adjusting the number of amplification stages and transistor resizing can potentially achieve the desired performance. The other possibility is that the specified (*i.e.*, desired) timing criteria are much more stringent than can possibly be achieved by local optimizations. In such a case, more global architectural redesign are needed. At this point, the design is backed up to the first point where the projected approximations are inconsistent with the observed timing behavior, and an alternative architecture is chosen that appears consistent with the desired performance in light of the new information that has been uncovered.

Alternative designs essentially lead to alternative sets of (in)equations that embody the temporal constraints relevant to that design. As a design evolves, the variables in the inequations that represent unknown complexity estimates may be ascribed concrete values that reflect an extrapolation (pessimistic or optimistic) of the final complexity. Thus has the result of either gracefully decreasing the number of variables to zero while leaving the constraints satisfied, or of leading to a stage where the constraints are violated. At this juncture, the choice of an alternative (*e.g.*, architecture) for a part of the design would lead to a new set of inequations that supplant the existing set of inequations *for this component*, while leaving the other constraints in place. This has the significant benefit of preserving the useable information relating to parts of the design that are not discarded. The iterative design process then continues until an implementation is obtained that satisfies the desired performance criteria, or until the user (or design tool, as the case may be), declares the specifications unsatisfiable. In practise, it is desirable to minimize the number of such major design iterations.

11. Summary

We have sketched a framework that allows the details of system timing to be gradually introduced as a design progresses. The main motivation for such a framework is to provide the basis for an agent that deals with timing issues in the context of very high level silicon compilers. In such a situation, the functionality of a system is a primary factor influencing the initial stages of design, although high level performance requirements may serve to critically influence the design strategy. While it is obvious that timing details such as clock widths and stability constraints on data signals need to be present in the final design, there is no standard paradigm for incorporating such decisions into the design process. We have examined in this paper how and when such timing details are introduced into the design description, how they may be specified, how they interact and with other design decisions, and how best to conceptually structure the introduction of these details such that their impact on the overall design is clearly understood.

Initially, no timing details need be explicit, and the only temporal constraints are due to causality, stability and functionality. As a design evolves, and decisions relating to the hardware resources used to implement functions are made, temporal constraints arise due to a need to share access to resources. The choice of timing and, subsequently, clocking disciplines further refines the overall temporal sequencing of events in the system. A subsequent level of timing detail is brought into play when circuit models (of varying degrees of sophistication) are used to improve the delays through parts of the circuit by using local logic optimizations, introducing additional stages of logic (buffers), and transistor resizing; considerations at this level can potentially influence some of the architectural details in a local sense.

It is important to stress that the framework introduced in this paper is preliminary, and much work remains to be done before the concepts enunciated herein can be considered fully developed.

Acknowledgements. I would like to thank Bryan Ackland for especially valuable discussions relating to previous drafts of this paper; Sid Ahuja, David Horn, and Gene Sampieri for conversations on how they design systems; and David Booth for comments on an earlier draft.

References

[1]. Lusk, E.L and Ross A. Overbeek, *The Automated Reasoning System ITP,* Argonne National Laboratory, Argonne, IL (1984).

[2]. Subrahmanyam, P. A., "The Algebraic Basis of an Expert System for VLSI Design," pp. 59-82 in *Formal Aspects of VLSI Design,* ed. G.Milne and P.A.Subrahmanyam , North-Holland, Amsterdam (1986).

[3]. Bryant, R., "A Switch-level model and Simulator for MOS Digital Systems," *IEEE Trans. Computers,* (February, 1984).

[4]. Davidson, S., Landskov, D., Shriver, B. D., and Mallett, P. W., "Some Experiments in local microcode compaction for horizontal machines," *IEEE Transactions on Computers* **C-30**(7) pp. 460-477 (July 1981).

[5]. Fishburn, J.P. and A. E. Dunlop, *TILOS: A Posynomial Programming Approach to Transistor Sizing,* Proc. Inter. Conf. on Computer-Aided Design (November, 1985).

[6]. Hanna, F. and G.Daeche, "Specification and Verification Using Higher-Order Logic: A Case Study," pp. 179-213 in *Formal Aspects of VLSI Design,* ed. G.Milne and P.A.Subrahmanyam, North-Holland, Amsterdam (1986).

[7]. Gordon, M., "Why Higher-Order Logic is a good formalism for specifying and verifying hardware," pp. 153-178 in *Formal Aspects of VLSI Design,* ed. G.Milne and P.A.Subrahmanyam, North-Holland, Amsterdam (1986).

[8]. Hayes, A.B., "Stored State Asynchronous Sequential Circuits," *IEEE Trans. on Computers* **C-30** pp. 596-600 (August 1981).

[9]. Horn, David, *Personal communication.* 1986.

[10]. Smith, R.J. II, "Fundamentals of Parallel Logic Simulation," *Proc. of 23rd Design Automation Conference,* pp. 2-11 (June 1986).

[11]. Krambeck, R.H., Charles M. Lee and H-F. S Law, "High-Speed Compact Circuits with CMOS," *IEEE Journal of Solid-State Circuits* **SC-17**(3) pp. 614-619 (June 1982).

[12]. Goncalves, N.F., and Hugo J. De Man, "NORA: A Race free Dynamic CMOS Technique for Pipelined Logic Structures," *IEEE Journal of Solid-State Circuits* **SC-18**(3) pp. 261-266 (June 1983).

[13]. G. De Micheli and A. Sangiovanni-Vincentelli, "Multiple Constrained Folding of Programmable Logic Arrays: Theory and Applications," *IEEE Transactions on CAD,* (July 1983).

[14]. Mishra, B. and E.Clarke, "Automatic and hierarchical verification of asynchronous circuits using temporal logic," Technical Report CMU-CS-83-155, Dept. of Computer Science, CMU (September 1983).

[15]. Moszkowski, B., "A temporal logic for multilevel reasoning about hardware," *IEEE Computer* **18**(2) pp. 10-19 (February 1985).

[16]. Nagel, L., *SPICE2: A Computer Program to Simulate Semiconductor Circuits,* Electronics Research Laboratory, University of California, Berkeley (1975).

[17]. Orton, D.W.R., *Clocked Dynamic Logic for CMOS,* Racal Research Internal Memo, Worton Drive, Worton Grange Industrial Est., Reading RG2 OSB, England (January 1984).

[18]. Marple, David P., *Performance Optimization of Digital VLSI Circuits,* Technical Report CSL-TR-86-308, Stanford University (October 1986).

[19]. P.Quinton, *The Systematic Design of Systolic Arrays,* IRISA Technical Report (May 1983).

[20]. Subrahmanyam, P. A. and S. Purushothaman, "An Algebraic Basis for Specifying and Reasoning about Protocols For Designing Self Timed Circuits," pp. 133-144 in *VLSI Design of digital systems,* ed. F. Anceau and E.J.Aas, North Holland (1983).

[21]. Seitz, C., "Self Timed Systems," *Caltech Conference on VLSI,* (1980).

[22]. Subrahmanyam, P.A., "Synapse: An Expert System for VLSI Design," *IEEE Computer* **19**(7) pp. 78-89 (July 1986).

[23]. Sutherland, Ivan, *Personal communication.* 1986.

[24]. Sutherland, Ivan, *A Theory of Logical Effort,* SSA 4679, Sutherland, Sproull and Associates, Inc. (1986).

[25]. Szymanski, T.G., *LEADOUT: A Static Timing Analyzer for MOS Circuits,* Proc. Inter. Conf. on Computer-Aided Design (November, 1986).

[26]. Weste, N. and Esraghian, K., *Principles of CMOS VLSI Design: A Systems Perspective,* Addison-Wesley, Ma. (1985).

6

BIDS: A Method for Specifying and Verifying Bidirectional Hardware Devices*

David R. Musser Paliath Narendran
William J. Premerlani

General Electric Company
Corporate Research and Development
Schenectady, New York 12301

Abstract

Previous methods of formally verifying hardware designs have been highly simplified in comparison to the physical operation of the hardware devices themselves, which is appropriate for many kinds of circuit design but not for others. In particular, the register-transfer model and the state-transition model have not been able to handle bidirectionality (allowing lines sometimes to be used for input and at other times for output). The predicate model can handle simple cases of bidirectionality, but breaks down in others, particularly when gate capacitance and/or charge-sharing are involved. In this paper we describe BIDS (*BI*directional *D*evice *S*pecification), a more elaborate model of hardware devices for formal verification, which we believe will be adequate for most techniques that designers commonly use, including bidirectionality, gate capacitance, and charge-sharing.

1 Introduction

In the literature on formal hardware verification several different methods of formally modelling hardware circuits have been used. Three of the most commonly used methods are:

- Register-transfer models—hardware devices are divided into combinational and sequential devices, with both being specified by functions, with the functions for sequential devices taking an extra argument representing an internal state that is updated when the circuit is clocked, e.g., [3], [7].

*Work partially supported by Contract F33615-85-C-1862, AFWAL, Wright-Patterson Air Force Base, Ohio

- State transition models—each type of device is modelled by a state machine, and connections between them are represented by equating some state variables of one with those of another, e.g., [2].
- Predicate models—hardware devices are specified using predicates that constrain the values present on lines connecting them, e.g., [8], [4].

In the previous literature these models have been highly simplified in comparison to the physical operation of the hardware devices themselves, which is appropriate in many kinds of circuit design but not for others. In particular the register-transfer models and state-transition models, as they have been employed in the literature, have not been able to handle bidirectionality—allowing lines sometimes to be used for input and at other times for output—because these models are typically defined in terms of functions. When using functions it has been necessary to identify their arguments as input lines and to associate the function value with an output line. A claimed advantage of the predicate model is that the predicates are relations on their arguments, and thus the lines that are identified with arguments do not have to be considered to be either input or output; i.e., they can be bidirectional.

However, the predicate model does not capture all relevant aspects of bidirectionality. To see this, let us consider a very simple example, that of a CMOS implementation of an inverter using the predicate model as in [8] or [4]. The circuit in Figure 1 can be shown to satisfy the specification

$$\mathbf{Inv}(i, o) \equiv (o = \neg i)$$

where i represents the input to the inverter and o the output, and $\neg$ is the standard "not" operator of predicate logic. Note, however, that since $(o = \neg i) \equiv (i = \neg o)$, we also have

$$\mathbf{Inv}(i, o) \equiv (i = \neg o)$$

implying that the device would still function as an inverter even if it were "turned around," or, looked at another way, it could used at different times as an inverter with i as its input line and at other times with o as its input line. But from Figure 1, it is evident that the actual circuit would not operate as an inverter if turned around. Thus the simple way in which Moszkowski, Gordon and others have used the predicate model is inadequate for some aspects of bidirectionality.

There are other aspects of circuit design that are also not handled by the commonly used methods of hardware device modelling for formal verification. Features that designers sometimes want to take advantage of include transistor gate capacitance (in NMOS devices) and charge-sharing. Gate capacitance has been dealt with in [3], for example, using the register-transfer model, but not in a way in which it could be combined with charge-sharing or bidirectionality.

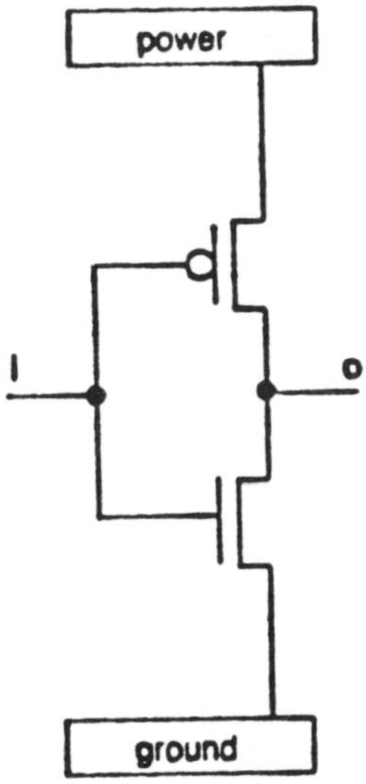

Figure 1: CMOS Implementation of an Inverter

In this paper we describe a more elaborate model of hardware devices for formal verification, which we believe will be adequate for most techniques that designers commonly use, including bidirectionality, gate capacitance, and charge-sharing. We call the method BIDS, for *BI*directional *D*evice *S*pecification.

The BIDS approach is based on (i) a combination, adaptation, and extension of several different approaches described in recent literature on hardware verification, and on (ii) previous experience with proof-based approaches to software verification using two proof-assistance systems developed at CRD, the Affirm-85 verification system and RRL, the Rewrite Rule Laboratory. The formulation of the method presented in this paper is based on the state-transition approach, but a similar extension could probably also be made to the predicate or register-transfer models.

The BIDS model is significantly more complex than models that have been used in previous research on formal hardware verification (though it is perhaps simpler than many models used for simulation). In the conclusion section we discuss possible ways of reducing this complexity.

2 Details of the method

The BIDS method uses the formalism of abstract data types [5], [6], as implemented in the Affirm-85 system [1], [9], [10]. All the concepts involved in specifying circuits are treated as abstract data types, which are used as a way of specifying state-transition systems [11].

Each abstract data type represents a set of states of a particular kind of hardware device, such as a wire, transistor, capacitor, etc., and the functions of the type define either

- *new states* of the device (transitions from states given as inputs to the the function); these are *constructor* functions of the type.
- *wires* connected to the device; these are *observer* functions of the type.

Type *wire* is defined more elaborately than in most other formal approaches to allow dealing with bidirectionality and other design features. A wire is represented as an aggregate type with four quantities associated with it, namely *value*, *strength*, *capacitance* and *charge*.

- *Value* is logically true or false and is associated with the Thevenin equivalent voltage of a node.
- *Strength* takes on integer values that rank the reciprocal of the Thevenin equivalent resistance of a node. A strength of 2 indicates a low resistance associated with (physical) wires, conducting transistors and power supplies. A strength of 1 indicates a moderate resistance associated with pullup resistance. A strength of 0 indicates an open circuit such as is associated with an nonconducting transistor.
- *Capacitance* takes on integer values that rank the total capacitance capable of delivering charge to the node.
- *Charge* takes on integer values that correspond to the sum of the charges of capacitors connected to the node.

Some thought was given to allowing the capacitance and charge parameters to take on real values but it was concluded that integers should be adequate. Actually, for trapped charge representation, it is only necessary to distinguish between zero and nonzero values. Integers are required for charge-sharing. The charge and capacitance of an output terminal model are important only if the strength is zero, representing an open circuit state.

A pair of wires, one input and one output, constitute a *port*. A port corresponds to an electrical terminal on a device. The pair of wires is used to model the interaction between the terminal and the rest of the circuit. The input wire describes the logical conditions that the rest of the circuit would impose if the connection were broken. The output wire describes the conditions the port would impose if the connection were broken. Intuitively, the port is supposed to be "receiving" data on its input wire and "sending" data on its output wire. For a device with several ports, the output wire for one of the ports can be determined from the behavior of the device and the conditions at the input wires at all of the other ports.

Two ports are connected in the model by making the output wire of one go to the input wire of the other (i.e., equating them) and vice versa, the rationale being the handling of bidirectionality. Two wires are equal when their values and strengths are the same. The charge and capacitance can be ignored when strength is greater than zero. When the strengths of two wires are zero, then they are equal if they have the same capacitance and charge.

The major circuit components such as "resistor," "power," "ground," "capacitor," and "transistor," are again specified as abstract data types. Nets in the circuit (i.e., places where three or more lines meet) are modelled using the concept called a *three-wire join* with three ports. This enables us to break more complicated nets down into simpler ones. In terms of the specification, the join is the most complicated primitive component, because it embodies the essence of the interaction of the circuit components.

The other primitive elements used in the specifications are *power*, *ground*, and *capacitor*, each of which has one port, *resistor*, which has two ports, and *transistor*, which has three ports.

1. *power* – provides an output line of strength 2, value true with no charge or capacitance.

2. *ground* – provides an output line of strength 2, value false with no charge or capacitance.

3. *resistor* – The strength of the output line of a port of a resistor is equal to the smaller of the the strength of the input line of the other port and 1. The rest of the components are the same as those of the input line of the other port.

4. *capacitor* – The capacitor is the memory element, and is important in nets that are not connected to sources with a strength greater than zero. The model remembers the previous value stored, until the input strength becomes greater than zero, at which time the new value is stored. Charge-sharing is modelled by combining the charge on the capacitor with the total charge attached to the total input capacitance. The strength of the output wire is always 0 and the capacitance associated with it is a constant, C, the capacitance of the element.

5. *transistor* – A MOS transistor is simply either a wire or an open circuit, depending on the value of its gate input. There are 3 ports, namely *gate*, *source* and *drain*. For an "n" type transistor, the *source* and the *drain* ports are connected when the value of the gate input is high and disconnected (i.e. there is an open circuit between them) when the value is low. The model of the "p" type is similar, except that the sense of the control due to the gate is inverted. Gate capacitance is not modelled in the transistor model itself, but can be handled by including a capacitor in the circuit. Similarly, a resistor can be included to model pullup resistance associated with an NMOS transistor.

In the next section we give the abstract data type specifications of these primitive circuit components.

3 Specifications of Primitive Elements

Since all the experiments were run using Affirm-85, we follow the Affirm-85 conventions for specifying each of the primitive elements as abstract data types.

3.1 Wire

As discussed in the last section, *wire* has four quantities, namely *strength*, *value*, *capacitance* and *charge*. Affirm-85 allows us to represent this as a record as in PASCAL and provides the notation **strength(w)**, **value(w)**, etc., where **w** is a wire.

```
type wire;
pascaltype record strength: Integer, value: Boolean,
                  capacit: Integer, charge: Integer end;
declare v1,v2: Boolean;
declare w1,w2: wire;
interface unc: Boolean;
interface dc,imm: Integer;
define w1 = w2 ==
       if (strength(w1) = imm) or (strength(w2) = imm)
         then true
         else
          if (strength(w1) = dc) or (strength(w2) = dc)
            then (value(w1) = value(w2))
            else
             if strength(w1) = 0
               then (strength(w2) = 0) and
                    (capacit(w1) = capacit(w2)) and
                    (charge(w1) = charge(w2))
               else
                 (strength(w1) = strength(w2)) and
                 (value(w1) = value(w2));
end;
```

A few words of explanation may be necessary here, about the (constant) values *unc*, *dc* and *imm*. The boolean value *unc* stands for "uncertain" and is important in a situation where we do not know whether the value on a wire is true or false. An instance is the output wire of the gate of a transistor which we shall encounter later on. Since we cannot assert whether the line is high or low, we thought it best to create a third "uncertain" value. If a circuit is (wrongly) designed in such a way that it depends crucially on such a value, this will be detected while doing the proofs. The integer values *dc* and *imm* stand for "don't-care" and "immaterial" respectively and play slightly different roles; *dc* takes care of the case where we only care to compare the boolean values of two wires and not the strengths and *imm* is for the situation where the whole line is unimportant. A case in point for the use of *imm* is a multiplexor, where we don't care about the connections made when the strength of the control input is 0.

3.2 Resistor

```
type resistor;
declare s:resistor;
interface in1(s), out1(s), in2(s), out2(s): wire;
axiom  out1(s) =
  if strength(in2(s)) = 2
    then wire(1,value(in2(s)),capacit(in2(s)),charge(in2(s)))
    else in2(s);
axiom  out2(s) =
  if strength(in1(s)) = 2
    then wire(1,value(in1(s)),capacit(in1(s)),charge(in1(s)))
    else in1(s);
end;
```

3.3 Ground

```
type ground;
declare s:ground;
interface in(s), out(s): wire;
axiom  out(s) = wire(2,false,0,0);
end;
```

3.4 Power

```
type power;
declare s:power;
interface in(s), out(s): wire;
axiom  out(s) = wire(2,true,0,0);
end;
```

3.5 Capacitor

The capacitor has only one port; the output line of the port has zero strength and its value depends on the stored charge. Since the charge that gets stored at any particular moment depends on the strength of the input and the charge that was stored *previously*, we have to model it using a state variable. This is implemented in Affirm-85 using the state transition approach. In the following definition, `s` is the state variable and the function `p(s)` is the transition function. Intuitively, `p(s)` stands for the state after a "pulse" is received. `Q(s)` stands for the stored charge in state `s`.

```
type capacitor;
pascaltype record c: Integer end;
declare s:capacitor;
interface p(s):capacitor;
interface in(s), out(s): wire;
interface Q(s): Integer;
axiom Q(p(s)) == if strength(in(s)) > 0
                 then if value(in(s)) = true then c(s) else 0
                 else if (2 * Q(s) + 2 * charge(in(s))  >
                             c(s) + capacit(in(s)) )
                      then c(s)
                      else 0;

axiom out(s) == if (Q(s) > 0) then wire(0,true,c(s),c(s))
                               else wire(0,false,c(s),0);

end;
```

3.6 Transistor

```
type transistor;
declare s:T;
interface p(s):T;
interface in1(s),out1(s),in2(s),out2(s),
       incntl(s),outcntl(s):wire;
axiom out2(s) ==
 if value(incntl(s)) = true then in1(s) else wire(0,unc,0,0);
axiom out1(s) ==
 if value(incntl(s)) = true then in2(s) else wire(0,unc,0,0);
axiom outcntl(s) == wire(0,unc,0,0);
end;
```

Note that `incntl` and `outcntl` stand for the wires in the gate port of the transistor.

3.7 Join

The join has three ports, and the output wire of a port depends on the input wires of the other two roughly as follows: if the strength of one input wire dominates over the other, then its strength and value are carried over to the output wire under consideration. If the strengths are both equal and non-zero, and the values are different, then the value of the output wire is uncertain. If the strengths are both zero then the value of the output wire is true if the combined charge of both inputs exceeds half the combined capacitance. The charge and capacitance of an output wire are, respectively, the sums of the charges and capacitances of the input wires of the other two ports.

```
type join;
declare s:join;
declare a,b: wire;
interface in1(s),out1(s),in2(s),out2(s),in3(s),out3(s): wire;
interface sumcap(a,b), sumchg(a,b): Integer;
interface joinout(a,b): wire;
axiom sumcap(a,b) == capacit(a) + capacit(b);
axiom sumchg(a,b) == charge(a) + charge(b);
axiom joinout(a,b) ==
 if strength(a) > strength(b)
 then wire(strength(a),value(a),sumcap(a,b),sumchg(a,b))
 else
   if strength(a) < strength(b)
   then wire(strength(b),value(b),sumcap(a,b),sumchg(a,b))
   else
     if strength(a) > 0
     then
       if value(a) = value(b)
       then wire(strength(a),value(a),sumcap(a,b),sumchg(a,b))
       else wire(strength(a),unc,sumcap(a,b),sumchg(a,b))
     else if 2 * sumchg(a,b) > sumcap(a,b)
          then wire(strength(a),true,sumcap(a,b),sumchg(a,b))
          else wire(strength(a),false,sumcap(a,b),sumchg(a,b));
axiom out1(s) == joinout(in2(s),in3(s));
axiom out2(s) == joinout(in3(s),in1(s));
axiom out3(s) == joinout(in1(s),in2(s));
end;
```

4 Verification

Verification is carried out in Affirm-85 by

1. specifying the circuit in terms of its desired properties as an abstract data type (the *behavioral specification*),

2. specifying the actual internals of the circuit as an "implementation" of the above type (the *structural specification*), and

3. verifying the implementation (showing that the properties of the behavioral specification are satisfied by the structural specification).

Affirm-85 has commands that enable the user to generate the *verification conditions* (i.e., theorems that must be proved for verification) and carry through the proof interactively. The environment in which the verification is carried out would typically include the primitive types discussed earlier and

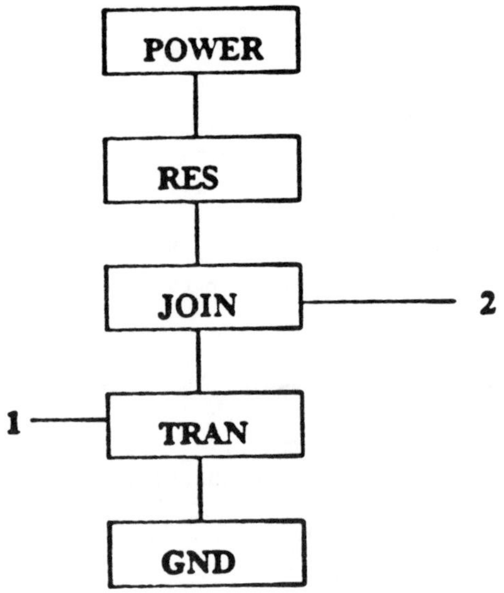

Figure 2: An Inverter Circuit

(possibly) other types that the user may have defined on his own. These user-defined types could in turn be behavioral specifications of circuits that were verified earlier. For large circuits, this allows us to break the problem down into several smaller ones in a structured manner; for once a circuit is verified, we only need its behavioral specification in later applications that involve this circuit as a component.

5 Examples

We now discuss the verification of three circuits, namely inverter, norgate and multiplexor ("mux"), using the BIDS model. We discuss them in that order because (a) they are of increasing complexity, and (b) the inverter is used in both norgate and multiplexor. Net capacitances are ignored in all three for simplicity.

5.1 Inverter

The inverter has two ports, say `A` and `B`, and can be behaviorally specified as follows: if the value of the input wire of port `A` is true then the output wire of `B` has strength 2 and value false; if the value is false, then the output wire of `B` has strength 1 and value true. The charge and capacitance values of the output wires of both ports are zero. Hence the type definition for the behavioral specification is

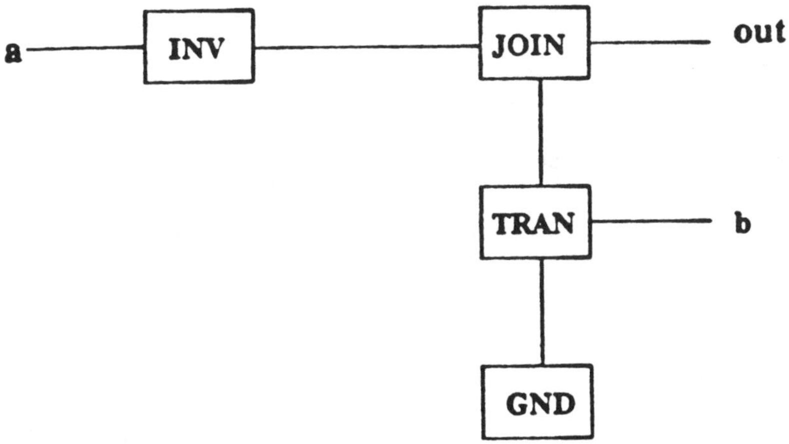

Figure 3: Norgate Circuit

```
type inverter;
declare s:T;
interface p(s):T;
interface in1(s),out1(s),in2(s),out2(s):wire;
axiom out2(s)  == if value(in1(s)) = false then
                     wire(1,true,0,0) else wire(2,false,0,0);

axiom out1(s)  == wire(0,unc,0,0);
end;
```

The circuit consists of a power source, a resistor, a transistor, a ground and a join (see Figure 2). The implementation is specified as

```
implement inverter;
interface tranA(s): transistor;
interface powerA(s): power;
interface resistorA(s): resistor;
interface groundA(s): ground;
interface joinA(s): join;
axiom  in1(tranA(s)) == out(groundA(s));
axiom  in(groundA(s)) = out1(tranA(s));
axiom  in2(tranA(s)) == out1(joinA(s));
axiom  in1(joinA(s)) == out2(tranA(s));
axiom  incntl(tranA(s)) == in1(s);
axiom  out1(s) = outcntl(tranA(s));
```

```
axiom  in2(joinA(s)) == in2(s);
axiom  out2(s) == out2(joinA(s));
axiom  in3(joinA(s)) == out2(resistorA(s));
axiom  in2(resistorA(s)) = out3(joinA(s));
axiom  in1(resistorA(s)) = out(powerA(s));
axiom  in(powerA(s)) = out1(resistorA(s));
axiom  good(s) = true;
end;
```

Once all the necessary primitive types (namely, wire, join, transistor, resistor, power and ground) and the above specifications are entered, we can generate the verification conditions using a *genimpvcs* command (*gen*erate *imp*lementation *v*erification *c*ondition*s*). The circuit is verified once these are proved correct.

5.2 Norgate

There are two input ports, say **A** and **B**, and one output port. The behavioral specification is as follows:

```
type norgate;
declare s:T;
interface p(s):T;
interface in(s),out(s),ina(s),outa(s),
  inb(s),outb(s):wire;
interface uncertain(s): Boolean;
axiom  outa(s) == wire(0,unc,0,0);
axiom  outb(s) == wire(0,unc,0,0);
axiom  out(s) == if value(ina(s)) = true
                 then wire(2,false,0,0)
                 else
                  if value(inb(s)) = true
                  then wire(2,false,0,0)
                  else wire(1,true,0,0);
end;
```

The structural specification consists of an inverter, a transistor, a ground and a join (see Figure 3).

```
implement norgate;
interface join(s):join;
interface tran(s): transistor;
interface inv(s):inverter;
interface ground(s):ground;
axiom in1(join(s)) == out2(inv(s));
axiom in2(inv(s)) == out1(join(s));
axiom in2(join(s)) == out2(tran(s));
axiom in2(tran(s)) == out2(join(s));
```

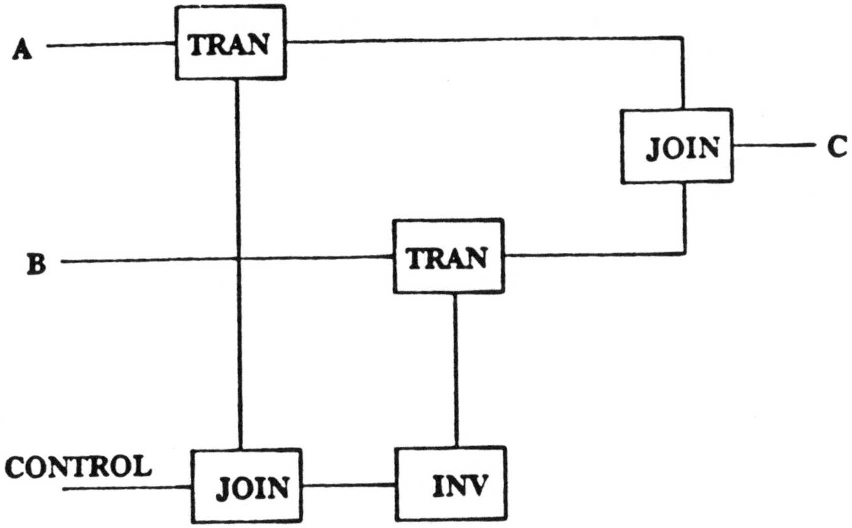

Figure 4: Multiplexor Circuit

```
axiom in3(join(s)) == in(s);
axiom out(s) == out3(join(s));
axiom in1(inv(s)) == ina(s);
axiom outa(s) == out1(inv(s));
axiom incntl(tran(s)) == inb(s);
axiom outb(s) == outcntl(tran(s));
axiom in1(tran(s)) == out(ground(s));
axiom in(ground(s)) == out1(tran(s));
axiom good(s) == true;
end;
```

Note that during the proof we only need the behavioral specification of the inverter.

5.3 Multiplexor

The circuit is shown in Figure 4. CONTROL is used only as input but A, B, and C are used bidirectionally. The main behavioral specification of the device is that if the value of the CONTROL input is true, then A and C are connected; if it is false, B and C are connected. However, we found that we cannot assert this in its full generality; the above characterization of the mux/demux device holds good only when the strength of the CONTROL input is non-zero. If the strength of the CONTROL input is zero, then all that we can say is that one of the lines A, B will be connected to line C. Thus our behavioral specification only specifies the case when the input strength is non-zero; when the strength is zero, it is "immaterial" which wire is connected to which.

```
type mux;
declare s:T;
interface p(s):T;
interface in(s),out(s),ina(s),outa(s),
  inb(s),outb(s),inc(s),outc(s):wire;
{''in'' and ''out'' are the input and
 output wires of the CONTROL port.}
axiom  outc(s) ==  if strength(in(s)) > 0
                   then if value(in(s)) = true
                        then ina(s)
                        else inb(s)
                   else wire(imm,unc,0,0);
axiom  outa(s) ==  if strength(in(s)) > 0
                   then if value(in(s)) = true
                        then inc(s)
                        else wire(0,unc,0,0)
                   else wire(imm,unc,0,0);
axiom  outb(s) ==  if strength(in(s)) > 0
                   then if value(in(s)) = false
                        then inc(s)
                        else wire(0,unc,0,0)
                   else wire(imm,unc,0,0);
axiom  out(s) == wire(0,unc,0,0);
end;
```

The case when the input strength is zero can be given as a separate theorem and proved, and this is what we actually did.

```
theorem t\\#1, all s ((strength(in(s)) = 0) imp
                    ((outc(s) = ina(s)) or (outc(s) = inb(s)))
                   );
```

(**imp** stands for implies.)

The structural description of the mux/demux circuit is in terms of two transistors and an inverter as shown in Figure 4.

```
implement mux;
interface tran1(s), tran2(s):transistor;
interface join1(s), join2(s):join;
interface inv(s):inverter;
axiom incntl(tran2(s)) == out2(inv(s));
axiom in2(inv(s)) == outcntl(tran2(s));
axiom in1(join1(s)) == in(s);
axiom out(s) == out1(join1(s));
```

```
axiom incntl(tran1(s)) == out2(join1(s));
axiom in2(join1(s)) == outcntl(tran1(s));
axiom in1(inv(s)) == out3(join1(s));
axiom in3(join1(s)) == out1(inv(s));
axiom in1(join2(s)) == out2(tran2(s));
axiom in2(tran2(s)) == out1(join2(s));
axiom in2(join2(s)) == out2(tran1(s));
axiom in2(tran1(s)) == out2(join2(s));
axiom in1(tran1(s)) == ina(s);
axiom outa(s) == out1(tran1(s));
axiom in1(tran2(s)) == inb(s);
axiom outb(s) == out1(tran2(s));
axiom in3(join2(s)) == inc(s);
axiom outc(s) == out3(join2(s));
axiom good(s) == true;
end;
```

We also found that the following assumptions are needed in the proofs:

```
assume a\#1, all w (strength(w) >= 0);
assume a\#2,
  all xx ((strength(in(xx)) = 0) imp ((capacit(in(xx)) = 0)
   and (charge(in(xx)) = 0)) );
```

where the types of **w** and **xx** are **wire** and **mux** respectively. The assumption **a#1** merely states the obvious fact that a wire cannot have negative strength. Since strengths of wires can be only 0, 1 or 2 in our model, we could have given a stronger assumption, but we found that the above weaker form is enough. The second assumption is that if the input wire of the **CONTROL** port has no strength, then it can have no capacitance or charge associated with it either. Assumption **a#1** is needed in the proofs of the verification conditions, whereas **a#2** is needed in the proof of theorem **t#1**.

It must be noted that the mux example needed more user interaction that the relatively simpler inverter and norgate examples. In larger examples even more user interaction may be needed. Finding ways of reducing the amount of interaction needed is a topic of ongoing research.

6 Conclusion

We have presented a method called BIDS for specifying and verifying low-level circuits. This method is distinguished from previous methods in that it allows relatively complete treatment of issues such as bidirectionality and stored charge. Many circuits can be dealt with using simpler approaches, as evidenced by a number of examples in the recent literature, without the rather high overhead of specification and proof complexity that the BIDS method incurs. However, we feel that it is important have available a method that can handle bidirectionality and stored charge in full generality for those cases in which a designer has good reasons to take advantage of them.

In any case, some of the complexity of the BIDS method could be reduced by having a preprocessor that constructs the set of connection equations from input in some simpler form. One of the most promising approaches is to construct the structural specifications from graphical input, and in fact such a facility has already been partially implemented as part of an Interactive VHDL Designers Workstation under development at General Electric Corporate Research and Development Center. A graphical editor has been developed to enter structural circuit specifications, from which equations and other formulae for the theorem prover are generated (and the proofs have been carried out in several examples). Thus far only a simpler version of the state-transition approach has been used, but we plan to experiment soon with automatic generation of BIDS model equations from graphical circuit representations.

Although Affirm-85 was used in the experiments described in this paper, we have since shifted our attention to RRL, a "Rewrite Rule Laboratory" currently being developed at CRD. RRL has a more powerful state-of-the-art theorem prover and is more naturally suited for reasoning with equations. The latter reason is quite important since equations play a central role in abstract data type specifications. Another feature of RRL is a better and more automatic way to prove theorems that require the use of induction. On the other hand, RRL currently lacks some of the proof interaction features of Affirm-85 that permit the user to provide guidance in cases where automatic proof attempts fail. We plan to adapt these Affirm-85 features for use in RRL.

References

[1] *Affirm Reference Manual* (D. H. Thompson and R. W. Erickson, eds.), USC Information Sciences Institute, Feb. 1981.

[2] Barrow, H. G., "Proving the Correctness of Digital Hardware Designs," *VLSI Design*, July 1984.

[3] Gordon, M., "A Very Simple Model of Sequential Behaviour of *n*MOS," *Proc. VLSI International Conference*, J. Gray (ed.), Academic Press, London and New York, 1981.

[4] Gordon, M. "Why Higher-Order Logic is a Good Formalism for Specifying and Verifying Hardware," Technical Report, University of Cambridge, Cambridge, U. K., 1985.

[5] Guttag, J. V., and Horning, J. J., "The Algebraic Specification of Abstract Data Types," *Acta Informatica* 10, 27-52 (1978).

[6] Guttag, J. V., Horowitz, E., and Musser, D. R., "Abstract Data Types and Software Validation," *CACM* 21(12), pp. 1048-1064, December 1978.

[7] Hunt, W. A., Jr., "FM8501: A Verified Microprocessor," Technical Report 47, University of Texas at Austin, December 1985.

[8] Moszkowski, B., "A Temporal Logic for Multilevel Reasoning about Hardware," *IEEE Computer*, February 1985.

[9] Musser, D. R., "Abstract Data Type Specification in the Affirm System," *IEEE Transactions on Software Engg.*, Vol. SE-6 (1), Jan. 1980.

[10] Musser, D. R., and Cyrluk, D. A., "Affirm-85 Reference Manual," General Electric Corporate Research and Development Center, Schenectady, NY 12301, August 1985.

[11] Sunshine, C. A., Thompson, D. H., Erickson, R. E., Gerhart, S. L., and Schwabe, D., "Specification and Verification of Communication Protocols in AFFIRM Using State Transition Models," *IEEE Transactions on Software Engg.*, Vol. SE-8, No. 5, September 1982.

7

HARDWARE VERIFICATION IN THE INTERACTIVE VHDL WORKSTATION *

Paliath Narendran
G.E. Corporate Research and Development Center
Schenectady, NY 12345

and

Jonathan Stillman
State University of NY at Albany
(Consultant to GE-CRD)
Albany, NY 12222

1. INTRODUCTION

Formal Hardware Verification has gained a lot of attention recently as a viable alternative to simulation. Several notable achievements have been reported, such as Hunt [4], Birtwistle et. al. [1]. In this document we describe one aspect of an ongoing project aimed at developing an integrated environment for hardware description, simulation, and verification. We report here on the current efforts toward developing tools for formal verification of hardware and how these tools interact with other parts of the IVW workstation being developed at General Electric Research and Development Center (GE CRD). This is of necessity a preliminary document, as our work in developing an environment for highly automatic verification of hardware is at an early stage.

There have been several ongoing activities on Software Specification, Verification, and Automated Reasoning at the CRD. Of these, our activities have centered on the Affirm-85 verification system and the Rewrite Rule Laboratory theorem prover. Most of the experimentation at the initial stages of the project was done with the Affirm-85 system and will be reported in another paper (Musser, Narendran, and Premerlani [12]). Here we report on the work currently being done with the Rewrite Rule Laboratory (RRL). This work is part of the Interactive VHDL workstation project (IVW) under Air Force contract number F33615-85-C-1862.

---

* Work partially supported by Contract F33615-85-C-1862, AFWAL, Wright-Patterson Air Force Base, Ohio.

2. THE IVW WORKSTATION

The **Interactive VHDL Workstation** project (IVW) is aimed at building a prototype VHDL workstation environment for design capture. The designs can be entered textually as well as graphically, although use of graphic editors is encouraged. The IVW integrates several graphical editors and specialized expert systems with a common object-oriented, design database, thus providing a more flexible medium of capturing design ideas than text entry systems. Once a design database is built, VHDL code can be automatically generated. Furthermore, a graphic editor, called the **structure editor**, is provided to maintain a correlation between several views of an object; changes made in one are reflected in the other views, both textually and graphically. (See Appendix C.) The design data base enables the user to store his designs and recall them later in designing higher level circuits.

The major goal of the hardware verification project is to provide the user with tools to (a) enter the behavioral description of the circuit, and (b) automatically verify whether the circuit satisfies the behavior. The connectivity information for the structural description is generated from the graphical input. Verification is done using the Rewrite Rule Laboratory, an experimental theorem prover currently being developed at GE CRD. Emphasis is placed on the following issues:

(a) The user should have a fairly powerful and easy-to-use front end for entering circuit descriptions. Furthermore, this should be language-independent, and should allow the user to build his circuit up hierarchically. The structure editor and the underlying design database are designed with this goal in view.

(b) The verification should be automatic, or as automatic as possible. Our current efforts have been directed toward freeing the user from any detailed knowledge of the workings of the theorem prover. This approach has certain limitations which will be explained later.

Note that currently we do not propose to build tools for the user to go directly from VHDL to RRL.

As outlined in an earlier paper, work has also been done on modeling CMOS/NMOS circuits at the transistor level. The circuits are specified as abstract data types. We have not incorporated this into the verification framework at present. This will be done in the near future.

3. THE REWRITE RULE LABORATORY

The **Rewrite Rule Laboratory** is a theorem proving program based on equational reasoning. The main approach is to treat axioms, given as equations, as rewrite rules; the loss of bidirectionality that this results in is compensated for by adding new rules. The goal is to find a 'complete' system where every two equivalent terms can be rewritten to a single term which is called a 'normal form' or 'canonical form.' Once such a complete system is found, every theorem in our theory can be proved - the proof involves rewriting the terms to their normal forms and checking if they are the same.

RRL works as follows: given a set of equational axioms, the program helps the user turn them into rewrite rules that are *finitely terminating* which essentially means that rewriting will never go on indefinitely. This part of the program is highly interactive since finite termination is established by setting certain precedences among the function symbols. The user is queried for these precedences. Searching for a complete set of rules is done using the Knuth-Bendix completion procedure (Knuth and Bendix [7]) which searches for ambiguities in rewriting by overlapping the left-hand sides of rules. Appendix A gives a typical RRL session where the input is the equational theory of groups.

RRL also has a theorem prover for the full first-order predicate calculus where quantified formulae are allowed. The approach is refutational - the conclusion is negated and a contradiction is searched for. The proof procedure is equational in nature (Kapur and Narendran [6]). The formulae are represented as polynomials over the boolean ring whose operators are "exclusive-or" and "and." Searching for contradiction is done using a procedure similar to the Knuth-Bendix completion procedure. For formulae without equality, this procedure is almost completely automatic.

RRL has a fairly powerful method for proving inductive theorems about equational specifications. This is called the "inductionless induction method" or "proof by consistency" (Kapur and Musser [5]). The approach is, in some sense, the dual of the "proof-by-contradiction" approach: the inductive theorem to be proved is added to the set of axioms, and if there is no inconsistency, we can assert the theorem to be true under certain conditions. However, this component of RRL needs more user interaction than the two mentioned before; in many cases the user has to coax the system into proving the theorem by proving several intermediate lemmas.

The typing facility of RRL is very simple and primitive and still under development. Currently, variables are not typed; therefore the user cannot "overload" his functions.

Apart from all this, RRL also gives the user the option to choose from several strategies in overlapping, rewriting, addition of new variables, etc. Statistics are provided about the times spent in most of the activities. The examples in the appendices illustrate many of these features.

4. A MODEL FOR CIRCUIT DESCRIPTION AND VERIFICATION

Our approach to circuit description has been to provide a set of predefined elements as part of the structure editor's library; these elements have ports and behavioral descriptions, but no "internals". For instance, an "and" gate might be considered primitive at some point, with 2 input ports, 1 output port, and a behavioral description defining the output value as the logical "and" of the 2 input values. (Thus far we have restricted our behavioral descriptions to first-order predicate calculus formulae in order to exploit the first-order capabilities of RRL, but this is not an inherent limitation.)

New circuits are built by specifying their ports, the components (from the library) that make up the internals of the new circuit, and how those components are connected to one another and to the ports of the circuit being created. Finally, a behavioral description must be entered to describe the input-output behavior of the new circuit. Each behavioral definition is broken up into three parts: **initialization assumptions, operational assumptions**, and the **behavioral statement. Initialization assumptions** are those properties which follow from the circuit having been properly initialized; that is, we may want to assume that the circuit is in some given state at the time we begin to examine it. **Operational assumptions** involve constraints placed on the circuit from without; they often involve restrictions on the possible states of the circuit, such as stability of input signals, constraints on the values that input signals may have, or the behavior of an external clock. The **behavioral statement** describes how the circuit is expected to behave given that the initalization and operational assumptions hold.

Once this information has been entered, the new circuit becomes part of the design library and can be used as a component of other circuits. Before we can be certain that the circuit behaves as specified, however, we must verify that the behavior specified indeed follows from the structure of the circuit. The structural specification of the circuit can be obtained by examining each of its components in turn, "instantiating" its behavioral description by substituting the actual names of the wires hooked up to the ports in the particular instance for the port names. This results in a set of axioms which describe how the components force the circuit to behave. Since we are using a refutational approach to verification, the behavioral description of the circuit is negated and entered as an axiom. The resulting set of axioms is submitted to RRL where the actual proof takes place.

5. EXAMPLES:

5.1 The RS latch

A simple example, the RS latch, is shown below:

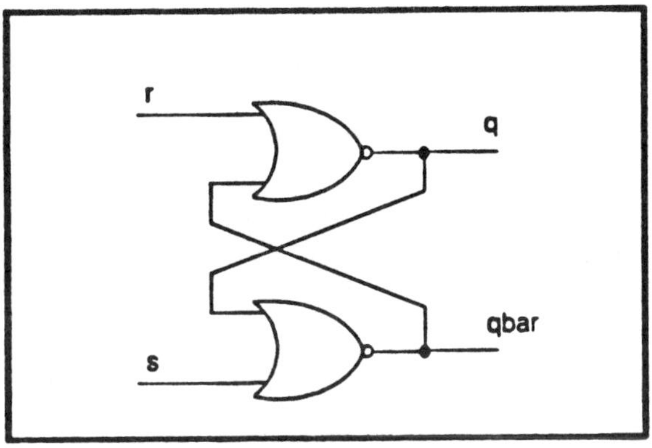

- Figure 1 -

The latch is made up of two "nor" gates, connected as shown. Assuming a unit delay in propagating the input values of a gate to the output value, the "nor" gate has the behavioral definition:

all x (output(f(x)) equ (not(input1(x) or input2(x))))

where the variable x and the function f are used to express time (for instance, x is thc initial time, f(x) is one time unit later, f(f(x)) is two time units later, etc.).

When instantiated to reflect the actual names of the wires connected to the input and output ports of each instance of a "nor" gate, the axioms derived from the structural description of the latch are:

1. **all x (q(f(x)) equ (not (r(x) or qbar(x))))**
and
2. **all x (qbar(f(x)) equ (not (s(x) or q(x))))**

In this example the only initialization assumption is that the q and qbar values start out complements of one another. The operational assumptions are that the r and s input lines are stable long enough for the circuit to stabilize and that the input lines are not simultaneously high. Finally, the behavior of the circuit given that the initialization and operational assumptions hold is given. Thus the behavioral definition of the latch is:

all x (((q(x) equ (not(qbar(x)))) and
(all x (not(r(x) and s(x)))) and
(r(x) equ r(f(x))) and (s(x) equ s(f(x))))
→ ((q(f(f(x))) equ (s(x) or (q(x) and (not (r(x)))))) and
(qbar(f(f(x))) equ (not (q(f(f(x)))))))

That is, if q and qbar start out complemented, and the r and s inputs are stable for two time units, then when the circuit stabilizes, the q and qbar outputs are still complements, and furthermore the value of q is high if and only if either the s input was high or q was already high and the r input wasn't high.

The input to RRL consists of the two derived axioms and the negation of the behavioral definition. RRL runs the Knuth-Bendix completion procedure on this set of axioms; if an inconsistent system results, we may conclude that the behavioral definition is correct.
(A full transcript of the proof is given in Appendix B.)

5.2 A CMOS full adder

In the above example, gates are considered to be primitive since no description of their architecture was entered. At another point, however, we might choose to use transistors as the primitive level, and would thus have to build gates from them. In the next example we show how one might use transistors as the primitive level, building a full adder circuit using transistors as the components. The two types of transistors we use are the n-transistor and the p-transistor. A description of these is given in the figure below:

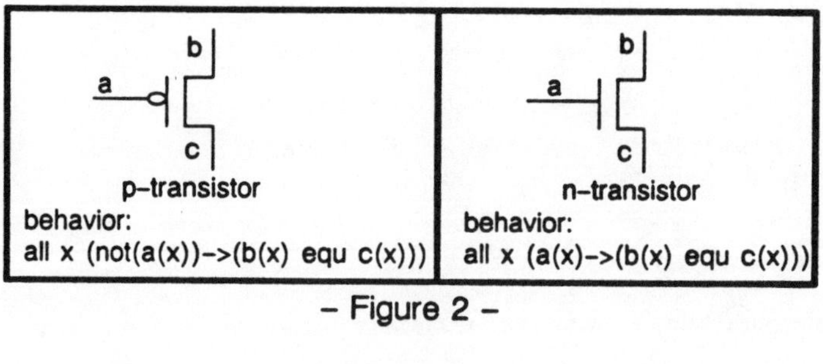

- Figure 2 -

One way of implementing a full adder in CMOS is shown in the figure below (Gordon [3]):

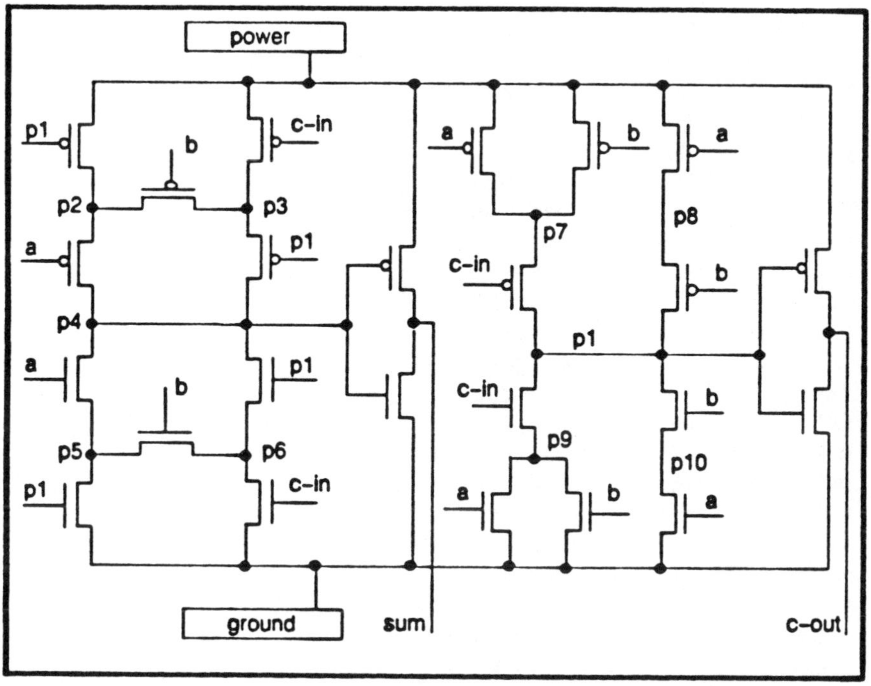

- Figure 3 -

Given such a circuit description we would like to ascertain whether it corresponds to our conception of what a full adder does. The behavioral definition follows:

(all x (all x (power(x) and (not(ground(x))))) →
((sum(x) equ (a(x) xor b(x) xor c-in(x))) and
(c-out(x) equ ((a(x) and b(x)) or
(a(x) and c-in(x)) or
(b(x) and c-in(x))))))

In other words, given that the power signal is always high and that the ground signal is always low, the the circuit being described behaves like a full adder (with the result available at the same instant that the input signals arrive). The axioms (both generated structural axioms and the negation of the behavioral definition) given to RRL are listed below:

```
(all x (((not (p1 (x)) → (power (x) equ p2 (x))))))
(all x (((not (a (x)) → (p2 (x) equ p4 (x))))))
(all x (((not (b (x)) → (p3 (x) equ p2 (x))))))
(all x (((not (c-in (x)) → (power (x) equ p3 (x))))))
(all x (((not (p1 (x)) → (p3 (x) equ p4 (x))))))
(all x (((not (p4 (x)) → (power (x) equ sum (x))))))
(all x (((not (a (x)) → (power (x) equ p7 (x))))))
(all x (((not (c-in (x)) → (p7 (x) equ p1 (x))))))
(all x (((not (b (x)) → (power (x) equ p7 (x))))))
(all x (((not (a (x)) → (power (x) equ p8 (x))))))
(all x (((not (b (x)) → (p8 (x) equ p1 (x))))))
(all x (((not (p1 (x)) → (power (x) equ c-out (x))))))
(all x (((a (x) → (p4 (x) equ p5 (x))))))
(all x (((p1 (x) → (p5 (x) equ ground (x))))))
(all x (((b (x) → (p6 (x) equ p5 (x))))))
(all x (((p1 (x) → (p4 (x) equ p6 (x))))))
(all x (((c-in (x) → (p6 (x) equ ground (x))))))
(all x (((p4 (x) → (sum (x) equ ground (x))))))
(all x (((c-in (x) → (p1 (x) equ p9 (x))))))
(all x (((a (x) → (p9 (x) equ ground (x))))))
(all x (((b (x) → (p9 (x) equ ground (x))))))
(all x (((b (x) → (p1 (x) equ p10 (x))))))
(all x (((a (x) → (p10 (x) equ ground (x))))))
(all x (((p1 (x) → (c-out (x) equ ground (x))))))
(not (all x ((all x (power (x) and not (ground (x)))) →
 (((sum (x) equ (a (x) xor b (x) xor c-in (x))) and
 (c-out (x) equ ((a (x) and b (x)) or
                 (a (x) and c-in (x)) or
                 (b (x) and c-in (x)))))))))
```

The verification is completely automatic and takes approximately 2 minutes.

6. OTHER VERIFIED CIRCUITS

At this point we have also automatically verified a number of other circuits, including circuits specified hierarchically and circuits such as clocked flip-flops with fairly complex timing patterns. We have also used this system to verify reasonably large combinational circuits at the gate level, even though the behavioral descriptions may be very different from the actual circuit description. This has proved useful in verifying sections of a systolic chip currently under development at GE. Work is presently underway on development of some methods for handling integer reasoning in RRL.

7. PROBLEMS AND ISSUES

Our experiments thus far have uncovered several problems in the Rewrite Rule Laboratory and with automatic theorem proving in general. We briefly discuss some of them here. One is the inability of the prover to handle integers efficiently. At present, the only recourse is to specify integers and operations on them by Peano Arithmetic. Since many of the circuits (such as ALUs and multipliers) operate on integers, this is a serious drawback. Thus an important research problem in automated reasoning would be to develop ways of handling fragments of integer arithmetic (using built-in decision procedures, for instance). Progress in this direction has been reported in (Boyer and Moore [2]) where linear arithmetic has been incorporated into the basic logic of the theorem prover.

We also feel that some interaction will be needed when the circuits grow larger in size. Some of this is reflected even in RRL, since proof of many theorems would involve induction and, as mentioned before, the induction component of RRL requires a certain amount of user interaction. Interactive capabilites would also be a useful option for knowledgeable users who could guide proofs that otherwise might either finish much more slowly or not finish at all.

One fundamental problem in using an automatic theorem prover such as RRL for verification of circuits is that it is unlikely that one's first description will be correct; when this happens, the prover may never halt. Since the formalism used will probably be at least as powerful as first order predicate calculus this problem cannot be avoided, but developing methods for obtaining meaningful diagnostic information about incorrect circuit specifications is critical if such a tool is to be of practical value.

8. CONCLUSION

As mentioned earlier, one of the issues that we had placed a lot of emphasis so far was on making the verification process as automatic as possible. We believe we have made some progress in that direction, though more research is needed to incorporate integer capabilities. We have also been able to successfully handle *feedback*, as the flip-flop example illustrates. The overall approach has been to utilize the full power of first-order predicate calculus. Absence of usable theorem provers has prevented us from using other paradigms such as temporal logic (Moszkowski [9]).

We have been able to provide the user with an easy-to-use front-end for entering his structural information. The graphic editor allows him to completely "draw" the circuit on the screen; the structural specification to be input to RRL is generated automatically. We do not have such facilities at present for *behavioral* specification. The obvious possibility is that of using VHDL itself for behavioral specification. Due to reasons that are beyond the scope of this paper, we have decided against it. We are currently working on designing a user interface for behavioral specification.

We would like to acknowledge the contributions of several of our co-workers whose efforts are making this work possible. Among them, notably, are Art Duncan, Wendy Haas-Murren, Kevin Spier and Rich Welty. We got considerable help in the use of RRL from Dave Cyrluk and Deepak Kapur. We would also like to thank Jim Caldwell, Bob Hardy, Dave Musser, Bill Premerlani, Dan Rosenkrantz, Joel Sturman and Fathy Yassa for contributing several ideas and suggestions. Above all, we would like to acknowledge Dave McGonagle, who is a strong motivating force for this work as well as a constant source of encouragement.

9. REFERENCES

[1] Birtwistle, G., Joyce, J., Liblong, B., Melham, T., and Schediwy, R. (1985). Specification and VLSI design. Research Report No. 85/220/33, Dept. of Computer Science, University of Calgary, Calgary, Alberta, Canada.

[2] Boyer, R. and Moore, J.S. (1985). Integrating Decision Procedures into Heuristic Theorem Provers: A Case Study of Linear Arithmetic. Technical Report ICSCA-CMP-44, Institute for Computing Science, University of Texas at Austin.

[3] Gordon, M. (1985). Why Higher-Order Logic is a good formalism for specifying and verifying hardware. Technical Report 77, Computer Laboratory, University of Cambridge, Cambridge, U.K.

[4] Hunt, W.A., Jr. (1985). FM8501: A Verified Microprocessor. Technical Report 47, Institute for Computing Science, Univ. of Texas at Austin.

[5] Kapur, D. and Musser, D.R. (1984). Proof by Consistency. To appear in *Artificial Intelligence*.

[6] Kapur, D. and Narendran, P. (1985). An Equational Approach to Predicate Calculus. In: Proceedings of IJCAI-85, Los Angeles.

[7] Kapur, D., Sivakumar, G., and Zhang, H. (1986). RRL: A Rewrite Rule Laboratory. In: Proceedings of *8th Conference on Automated Deduction (CADE-86)*, Oxford, U.K.

[8] Knuth, D.E. and Bendix, P.B. (1970). Simple Word Problems in Universal Algebras. in *Computational Problems in Abstract Algebras* (ed. J. Leech), Pergamon Press, 263-297.

[9] Moszkowski, B. (1985). A Temporal Logic for Multilevel Reasoning about Hardware. *IEEE Computer*.

[10] Musser, D.R. (1980). Proving Inductive Properties of Abstract Data Types. In: *Proc. of 7th Principles of Programming Languages Conf.*, Las

Vegas, Nevada.

[11] Musser, D.R., and Cyrluk, D.A. (1985). Affirm-85 Installation Guide and Reference Manual Update. GE Corporate R&D Center, Schenectady, NY 12345.

[12] Musser, D.R., Narendran, P. and Premerlani, W. (1987). BIDS: A Method for Specifying and Verifying Bidirectional Hardware Devices. Appears elsewhere in this volume.

10. APPENDIX A

```
WELCOME TO REWRITE RULE LABORATORY (RRL)
(c) Copyright 1985,1986 G.Sivakumar. All rights reserved.
(c) Copyright 1985,1986 Hantao Zhang. All rights reserved.
(c) Copyright 1985,1986 General Electric Company. All rights reserved.

Type Add, Akb, Auto, Break, Clean, Delete, Dump, Grammar, Init, Kb, List,
   Log, Makerule, Narrow, Norm, Order, Option, Operator, Prove, Quit,
   Read, Refute, Stats, Suffic, Undo, Unlog, Write or Help.
RRL-> add
Type your equations, enter a ']' when done.
x + (y + z) == (x + y) + z
0 + x == x
i(x) + x == 0
]

Equations read in are:
 1. (X + (Y + Z)) == ((X + Y) + Z)  [USER, 1]
 2. (0 + X) == X  [USER, 2]
 3. (I(X) + X) == 0  [USER, 3]

New constant set is: { 0 }
Time = 62.75 sec

Type Add, Akb, Auto, Break, Clean, Delete, Dump, Grammar, Init, Kb, List,
   Log, Makerule, Narrow, Norm, Order, Option, Operator, Prove, Quit,
   Read, Refute, Stats, Suffic, Undo, Unlog, Write or Help.
RRL-> kb                    {stands for "Knuth-Bendix"}
----- Step 1 -----

Trying to orient equation:
  (X + (Y + Z)) == ((X + Y) + Z)  [USER, 1]

  To prove:  (X + (Y + Z))  >  ((X + Y) + Z)
```

or: (X + (Y + Z)) < ((X + Y) + Z)
I have no precedence suggestions.
Try doing Equiv or Status.

Type Abort, Display, Equiv, Twoway, Operator, Status, LR, RL, Postpone, MakeEq, Quit, Undo or Help.
RRL>KB> **status**
Type operator you wish to set status for: **+**
Type LR, RL, HELP --> **lr** {stands for "left-to-right"}
Operator, +, given status: LR

----- Step 2 -----

Adding rule: [1] ((X + Y) + Z) ---> (X + (Y + Z)) [USER, 1]

Adding rule: [2] (0 + X) ---> X [USER, 2]

Trying to orient equation:
(I(X) + X) = = 0 [USER, 3]

To prove: (I(X) + X) > 0
Here are some precedence suggestions:
1. + > 0
2. I > 0
Either type a list of numbers or
Type Abort, Display, Equiv, Twoway, Operator, Status, LR, RL, Postpone, MakeEq, Quit, Undo or Help.
RRL>KB> **1 2** {+ and I are given higher precedence over 0}

----- Step 3 -----

Adding rule: [3] (I(X) + X) ---> 0 [USER, 3]

Computing critical pairs with:
[2] (0 + X) ---> X [USER, 2]

Computing critical pairs with:
[3] (I(X) + X) ---> 0 [USER, 3]

Computing critical pairs with:
[1] ((X + Y) + Z) ---> (X + (Y + Z)) [USER, 1]

Adding rule: [4] (I(Y) + (Y + Z)) ---> Z [3, 1]

Computing critical pairs with:
[4] (I(Y) + (Y + Z)) ---> Z [3, 1]

Adding rule: [5] (I(0) + Z) ---> Z [2, 4]

Adding rule: [6] (I(I(Z)) + 0) ---> Z [3, 4]

Adding rule: [7] (I((X + Y)) + (X + (Y + Z))) ---> Z [1, 4]

Adding rule: [8] (I(I(Y)) + Z) ---> (Y + Z) [4, 4]

Deleting rule: [6] (I(I(Z)) + 0) ---> Z [3, 4]

Adding rule: [9] (Z + 0) ---> Z [DELETED, 6]

Computing critical pairs with:
[9] (Z + 0) ---> Z [DELETED, 6]

Adding rule: [10] I(0) ---> 0 [3, 9]

Deleting rule: [5] (I(0) + Z) ---> Z [2, 4]

Computing critical pairs with:
[10] I(0) ---> 0 [3, 9]

Computing critical pairs with:
[8] (I(I(Y)) + Z) ---> (Y + Z) [4, 4]

Adding rule: [11] I(I(Y)) ---> Y [9, 8]

Deleting rule: [8] (I(I(Y)) + Z) ---> (Y + Z) [4, 4]

Computing critical pairs with:
[11] I(I(Y)) ---> Y [9, 8]

Adding rule: [12] (Y + I(Y)) ---> 0 [11, 3]

Adding rule: [13] (Y + (I(Y) + Z)) ---> Z [11, 4]

Computing critical pairs with:
[12] (Y + I(Y)) ---> 0 [11, 3]

Adding rule: [14] (X + (Y + I((X + Y)))) ---> 0 [1, 12]

Computing critical pairs with:
[13] (Y + (I(Y) + Z)) ---> Z [11, 4]

Adding rule: [15] (X + (Y + (I((X + Y)) + Z))) ---> Z [1, 13]

```
Computing critical pairs with:
  [14] (X + (Y + I((X + Y)))) ---> 0  [1, 12]

Adding rule:  [16] (X + (Y + (Y1 + I((X + (Y + Y1)))))) ---> 0  [1, 14]

Adding rule:  [17] (Y + I((Y1 + Y))) ---> I(Y1)  [14, 4]

  Deleting rule: [14] (X + (Y + I((X + Y)))) ---> 0  [1, 12]

Computing critical pairs with:
  [17] (Y + I((Y1 + Y))) ---> I(Y1)  [14, 4]

Adding rule:  [18] (X + (Y + I((Y1 + (X + Y))))) ---> I(Y1)  [1, 17]

  Deleting rule: [16] (X + (Y + (Y1 + I((X + (Y + Y1)))))) ---> 0  [1, 14]

Adding rule:  [19] (Y + I((X + (Y1 + Y)))) ---> I((X + Y1))  [1, 17]

  Deleting rule: [18] (X + (Y + I((Y1 + (X + Y))))) ---> I(Y1)  [1, 17]

Adding rule:  [20] (X + (I((Y + X)) + Z)) ---> (I(Y) + Z)  [17, 1]

  Deleting rule: [15] (X + (Y + (I((X + Y)) + Z))) ---> Z  [1, 13]

Trying to orient equation:
  (I(Y) + I(Y1)) == I((Y1 + Y))  [17, 4]

  To prove:  (I(Y) + I(Y1))  >  I((Y1 + Y))
         or:  (I(Y) + I(Y1))  <  I((Y1 + Y))
  Here are some precedence suggestions:
     1.  + > I                    2.  I > +
Either type a list of numbers or
Type Abort, Display, Equiv, Twoway, Operator, Status, LR, RL, Postpone,
   MakeEq, Quit, Undo or Help.
RRL>KB> 2
----- Step 4 -----

Adding rule:  [21] I((Y + Y1)) ---> (I(Y1) + I(Y))  [17, 4]

  Deleting rule:  [7] (I((X + Y)) + (X + (Y + Z))) ---> Z  [1, 4]

  Deleting rule: [17] (Y + I((Y1 + Y))) ---> I(Y1)  [14, 4]

  Deleting rule: [19] (Y + I((X + (Y1 + Y)))) ---> I((X + Y1))  [1, 17]

  Deleting rule: [20] (X + (I((Y + X)) + Z)) ---> (I(Y) + Z)  [17, 1]
```

```
Computing critical pairs with:
  [21] I((Y + Y1)) ---> (I(Y1) + I(Y)) [17, 4]

Your system is canonical.

  [1] ((X + Y) + Z) ---> (X + (Y + Z)) [USER, 1]
  [2] (0 + X) ---> X [USER, 2]
  [3] (I(X) + X) ---> 0 [USER, 3]
  [4] (I(Y) + (Y + Z)) ---> Z [3, 1]
  [9] (Z + 0) ---> Z [DELETED, 6]
 [10] I(0) ---> 0 [3, 9]
 [11] I(I(Y)) ---> Y [9, 8]
 [12] (Y + I(Y)) ---> 0 [11, 3]
 [13] (Y + (I(Y) + Z)) ---> Z [11, 4]
 [21] I((Y + Y1)) ---> (I(Y1) + I(Y)) [17, 4]

Processor time used                   = 31.375 sec
Number of rules generated             = 21
Number of critical pairs              = 73
Time spent in normalization           = 0.25 sec
 (0.7968128 percent of time)
Time spent while adding rules         = 2.625 sec
 (8.366534 percent of time)
 (keeping rule set reduced)
Time spent in unification             = 0.5 sec
 (1.5936255 percent of time)
Time spent in ordering                = 27.375 sec
(87.25099 percent of time)

Total time (including 'undo' action)  = 31.75 sec
Time = 38.25 sec


Type Add, Akb, Auto, Break, Clean, Delete, Dump, Grammar, Init, Kb, List,
   Log, Makerule, Narrow, Norm, Order, Option, Operator, Prove, Quit,
   Read, Refute, Stats, Suffic, Undo, Unlog, Write or Help.
RRL-> list
No equations in current system.

Rules:
  [1] ((X + Y) + Z) ---> (X + (Y + Z)) [USER, 1]
  [2] (0 + X) ---> X [USER, 2]
  [3] (I(X) + X) ---> 0 [USER, 3]
  [4] (I(Y) + (Y + Z)) ---> Z [3, 1]
  [9] (Z + 0) ---> Z [DELETED, 6]
 [10] I(0) ---> 0 [3, 9]
```

```
 [11] I(I(Y)) ---> Y  [9, 8]
 [12] (Y + I(Y)) ---> 0  [11, 3]
 [13] (Y + (I(Y) + Z)) ---> Z  [11, 4]
 [21] I((Y + Y1)) ---> (I(Y1) + I(Y))  [17, 4]
Time = 12.875 sec

Type Add, Akb, Auto, Break, Clean, Delete, Dump, Grammar, Init, Kb, List,
   Log, Makerule, Narrow, Norm, Order, Option, Operator, Prove, Quit,
   Read, Refute, Stats, Suffic, Undo, Unlog, Write or Help.
RRL-> prove
Type equation to prove in the format:  L == R (if C)
Enter a ']' to exit when no equation is given.

I(I(I(z))) == I(z)

Yes, it is equational theorem.
Time = 75.75 sec

Type Add, Akb, Auto, Break, Clean, Delete, Dump, Grammar, Init, Kb, List,
   Log, Makerule, Narrow, Norm, Order, Option, Operator, Prove, Quit,
   Read, Refute, Stats, Suffic, Undo, Unlog, Write or Help.
RRL-> quit
Good bye .
```

11. APPENDIX B

```
          WELCOME TO REWRITE RULE LABORATORY (RRL)
         (c) Copyright 1985,1986 G.Sivakumar.  All rights reserved.
         (c) Copyright 1985,1986 Hantao Zhang.  All rights reserved.
   (c) Copyright 1985,1986 General Electric Company.  All rights reserved.

Type Add, Akb, Auto, Break, Clean, Delete, Dump, Grammar, Init, Kb, List,
   Log, Makerule, Narrow, Norm, Order, Option, Operator, Prove, Quit,
   Read, Refute, Stats, Suffic, Undo, Unlog, Write or Help.
RRL-> read latch.eqn
 Equations read in are:
 1. ALL X ((Q(F(X)) EQU NOT((R(X) OR QBAR(X)))))  [USER, 1]
 2. ALL X ((QBAR(F(X)) EQU NOT((S(X) OR Q(X)))))  [USER, 2]
 3. NOT(ALL X ((((Q(X) EQU NOT(QBAR(X))) AND
                (ALL X (NOT((R(X) AND S(X)))) AND
                   ((R(X) EQU R(F(X))) AND (S(X) EQU S(F(X)))))) ->
                   ((Q(F(F(X))) EQU (S(X) OR (Q(X) AND NOT(R(X))))) AND
                   (QBAR(F(F(X))) EQU NOT(Q(F(F(X))))))))  [USER, 3]
Time = 6.3125 sec
```

```
Type Add, Akb, Auto, Break, Clean, Delete, Dump, Grammar, Init, Kb, List,
   Log, Makerule, Narrow, Norm, Order, Option, Operator, Prove, Quit,
   Read, Refute, Stats, Suffic, Undo, Unlog, Write or Help.
RRL-> Kb
----- Step 1 -----

Adding rule:  [1] Q(F(X)) & QBAR(X) & R(X) --->
                  (Q(F(X)) & QBAR(X) + Q(F(X)) & R(X))  [USER, 1]

Adding rule:  [2] QBAR(X) & R(X) --->
                  (TRUE +
                  Q(F(X)) +
                  QBAR(X) +
                  R(X))  [USER, 1]

 Deleting rule:  [1] Q(F(X)) & QBAR(X) & R(X) --->
                     (Q(F(X)) & QBAR(X) + Q(F(X)) & R(X))  [USER, 1]

Adding rule:  [3] Q(X) & QBAR(F(X)) & S(X) --->
                  (Q(X) & QBAR(F(X)) + QBAR(F(X)) & S(X))  [USER, 2]

Adding rule:  [4] Q(X) & S(X) --->
                  (TRUE +
                  Q(X) +
                  QBAR(F(X)) +
                  S(X))  [USER, 2]

 Deleting rule:  [3] Q(X) & QBAR(F(X)) & S(X) ---> (Q(X) & QBAR(F(X)) +
                     QBAR(F(X)) & S(X))
 [USER, 2]

 S1 is a skolem function for X
   in the assertion: NOT(ALL X ((((Q(X) EQU NOT(QBAR(X))) AND
 (ALL X (NOT((R(X) AND S(X))))
 AND ((R(X) EQU R(F(X))) AND (S(X) EQU S(F(X)))))) ->
 ((Q(F(F(X))) EQU (S(X) OR (Q(X) AND NOT(R(X))))) AND
 (QBAR(F(F(X))) EQU NOT(Q(F(F(X))))))))

Adding rule:  [5] Q(S1) & QBAR(S1) ---> FALSE  [USER, 3]

Adding rule:  [6] QBAR(S1) ---> (TRUE + Q(S1))  [USER, 3]

 Deleting rule:  [5] Q(S1) & QBAR(S1) ---> FALSE  [USER, 3]

Adding rule:  [7] R(X) & S(X) ---> FALSE  [USER, 3]
```

```
Adding rule:  [8] R(S1) & R(F(S1)) ---> R(S1)  [USER, 3]

Adding rule:  [9] R(F(S1)) ---> R(S1)  [USER, 3]

  Deleting rule:  [8] R(S1) & R(F(S1)) ---> R(S1)  [USER, 3]

Adding rule:  [10] S(S1) & S(F(S1)) ---> S(S1)  [USER, 3]

Adding rule:  [11] S(F(S1)) ---> S(S1)  [USER, 3]

  Deleting rule: [10] S(S1) & S(F(S1)) ---> S(S1)  [USER, 3]

Adding rule:  [12] QBAR(F(S1)) & QBAR(F(F(S1))) & R(S1) --->
                   (Q(F(F(S1))) +
                   Q(F(F(S1))) & QBAR(F(S1)) +
                   Q(F(F(S1))) & QBAR(F(F(S1))) +
                   QBAR(F(S1)) & QBAR(F(F(S1))) +
                   Q(F(F(S1))) & R(S1) +
                   Q(F(F(S1))) & QBAR(F(S1)) & R(S1) +
                   QBAR(F(F(S1))) & R(S1))  [USER, 3]

Computing critical pairs with:
  [6] QBAR(S1) ---> (TRUE + Q(S1))  [USER, 3]

Adding rule:  [13] Q(S1) & R(S1) ---> (Q(S1) + Q(F(S1)))  [2, 6]

Computing critical pairs with:
  [11] S(F(S1)) ---> S(S1)  [USER, 3]

Adding rule:  [14] Q(F(S1)) & S(S1) --->
                   (TRUE +
                   Q(F(S1)) +
                   QBAR(F(F(S1))) +
                   S(S1))  [4, 11]

Computing critical pairs with:
  [9] R(F(S1)) ---> R(S1)  [USER, 3]

Adding rule:  [15] QBAR(F(S1)) & R(S1) --->
                   (TRUE +
                    Q(F(F(S1))) +
                    QBAR(F(S1)) +
                    R(S1))  [2, 9]

  Deleting rule: [12] QBAR(F(S1)) & QBAR(F(F(S1))) & R(S1) --->
                      (Q(F(F(S1))) +
```

```
Q(F(F(S1))) & QBAR(F(S1)) +
Q(F(F(S1))) & QBAR(F(F(S1))) +
QBAR(F(S1)) & QBAR(F(F(S1))) +
Q(F(F(S1))) & R(S1) +
Q(F(F(S1))) & QBAR(F(S1)) & R(S1) +
QBAR(F(F(S1))) & R(S1)) [USER, 3]

Adding rule: [16] QBAR(F(F(S1))) ---> Q(F(F(S1))) [DELETED, 12]

 The right hand side is reducible:
  [14] Q(F(S1)) & S(S1) --->
       (TRUE +
       Q(F(S1)) +
       QBAR(F(F(S1))) +
       S(S1)) [4, 11]

Computing critical pairs with:
 [16] QBAR(F(F(S1))) ---> Q(F(F(S1))) [DELETED, 12]

Adding rule: [17] Q(F(F(S1))) & R(F(F(S1))) --->
                  (TRUE +
                  Q(F(F(S1))) +
                  Q(F(F(F(S1)))) +
                  R(F(F(S1)))) [2, 16]

Computing critical pairs with:
 [13] Q(S1) & R(S1) ---> (Q(S1) + Q(F(S1))) [2, 6]

Derived: [18] TRUE ---> FALSE [4, 13]

Your system is not consistent.

Processor time used                     = 7.0 sec
Number of rules generated               = 18
Number of critical pairs                = 8
Time spent in normalization             = 0.9375 sec
 (13.392857 percent of time)
Time spent while adding rules           = 2.625 sec
 (37.5 percent of time)
 (keeping rule set reduced)
Time spent in unification               = 0.0 sec
 (0.0 percent of time)
Time spent in ordering                  = 0.3125 sec
 (4.464286 percent of time)
```

Time spent in boolean translation = 1.0625 sec
(15.178572 percent of time)

Total time (including 'undo' action) = 7.03125 sec
Time = 8.9375 sec

12. APPENDIX C

Structure editor representation of the full adder: shown are both a hierarchical view of the circuit and a CMOS implementation.

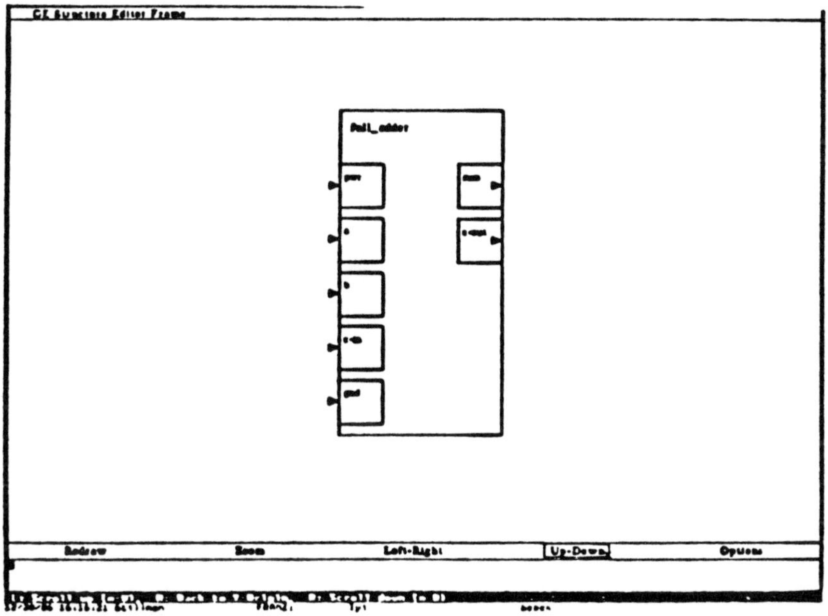

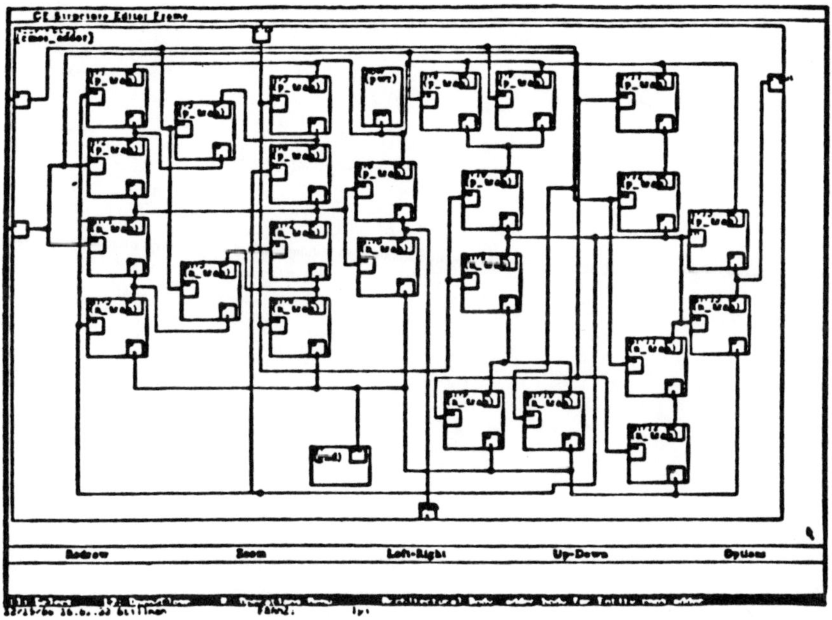

8

Contextual Constraints for Design and Verification

Bruce S. Davie
Department of Computer Science
University of Edinburg, U.K.

George J. Milne
Texas Instruments Ltd.
Manton Lane, Bedford MK41 7PA, U.K.

Abstract. A new approach to the design and verification of VLSI structures capitalizing on knowledge of the surrounding environment is proposed. This approach formalizes the use of contextual information as a means of simplifying the design and verification process and provides us with a new way of viewing the hierarchical evolution of a design. The proposed design process illustrates the different uses of contextual constraints to aid and guide the designer. In particular, constraints may be introduced to simplify verification and must in turn be satisfied by the design itself. Hence we find that design choices may be made to aid validation, an interesting reversal in the usual post-design role of simulation and formal verification.

1. INTRODUCTION

Design may be viewed as the repeated transformation of an abstract functional description through progressively more detailed levels of description until a concrete level of design is reached. Design is the process where information is added to a continuously developing design description, introducing more and more detail, while preserving the functionality of the design. The design evolution involves transforming a design description at a given level of abstraction (level of design detail) into a more detailed description which captures implementation of 'internal' properties belonging to an adjacent level. The behaviour of the design representations at adjacent levels should be closely related if not equivalent and this property should be demonstrable by mathematical proof, called verification, or by simulation.

The design and validation of a circuit will generally require knowledge of more than just the intended device specification; it will require knowledge of the context in which the device is intended to function. Some contextual information will be necessarily detailed, where a particular context in which the device is to operate in is known. Alternatively some contexts will be unknown and in this case the device must be designed to function correctly under all environments. Between the two, varying amounts of contextual information may be specified.

A context is similar to a specification in that both are used to define information relating to the design. While for verification purposes we are primarily concerned with behavioural information, contexts and specifications may also refer to structural and geometrical properties. In an integrated design and verification framework we therefore require a rigorous way in which to represent specifications, contexts and evolving designs. We will propose design and verification within contexts as a new approach to VLSI design. The essence of this approach will be illustrated using a flowchart to picture the design cycle together with a simple example to indicate where contextual constraints fit into this design cycle.

2. THE DESIGN PROCESS

We will begin by considering a hierarchical design methodology for VLSI in which design and validation are carried out in an integrated way. The advantages of such an approach have been discussed in detail in [1], [2]. The key advantage is the ability to detect errors as soon as they are made in the design process rather than needing to design a circuit fully before finding that it may be incorrect.

Figure 1 is a flow chart which shows how a designer might process through a design exercise in a hierarchical way, validating his design decisions at each successive level in the hierarchy before moving on to the next lowest level. We assume that at any stage his aim is to design a box; 'design' here means to partition the box into a number of smaller cells, specifying their interconnection and respective behaviours. A box is simply an abstract circuit whose internal details are not yet specified. The required behaviour of the box will be its spec; the behaviour that is attached to a cell is referred to in the diagram as a cellspec.

When the designer initially sets out to design a circuit, that whole circuit will be the box at level 1 in the hierarchy. As the design proceeds, the cells in turn become the boxes, which must also be designed. We have assumed for the purposes of our illustration that this repated partitioning terminates when a level is reached in the hierarchy where the cells are parts which already exist in a library of predefined parts. This would be the case in a standard cell design system. In a full custom design it might terminate when the cells are of suitable complexity to be laid out as leaf cells.

There are various points worth noting in this somewhat idealized representation of the design process. The partitioning of a box into cells (which includes specifying their interconnections) is presented as completely separate operation from the description of cell behaviours. The former operation is purely structural, the latter purely behavioural. This distinction may be considered artificial, as a designer will almost certainly have some idea of the intended behaviour of the cells when he does the partitioning. However, this association of behaviour with structure is not made formal until the specification phase, and so we consider the two operations separately here.

It should also be noted that we are not by any means proposing a fully automated design process here. While we would hope that some machine assistance would be available for many of the tasks, and the role of constraints in providing some of this assistance will be discussed below, operations such as partitioning will certainly require some amount of skill on the part of the designer.

There are various possibilities for *iterative* design here. If the behaviour of the box's implementation **imp**, which is obtained by composing the cellspecs, is not in some way equivalent[1] to the required behaviour spec, then it is necessary to make a new partitioning of the box. If it is eventually found that the box cannot be implemented successfully with any partitioning then we must move back up through the design hierarchy; the 'parent' box will then have to be redesigned.

At the end of the design the device which we set out to implement, and perhaps several intermediate devices as well, will have been fully designed down to same level of 'primitive' parts. So that these implementations may be used in later designs, they should be added to a library, as shown in the last stage on the chart.

Having now outlined a formal design and verification methodology, we are in a position to discuss the ways in which contextual constraints may be of assistance in such a methodology. In the following Section the concept of contextual constraint will be presented and its role in design and validation outlined.

3. CONTEXTUAL CONSTRAINTS

Knowledge of the environment or context in which a design is to be used can drastically aid the design and verification process. This is an approach which is informally taken by designers at present, although the lack of a formal design/verification framework prevents the technique from being elevated into a powerful design methodology. As a example consider

1 Note that the notion of equivalence is not defined here. It is dealt with in detail in [5].

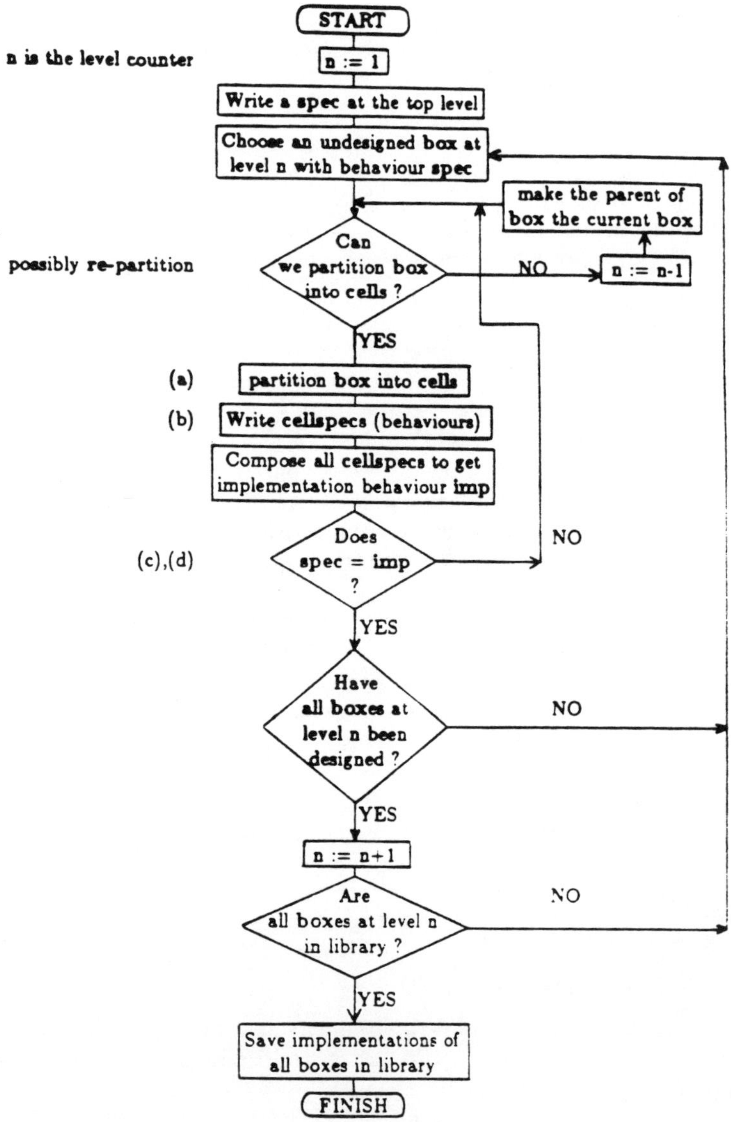

Figure 1: The Design Process

synchronous devices which use certain clocking regimes to simplify their design. While a context involving a specific system clock simplifies the design process by restricting the class of environments in which the circuit must work correctly, it also greatly simplifies the verification process. The circuit behaviour is restricted by the limited interactions which the environment or context will permit; it will evolve into a much reduced set of states through time while in any state only a limited set of possible actions will be permitted by the environment.

We propose a technique where specifications, designs and contextual information are all described in a common language. Contextual knowledge is captured as just another expression in the language and may be used to simplify the representation of a design in a rigorous manner. This uses properties of the mathematical model which underlies the descriptive language, so also simplifying design verification.

To illustrate the central ideas behind using contextual knowledge to constrain parts of the design process consider the task of verifying that a flip-flop constructed out of a number of gates is correct. While this at first sight appears to be a trivial example due to the structural simplicity of a flip-flop it should be noted that it has sophisticated time-related behaviour due to feedback. Describing the flip-flop behaviour, a necessary prerequisite for verification, requires that all response to all possible patterns of stimuli on its set and reset ports be represented. Certain patterns are known to cause a flip-flop to react in strange ways; 'spikes' on the inputs and simultaneous set and reset operations are two such examples. But we will frequently know something about the context in which the flip-flop is to be used. A system clock can be used to control the latching of signals such that spikes and simultaneous set and reset actions will never be generated. What is then required is to verify the flip-flop design only under the legitimate patterns of set and reset actions which we assume that the context will generate. At some stage the context must also be verified to assure ourselves that it generates the restricted behaviour as intended but this is a separate part of the design process and may be considered in isolation. The significance of contextual restriction is illustrated in the Section 4 where a D-type flip-flop is designed and verified using contextual constraints where possible.

The Role of Constraints

Four distinct uses of constraints may be isolated and their role in the total design process can be indicated by relating them to the flowchart in Section 2.

(a) They dictate, at least partially, the context in which a design component may be used and in so doing they restrict the design choices available when designing that component.

(b) They aid the component specification task by allowing us to ignore some of the possible behavioural patterns which an unconstrained environment might generate.

(c) They may be introduced to aid component verification by reducing the state-space under consideration as indicated in the above discussion on the design of a flip-flop.

(d) They may be generated when a verification proof is performed and will then be necessary to ensure design correctness. The context must then be modified to reflect this additional constraint and we have here an example of 'design for verification'.

The first two uses of constraints are primarily for design reasons while the latter two fulfill a validation role. However the use of constraint information always applies to both design and validation; what aids one will generally aid the other.

Constraints for Design

As mentioned previously we imagine design to center around the task of creating and validating interconnected design components called boxes, as indicated by the flowchart of Section 2. We can assume that each box representation held in the design library should include the constraint information pertaining to that box. Hence boxes are represented by an identifier name, structural information which is captured by the sort, a behavioural specification written as a CIRCAL expression, possibly some geometrical layout information and finally a (possibly null) contextual constraint expression which is also represented in CIRCAL. If no constraints are to be imposed on the box, i.e., if it is designed to function correctly in any context, then the constraint is null. The reader is referred to [3] for details of the representation of circuit structure and behaviour by CIRCAL expressions.

Property (a) above aids design partitioning which is the creative step in the design process. It therefore fits into the flow diagram at the partitioning stage as indicated by the (a) in the flowchart. Property (b) makes the specification task more straightforward and relates to the fundamentally important behavioural description stage of the design of a design component.

Constraints for Verification

Constraints may be imposed on a design component in one of two ways; the context that the component is to be used within may be known prior to design, for instance a certain clocking scheme may be predetermined. Alternatively constraints may be found to be necessary to allow component verification to proceed by imposing restrictions on the possible environmental interactions that the component can respond to. This constraint must then be the surrounding context, i.e., the other components which will interact with it.

Property (c) above relates to constraints imposed prior to verification, introduced either in anticipation of the verification to come or else by the given design partitioning. It fits into the flowchart int he partitioning stage or in the verification stage respectively. Property (d), the discovery of the need for constraints to aid the verification process obviously fits into the verification stage of the flow chart, where we attempt to construct a proof that the component implementation meets its specification. Various CIRCAL validation techniques are presented by Milne [4] and Traub [5].

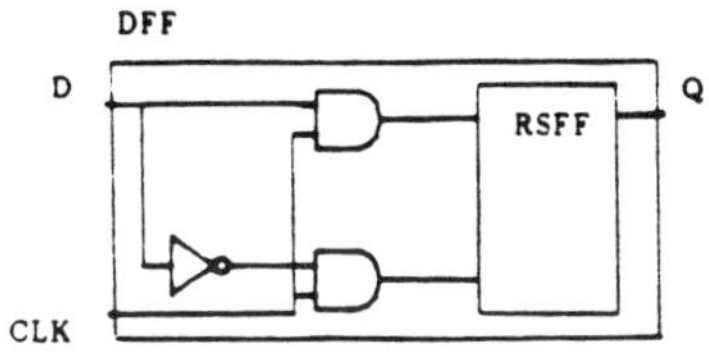

Figure 2: D flip-flop implementation

4. AN EXAMPLE

We will consider the situation where a designer is attempting to design a simple D flip-flop. While this example may seem too simple or may appear contrived since the design flip-flop is already well understood, it does represent a situation which could perhaps arise if designing in some new technology or in an attempt to fit the device into some awkwardly shaped or sized area of a chip. Furthermore, it is hoped that the ideas illustrated here could be applicable to one or more general and complex design situations. It should also be noted that this example is intended to illustrate only some of the various ways in which contextual constraints may be of assistance to a designer.

In the terminology of the above flowchart, the D flip-flop is the box which the designer wishes to design. Support he wishes to partition it into an RS flip-flop and some gates. As mentioned in Section 2, this is an essentially structural operation, and must be followed by formally associating behaviours with each of the cells. In specifying the behaviour of the RS flip-flop, it is necessary either to specify its behaviour fully (i.e., taking account of all possible patterns of inputs) or partially. As stated in Section 3, writing a full specification of an RS flip-flop is actually a non-trivial task, as its behaviour when presented with simultaneous ones on the set and reset inputs is quite complex and is usually undesirable. Thus we may wish to specify its behaviour without taking account of such undesirable input patterns. If this approach is to be taken, then steps must also be taken to ensure that the context in which the RS flip-flop will be placed will never generate the pattern of inputs. Thus in addition to the flip-flop's partial specification, we would define the constraint in which it may legitimately be used. The partial specification and the corresponding constraint on the environment are as follows:

$$\text{RSFF0}\{\text{res,set,q}\} \leq \text{res1 res0 RSFF0 set1 q1 set0 RSFF1}$$

$$\text{RSFF1}\{\text{res,set,q}\} \leq \text{res1 q0 res0 RSFFO} + \text{set1 set0 RSFF1}$$

$$\text{ENV}\{\text{res,set}\} \leq \text{res1 res0 ENV} + \text{set1 set0 ENV}$$

A full discussion of the CIRCAL syntax and its semantics is presented in [3]. In essence, the contextual constraint specified in the last line of the above description states that the context or environment (ENV) must only present certain patterns of stimuli on the ports 'res' and 'set'. These are specified in such a way that res and set are never simultaneously presented with ones. Note that for the sake of clarity in this illustration we have not dealt with other constraint mentioned, that of preventing input 'spikes'.

So we have shown that a contextual constraint may make the task of writing specifications simpler by allowing us to ignore certain patterns of inputs. Similarly, a part in a standard cell library, for example, might have a constraint associated with it, such as the set-up time required on its inputs. In this situation, as in the above, we see the need for a formal way of both specifying and manipulating constraints.

Having writing the constraints on the context of a device in a formal way, a designer must then ensure that the context does in fact meet those constraints. In the case where the context is not yet designed, this means that knowledge of the constraints will influence the designer's choices. In the case of the simple D flip-flop design, knowledge of the constraint that set and reset lines must never be simultaneously one might lead the designer to place an inverter between the two inputs; also, knowing that spikes on the inputs are undesirable, he might choose to clock the inputs using the arrangement of and gates shown in Figure 2. Having chosen to implement the context of the RS flip-flop in this way, the designer would formally associate behaviours with the components which make up the context and then verify that they do in fact meet the contextual constraint specified above.

In this example there is still another way in which constraints can be seen to be of assistance to the designer. This is the area of validation of the design. When the D flip-flop is the box in the terminology of Figure 1, the designer must verify that the composition of the behaviours of the RS flip-flop and the three gates satisfies the behavioural specification of the D flip-flop. The greatly simplified specification of the RS flip-flop which was allowed by the formal specification of a constraint on its context leads directly to a simpler proof of this behavioural equivalence.

Furthermore, in the next level down the hierarchy, when the RS flip-flop becomes the box and is in turn designed using a number of cells (e.g., nor gates), the verification process will also be simplified. This is again a direct result of the use of a simpler specification where certain patterns of input stimuli are disallowed by the use of a contextual constraint.

5. Concluding Remarks

We have informally presented contextual techniques for design and verification and indicated their utility as a practical tool for circuit design. This exercise has also illustrated the necessity of using a rigorous description and proof framework when we wish to take a design rule-of-thumb and extend and develop it into a powerful and far-reaching methodology. Here we have taken the informally used concept of contextual design and outlined how it may be used as a central feature of a design system.

REFERENCES

[1] Davie, B. S. *Hardware Description Languages: Some Recent Developments,* Departmental Report CSR-198-86, Edinburgh University, 1986.

[2] Davie, B. S. and Milne, G. J. *The Role of Behaviour in VLSI Design Languages* in Proc. IFIP WB 10.2 Conference, From HDL Descriptions to Guaranteed Correct Circuit Designs, Grenoble, Sept. 1986, (North-Holland).

[3] Milne, G. J. *CIRCAL. A Calculus for Circuit Description,* Integration, the VLSI Journal, vol. 1, nos. 2&3, p. 121-160, October 1983.

[4] Milne, G. J. *Simulation and Verification: Related Techniques for Hardware Analysis* in Proc. 7th. Int. Symposium on Computer Hardware Description Languages and their Applications, CHDL '85, Tokyo, Koomen and Moto-oka (eds.), (North-Holland), 1985.

[5] Traub, N. G. *A Formal Approach to Hardware Analysis,* Phd. thesis, Edinburgh University, 1986.

9

Abstraction Mechanisms for Hardware Verification

Thomas F. Melham

University of Cambridge
Computer Laboratory
Corn Exchange Street
Cambridge, CB2 3QG
England

Abstract: It is argued that techniques for proving the correctness of hardware designs must use abstraction mechanisms for relating formal descriptions at different levels of detail. Four such abstraction mechanisms and their formalisation in higher order logic are discussed.

Introduction

Recent advances in microelectronics have given designers of digital hardware the potential to build electronic devices of unprecedented size and complexity. With increasing size and complexity, however, it becomes increasingly difficult to ensure that such systems will not malfunction because of design errors.

This problem has prompted some researchers to look for a firm theoretical basis for correct design of hardware systems. Mathematical methods have been developed to model the functional behaviour of electronic devices and to verify, by formal proof, that their designs meet rigorous specifications of intended behaviour. Proving the correctness of a design using such hardware verification techniques typically involves:

- Writing a set of formulas S which express the specification of the device whose design is to be proven correct.
- Describing the design or implementation of the device by a second set of formulas I.
- Proving that a 'satisfaction' or correctness relation holds between the sets of formulas I and S—i.e. that the implementation satisfies the specification.

This process is usually carried out at each level of the structural hierarchy of the design. The top level specification is shown to be satisfied by some connection of components; the specifications of these components are in turn shown to be satisfied by their implementations, and so on—until the level of primitive components is reached.

If this technique is to be used to verify large and complex designs, it is clear that the 'satisfaction relation' that is used cannot be strict equivalence. Otherwise, at each level of the design hierarchy, the specifications will contain all the information present in the design descriptions at the level below. For large and complex designs this means that the specifications at the upper levels of the hierarchy will themselves become so large and complex that they can no longer be seen to reflect the intended behaviour of the device. Furthermore, proofs of equivalence between such complex formulas will become unmanageable, even when automated theorem proving tools are used.

A satisfaction or correctness relation based on the idea of *abstraction*, rather than equivalence, is the key to making formal verification of large designs tractable. The aim of this paper is to discuss the role of abstraction in hardware verification and to show how various types of abstraction can be formalised and used to control the complexity of specifications and correctness proofs.

The organisation of the paper is as follows. Section 1 introduces the idea of abstraction and its role in controlling the complexity of hardware verification. In section 2 an informal description is given of four basic abstraction mechanisms that can be used in proofs of hardware correctness. Section 3 contains a brief introduction to the logic that will be used to prove devices correct, and reviews the standard techniques for formally specifying the behaviour and structure of hardware in logic. Sections 4 through 7 show how each of the four basic abstraction mechanisms can be formalised in logic; in each section, an example is given to illustrate the use of the abstraction mechanism being discussed.

1 Abstraction

Abstraction involves the suppression of irrelevant detail or information, in order to concentrate on the things of interest. In hardware verification, the larger and more complex the system we are describing, the more irrelevant detail must be ignored to keep specifications small. Thus, in general, specifications will be abstractions of the devices actually implemented. This means that the satisfaction or correctness relation that is used must involve abstraction mechanisms that, by removing information, relate the more detailed implementation descriptions to their abstract specifications.

The specification of a microprocessor, for example, would include a description of the effect of each instruction on the machine registers. The description of a microcoded implementation of the system would contain far more information, including the exact sequence of microinstructions which implements each macroinstruction. To show that this design is correct, we would have to prove that each microcode sequence produces the effect on the registers that is required by the specification of the corresponding macroinstruction. Here, the abstraction mechanism used to relate the levels of description is the composition of state changes; by composing a sequence of low level state changes to get a single transition, we 'hide' from the top level specification information about the intermediate states that occur during macroinstruction execution. At the abstract level, we only know *what* each instruction does; at the more detailed implementation level, we also know *how* the instruction does it.

Another example is the correctness proof of a multiplier circuit. A multiplier is constructed from hundreds of transistors, each of which exhibits quite complex behaviour. Furthermore, this device performs multiplication by some particular method, e.g. repeated addition. All this information will be explicit in the formulas that describe the multiplier's design but will be irrelevant to its abstract specification, which will simply state that it correctly performs multiplication.

The correctness proof of this device will show that the logical formulas expressing its specification represent a valid abstract view of the formulas describing its design. The design description will include, for example, assertions about voltages present at certain points in the device; these must be translated into assertions in the specification about the numbers being multiplied. Such a translation is an abstraction mechanism for relating *numbers* in the specification to the *voltages* in the design which represent them.

If we have proved that a formal specification represents a valid abstract view of a device, we can use it in place of the more detailed implementation description when proving the correctness of a larger system which contains the device as a component. This will allow us to derive a more clear and concise behavioural description of the larger system's implementation than would otherwise be possible. If we do this at each level of the proof of a hierarchically structured design, we can control the size and complexity of the entire proof. In this way, abstraction mechanisms combine with hierarchical structuring to make it possible to handle proofs of large systems.

2 Four Types of Abstraction

The type of abstraction most fundamental to hardware verification is *structural* abstraction—the suppression of information about a device's internal structure. The idea of structural abstraction is that the specification of a device should not reflect its internal construction, but only its externally observable behaviour. The description of a device's implementation, however, must contain explicit information about its structure; the mechanism of structural abstraction therefore involves formalising the idea that such information concerns 'internal' structure.

A second type of abstraction, *behavioural* abstraction, concerns specifications that only partially define a device's behaviour, leaving unspecified its behaviour in certain states or for certain input values. Such partial specification is appropriate, for example, when it is known that a device will never have to operate in certain environments and it is therefore unnecessary to specify its expected behaviour in *all* environments. The mechanism of behavioural abstraction serves to relate partial specifications to implementation descriptions that fully define the device's behaviour, by showing that they agree on the device's behaviour for all states and inputs that are of interest, i.e. that are defined by the specification.

The concept of *data* abstraction is well known from programming language theory. There are several abstract data types which are useful for formal specification of hardware; a simple example is the type of boolean truth values, a data abstraction from analog signals routinely used by hardware designers to reason about circuits. Another example is the type of n-bit integers, represented at the less abstract level by vectors of booleans. A data abstraction step consists of constructing a mapping from the data types of an implementation description to the more abstract data types of the specification. This mapping is then used to show that the operations carried out on the low level data types correctly implement the desired operations on the high level types.

A final type of abstraction is *temporal* abstraction, in which the sequential or time-dependent behaviour of a device is viewed at different 'grains' of discrete time. An example of temporal abstraction is the unit delay or register, implemented using an edge triggered flip flop. At the abstract level of description the device is specified as a unit delay, one 'unit' of discrete time corresponding to the clock period; at the detailed level of description the grain of time is finer, several units of time corresponding to the delay through a gate. The proof of a microcoded computer design, mentioned above, also involves temporal abstraction; at the low level there is one microcode step per unit of discrete time, and at the higher level there is one machine instruction step per unit of discrete time. Relating the levels of temporal abstraction involves mapping points or periods of low level time to points or periods of high level time and showing that a many-step low level computation implements a one-step high level computation.

3 Specification using Higher Order Logic

The formalism that we use to specify and verify hardware is a variety of higher order logic, adapted for this purpose by Mike Gordon at the University of Cambridge. In this section, we first give a very brief introduction to this logic; a more complete and formal presentation is given by Gordon in [2,3]. We then review how the behaviour and structure of digital hardware can be specified using higher order logic.

3.1 Introduction to Higher Order Logic

Higher order logic includes terms corresponding to the standard notation of predicate calculus. Thus, '$P\,x$' expresses the proposition that x has property P and '$R(x, y)$' means that the relation R holds between x and y. Higher order logic has the usual logical operators $\neg$, $\wedge$, $\vee$, $\supset$ and $\equiv$ denoting negation, conjunction, disjunction, implication and equivalence respectively. The universal and existential quantifiers $\forall$ and $\exists$ are used to express the concepts of *every* and *some*; '$\forall x.P\,x$' means that P holds for every value of x and '$\exists x.P\,x$' means that P holds for some (i.e. at least one) value of x. Terms of the form '$(c \Rightarrow t_1 \mid t_2)$' denote the conditional 'if c then t_1 else t_2'. The term '$f \circ g$' denotes the composition of the functions f and g defined by $(f \circ g)(x) = f(g(x))$. The constants T and F denote the truth values *true* and *false* respectively.

The logic we use is *higher-order*. That is, variables are allowed to range over functions and predicates. The well ordering property of the natural numbers, for example, can be expressed using a variable P ranging over predicates:

$$\forall P.\,(\exists n.\,P\;n) \supset (\exists n.\,P\;n \wedge \forall n'.\,n'{<}n \supset \neg P\;n)$$

This states that if any predicate P is true of at least one number then there is a smallest number for which P is true.

In higher order logic, functions and predicates can take other functions and predicates as arguments. Functions can also yield functions as results. Consider, for example, the function Rise defined as follows:

$$(\mathsf{Rise}\; ck)(t) \equiv \neg ck(t) \wedge ck(t{+}1)$$

Rise is intended to express the notion that a clock ck rises at time t. The clock, modelled by a function ck from natural numbers to booleans, is given as the argument to Rise. The result of applying Rise to ck is also a function from numbers to booleans. Thus Rise is an example of a function that both takes a function as an argument and yields a function as a result.

Higher order logic is a typed logic; every term of the logic has a *type*, though these are usually omitted when a term is written down. Informally, types can be thought of as denoting sets of values and the value of a term as an element of the set denoted by the term's type. The basic types of the logic include the type of natural numbers, *num* and the type of boolean truth values *bool*. Types can be built from other types using *type operators*. For example, the type of functions from *num* to *bool* is denoted by '*num*→*bool*', using the infix type operator '→'.

Writing '*tm*:*ty*' indicates explicitly that a term *tm* has type *ty*. For example, the term 'Rise ck' can be written with explicit type information as

$$(\mathsf{Rise}\; ck){:}num{\to}bool$$

or even as

$$((\mathsf{Rise}{:}(num{\to}bool){\to}(num{\to}bool))\;(ck{:}num{\to}bool)){:}num{\to}bool$$

though it is seldom necessary to give this much type information.

An important primitive constant in the logic we are using is the ε-operator, the informal semantics of which are as follows. If $\mathsf{P}[x]$ is a predicate involving a variable x of type ty then $\varepsilon x.\,\mathsf{P}[x]$ denotes some value of type ty such that P is true of that value. If there is no such value (i.e. $\mathsf{P}[v]$ is false for each value v of type ty) then $\varepsilon x.\,\mathsf{P}[x]$ denotes some fixed but arbitrarily chosen value of type ty. Thus, for example, '$\varepsilon n.\,4{<}n \wedge n{<}6$' denotes the value 5, '$\varepsilon n.\,(\exists m.\, n = 2{\times}m)$' denotes an unspecified even natural number, and '$\varepsilon n.\, n{<}n$' denotes an arbitrary natural number. For further discussion of the ε-operator, see [2] or [10].

3.2 Specifying Behaviour in Logic

The technique for specifying hardware behaviour with higher order logic is well established; see, for example, Gordon's paper [4] or the work done by Hanna and Daeche [7]. Hardware devices can be specified by predicates that describe their behaviour in terms of the values on their external ports. Consider, for instance, a device with external ports a, b, c and d:

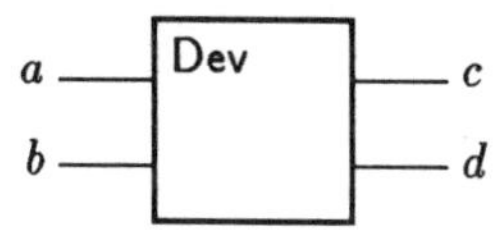

Such a device is specified using a four-place predicate Dev, defined such that '$\mathsf{Dev}(a,b,c,d)$' is true exactly when the combination of the values of a, b, c and d is one that could occur on the corresponding ports of the device.

Devices that exhibit time-dependent behaviour can be specified by allowing the values on the external ports to vary over time. An inverter with δ time units of delay, for example, can be specified by the predicate Inv defined as follows.

$$\mathsf{Inv}(i,o) \equiv \forall t.\, o(t{+}\delta) = \neg i(t)$$

In this specification, instants of discrete time are modelled by the natural numbers; i and o are functions from *num* to *bool* giving the sequence values on the input and output ports at successive moments of time. Such functions are sometimes called 'signals'. The predicate Inv specifies an inverter with δ delay by asserting that for all times t, the value of the output signal at time $t{+}\delta$ is the negation of the value of the input signal at time t.

3.3 Specifying Structure in Logic

Given predicates that specify the behaviours of the components of a design, we wish to derive a logical expression that describes both how these components are interconnected and what the net behaviour of the connected components will be. In logic, this is done simply by forming a conjunction of the predicates that specify the components, identifying variables that represent connected ports (see Gordon's paper [4]).

Suppose, for example, $\mathsf{P}_1(a, b, c, d)$ and $\mathsf{P}_2(e, f, g)$ are the predicates that specify the two components P_1 and P_2 of a design:

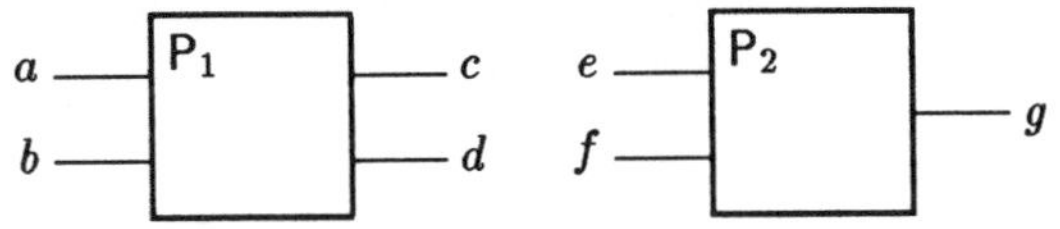

and suppose that d is connected internally to b, and c to both e and f:

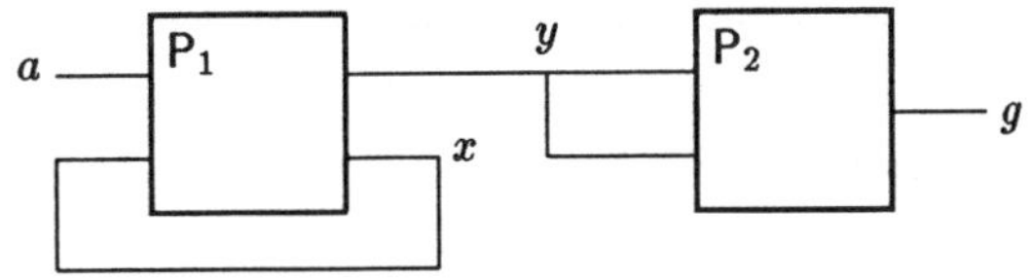

The resulting structure can be specified in logic by the predicate Imp, defined as follows:

$$\mathsf{Imp}(a, x, y, g) \equiv \mathsf{P}_1(a, x, y, x) \wedge \mathsf{P}_2(y, y, g)$$

The predicates P_1 and P_2 define the allowable values on the ports of the two components; the predicate Imp simply asserts that the values allowed on the ports of the connected components are just those allowed by *both* P_1 and P_2.

4 Formalising Structural Abstraction

We now consider how the mechanism of structural abstraction can be formalised in higher order logic. The abstract specification of a device's intended behaviour will, in general, be given in terms that do not make reference to any particular structure that realizes the behaviour. The formal description of some particular implementation, however, will contain information about its structure, in the form of constraints on signals not present in the specification. The correctness relation that we formulate must express the idea that these constraints are 'internal' to the implementation description.

Consider, for example, the design described by the predicate Imp, defined above. This predicate defines a logical relation that involves the external ports a and g as well as the internal signals x and y. An abstract specification of this device, however, will be given only in terms of the external ports a and g, and will not include information about the values of x and y. Suppose that such a specification is given by $\mathsf{Spec}(a, g)$. The correctness relation for this device must state that Spec and Imp agree on what values are allowed to occur on the external ports a and g, given that x and y are internal signals.

We begin by considering what it means for x and y to be 'internal' signals, hidden from the device's environment. If x and y are internal signals, then their values are determined solely by the constraints imposed internally by the components of the device. Thus, if '$\mathsf{Imp}(a, x, y, g)$' is true, we can interpret this as meaning that x and y are the values generated internally when the values a and g are present on the external ports.

Now, two values a and g are allowed to occur on the external ports just when it is *possible* for internal signals to be generated such that $\mathsf{Imp}(a, x, y, g)$ comes out true. Therefore two values a and g are allowable on the external ports just when there are *some* values x and y such that $\mathsf{Imp}(a, x, y, g)$ holds, i.e. when it is the case that $\exists\, x\, y. \mathsf{Imp}(a, x, y, g)$.

A correctness relation involving structural abstraction can therefore be formulated as follows. The design described by Imp is correct with respect to the abstract specification Spec if we can prove that:

$$\forall\, a\, g.\ \mathsf{Spec}(a, g) \equiv \exists\, x\, y. \mathsf{Imp}(a, x, y, g)$$

That is, Spec and Imp agree on what values are allowed on the external ports a and g, given that the signals x and y are internal to the design described by Imp.

Thus, correctness relations involving structural abstraction can be expressed formally in higher order logic by 'hiding' internal signals in the design description using the existential quantifier and proving that the resulting term is logically equivalent to the specification. This technique of using '$\exists$' to hide internal structure is fairly common; see, for example, [1,4] or [9].

4.1 An Example

This example illustrates the use of structural abstraction in a very simple correctness proof. A delay device with δ time units of delay can be specified by the predicate Delay, defined as follows:

$$\mathsf{Delay}(i, o) \equiv \forall t.\ o(t{+}\delta) = i(t)$$

A series connection of two such devices can be specified by the predicate Imp defined by:

$$\mathsf{Imp}(i, x, o) \equiv \mathsf{Delay}(i, x) \land \mathsf{Delay}(x, o)$$

If the internal signal x is hidden, this connection of devices should satisfy the specification given by:

$$\mathsf{Spec}(i, o) \equiv \forall t.\ o(t{+}2\delta) = i(t)$$

Formally, the correctness relation that we want to prove is:

$$\forall\, i\, o.\ \mathsf{Spec}(i, o) \equiv \exists\, x.\ \mathsf{Imp}(i, x, o)$$

The proof of this correctness statement proceeds as follows:

1. By the definition of Delay:

$$\exists\, x. \mathsf{Imp}(i, x, o) \equiv \exists\, x.(\forall t.\ x(t{+}\delta){=}i(t) \land \forall t.\ o(t{+}\delta){=}x(t))$$

2. By the symmetry of equality:

$$\exists\, x.\mathsf{Imp}(i,x,o) \equiv \exists\, x.(\forall t.\, x(t{+}\delta){=}i(t) \wedge \forall t.\, x(t){=}o(t{+}\delta))$$

3. Selective rewriting with the equation '$\forall t.\, x(t){=}o(t{+}\delta)$' yields:

$$\exists\, x.\mathsf{Imp}(i,x,o) \equiv \exists\, x.(\forall t.\, o((t{+}\delta){+}\delta){=}i(t) \wedge \forall t.\, x(t){=}o(t{+}\delta))$$

4. Since x does not occur in the left hand conjunct, the scope of the existential quantifier can be limited to the right hand conjunct:

$$\exists\, x.\mathsf{Imp}(i,x,o) \equiv \forall t.\, o((t{+}\delta){+}\delta){=}i(t) \wedge \exists\, x.(\forall t.\, x(t){=}o(t{+}\delta))$$

5. The right hand conjunct '$\exists x.\forall t.\, x(t){=}o(t{+}\delta)$' is a tautology; the equivalence therefore reduces to:

$$\exists\, x.\mathsf{Imp}(i,x,o) \equiv \forall t.\, o((t{+}\delta){+}\delta) = i(t)$$

6. Simplifying the expression '$(t{+}\delta){+}\delta$', yields:

$$\exists\, x.\mathsf{Imp}(i,x,o) \equiv \forall t.\, o(t{+}2\delta) = i(t)$$

7. From which the desired correctness statement follows immediately:

$$\exists\, x.\mathsf{Imp}(i,x,o) \equiv \mathsf{Spec}(i,o)$$

This proof, although trivial in itself, illustrates the general approach to proving correctness statements that use structural abstraction. This commonly involves rewriting with the equations defining internal signals (step 3 above) and eliminating these equations by restricting the scope of the existential quantifiers (steps 4 and 5). For other examples of the use of this technique, see [1] or [4].

5 Formalising Behavioural Abstraction

Behavioural abstraction involves proving the correctness of designs with respect to *partial* specifications of intended behaviour, i.e. with respect to specifications that do not completely define the full range of behaviour that a device can exhibit, but only define its behaviour in environments or states that are of particular interest.

In logic, specifications are expressed by constraints on the values allowed on the external signals of a device; the 'range of behaviour' defined by a specification is given by the set of values that satisfy these constraints. A *partial* specification constrains a device's signals to have certain values in situations that are significant or relevant, but leaves unconstrained the signal values in all other situations. This means that, in the situations of 'undefined' behaviour, the predicate defining a partial specification will be satisfied by signal values that would not be allowed by a more complete specification of behaviour. Thus, the partial specification of a device imposes weaker constraints on its signal values than a complete specification would.

The formalisation of behavioural abstraction in logic is straightforward. Suppose that $\mathsf{Spec}(a,b)$ is the partial specification of a device and $\mathsf{Imp}(a,b)$ is the design description. The implementation Imp is correct with respect to the partial specification Spec if the following correctness relation holds:

$$\forall a\, b.\ \mathsf{Imp}(a,b) \supset \mathsf{Spec}(a,b)$$

This correctness statement asserts that whenever signals a and b satisfy Imp, they will also satisfy Spec. The implication may also be true, however, of signals for which Spec is true but Imp is false, i.e. there may be values that are allowed by the weaker constraints of Spec but not allowed by the constraints given by Imp. Thus, logical implication can be used to express the idea that the specification Spec is a behavioural abstraction, imposing weaker constraints on the signals a and b than those of the design description Imp. This use of logical implication for behavioural abstraction is natural and well known; see for example [1,7].

5.1 An Example

We now consider an example of the use of behavioural abstraction: the correctness proof of an RS latch, implemented using two cross coupled nor-gates. If a nor-gate with unit delay is specified by the three-place predicate Nor defined by:

$$\mathsf{Nor}(i_1, i_2, o) \equiv \forall t.\ o(t{+}1) = \neg(i_1\, t \vee i_2\, t)$$

then the implementation description of an RS latch is given by:

$$\mathsf{RS\_imp}(s, r, q, \overline{q}) \equiv \mathsf{Nor}(r, \overline{q}, q) \wedge \mathsf{Nor}(s, q, \overline{q})$$

The abstract specification of the latch will use the auxiliary predicates store and stable, defined as follows. The term '$\mathsf{store}\ s\ r\ t\ v$' will have the meaning 'the signals s and r take on values at time t such that the value v will be stored in the latch'. The formal definition of store is:

$$\mathsf{store}\ s\ r\ t\ v \equiv s(t){=}v\ \wedge\ s(t{+}1){=}v\ \wedge\ r(t){=}\neg v\ \wedge\ r(t{+}1){=}\neg v$$

This definition states, for example, that to store the value T in the latch the set and reset signals must have the values $s{=}\mathsf{T}$ and $r{=}\mathsf{F}$ for two units of time.

The term '$\mathsf{stable}\ sig\ v\ t\ n$' will have the meaning 'the signal sig is stable with value v during the period from time t to time $t{+}n$'. Formally, stable is defined as follows:

$$\mathsf{stable}\ sig\ v\ t\ n \equiv \forall t'.\ t{\leq}t' \wedge t'{\leq}t{+}n \supset sig(t'){=}v$$

Using store and stable, the partial specification of the RS latch is given by:

$$\begin{array}{l} \mathsf{RS\_spec}(s,r,q,\overline{q}) \equiv \\ \quad \forall t\, v.\ \mathsf{store}\ s\ r\ t\ v \supset \\ \qquad \forall n.\ \mathsf{stable}\ s\ \mathsf{F}\ (t{+}2)\ n \wedge \mathsf{stable}\ r\ \mathsf{F}\ (t{+}2)\ n \supset \\ \qquad\quad \mathsf{stable}\ q\ v\ (t{+}2)\ n \wedge \mathsf{stable}\ \overline{q}\ \neg v\ (t{+}2)\ n \end{array}$$

This abstract specification states that if the value v is stored in the latch at time t then the values v and $\neg v$ will be held stable on the outputs q and $\overline{q}$ from time $t{+}2$, for as long as both s and r are held low.

This specification defines the behaviour of the latch only for well behaved set and reset inputs, s and r. The predicate RS_spec does not state, for example, what the behaviour of the device will be if s and r are high simultaneously for some period of time. While s and r are both high, the antecedant 'store s r t v' will be false; the specification will therefore be satisfied during this period, regardless of the values of q and $\overline{q}$. Thus, the partial specification RS_spec leaves unspecified—i.e. unconstrained—the values of the outputs q and $\overline{q}$ during periods of time when both s and r are high.

The design description RS_imp, however, does *not* leave unconstrained the values of q and $\overline{q}$ during periods when both s and r are high; the values of these outputs are well defined at all times by the equations for the two nor-gates that implement the latch. To reflect this difference between RS_imp and RS_spec, the correctness statement is formulated as an implication:

$$\forall s\, r\, q\, \overline{q}.\ \mathsf{RS\_imp}(s,r,q,\overline{q}) \supset \mathsf{RS\_spec}(s,r,q,\overline{q})$$

That is, RS_spec is a behavioural abstraction of RS_imp that agrees with RS_imp on the values of q and $\overline{q}$ whenever the inputs s and r are 'well behaved'. The proof of this correctness statement is straightforward; the main step is an induction on the variable n in the definition of RS_spec.

This simple example, in which a partial specification defines the device's output signals for only well behaved input signals, is typical of correctness proofs involving behavioural abstraction, and shows how using logical implication to formulate correctness makes it possible relate such simplified specifications to more detailed design descriptions.

6 Formalising Data Abstraction

In the examples given above, values present on the ports of the devices have been modelled by booleans—i.e. values of type '*bool*'. Other logical types, however, can also be used to model the range of values on the ports of a device. For example, a type of 8-bit words, '*word8*', can be introduced into the logic and used to specify the behaviour of the serial-in parallel-out shift register shown below.

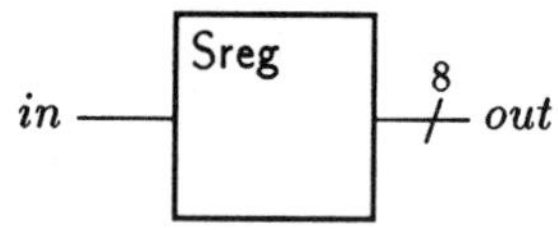

The behaviour of this device can be specified by a predicate Sreg(in, out), where in is a function of type '$num \rightarrow bool$' and out is a function of type '$num \rightarrow word8$'. Here, the logical type '$word8$' is used as an abstract data type in the specification of the shift register.

With such data types, correctness statements involving data abstraction can be formulated by using abstraction functions that map the data types of implementation descriptions to those of abstract specifications. There are two distinct tasks involved in using data abstraction in the correctness proof of a device:

- Constructing the logical types that will be used as abstract data types to describe the behaviour of the design and its specification.
- Defining abstraction mappings from the data types of the design description to those of the specification, and proving that the operations performed by the device on the detailed level data types correctly implement the specified operations on the abstract level data types.

The first of these will not be discussed in detail here. New logical types can be consistently introduced into the logic by constructing them from 'subsets' of existing types. The newly introduced types can then be characterised by a set of theorems—derived from their construction—that serve as 'axiomatizations' of the types. For details about introducing new types into the logic, see [2,3].

Once logical types have been defined for both the specification and the design description of a device, a correctness statement using data abstraction can be formulated. Consider, for example, a predicate Imp(a, b) describing the design of a device having values of type $ty1$ on its external ports. If Imp is a sequential device, the variables a and b will be signals of type '$num \rightarrow ty1$'. Suppose that the abstract specification for this device is given by a predicate Spec(a', b'), where a' and b' are signals of type '$num \rightarrow ty2$'. To prove Imp correct with respect to Spec, we must define an appropriate abstraction function f: $ty1 \rightarrow ty2$ to map values of type $ty1$ on the ports of Imp to values of type $ty2$ on the ports of Spec. Using f, the correctness statement that we want to prove is:

$$\forall a\, b.\ \mathsf{Imp}(a, b) \supset \mathsf{Spec}(\mathsf{f} \circ a, \mathsf{f} \circ b)$$

This states that if signals a and b satisfy Imp then the abstracted signals f $\circ\, a$ and f $\circ\, b$ will satisfy Spec. The function f maps values of type $ty1$ to values of type $ty2$; function composition with f simply translates the *sequences* of values of type $ty1$, given by a and b, to the corresponding sequences of values of type $ty2$.

In this simple example there is a one-to-one correspondence between the signals of the design description Imp and those of the specification Spec. In general, however, this may not be the case. An abstract signal of the specification of a device may be constructed from the values of several signals of its implementation description. An implementation of the Sreg device above, for example, could have eight boolean outputs that are mapped by an abstraction function to the single '*word8*' output of Sreg.

6.1 An Example

The following example illustrates the use of data abstraction in the correctness proof of a CMOS inverter. We first give the specifications of the components of the inverter, beginning with the N-type transistor:

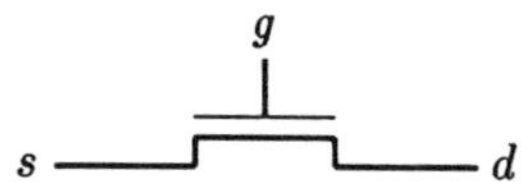

which can be specified by the predicate Ntran, defined as follows:

$$\mathsf{Ntran}(g, s, d) \equiv \forall t.\, g(t){=}\mathsf{H} \vee (g(t){=}\mathsf{Z} \wedge g(t{-}1){=}\mathsf{H}) \supset s(t){=}d(t)$$

Here, we are using the three-valued type '*tri*', with values H (*high*), L (*low*) and Z (*floating*), to model the values present on the lines g, s and d. The predicate Ntran asserts that the values at the source s and the drain d will be equal whenever 1) the gate g is high or 2) the gate is floating and the gate was high at the previous moment of time. Thus, Ntran specifies a transistor that 'stores' the value on its gate for one unit of discrete time.[1]

The specification of a P-type transistor is similar to Ntran. It is:

$$\mathsf{Ptran}(g, s, d) \equiv \forall t.\, g(t){=}\mathsf{L} \vee (g(t){=}\mathsf{Z} \wedge g(t{-}1){=}\mathsf{L}) \supset s(t){=}d(t)$$

This specification asserts that the source and drain are connected if the gate is driven low or the gate is floating and was low at the previous instant of time.

Two other components are used in the inverter: Pwr and Gnd. These model the power and ground nodes respectively, and are defined by:

$$\mathsf{Pwr}(w) \equiv \forall t.\, w(t) = \mathsf{H} \qquad \text{and} \qquad \mathsf{Gnd}(w) \equiv \forall t.\, w(t) = \mathsf{L}$$

Using these parts, the standard design of an inverter can be described by the predicate Imp, defined as follows:

$$\mathsf{Imp}(i, o) \equiv \exists p\, g.\, \mathsf{Pwr}(p) \wedge \mathsf{Gnd}(g) \wedge \mathsf{Ptran}(i, p, o) \wedge \mathsf{Ntran}(i, g, o)$$

[1] This model of transistor behaviour is, of course, very much simplified—but it will serve for the purposes of this example. For a better transistor model see, for example, [11]

The predicate Imp defines the behaviour of the inverter in terms of values of type '*tri*'. Suppose, however, that we wish to show that this implementation satisfies a specification Inv, defined in terms of boolean values on the external ports:

$$\mathsf{Inv}(i, o) \equiv \forall t.\, o(t) = \neg i(t)$$

Here, the signals i and o have type '*num→bool*'; in the implementation, however, they have type '*num→tri*'. To abstract from three-valued to boolean signals, we will define a data abstraction function f:*tri→bool* as follows:

$$\mathsf{f}(v) = \varepsilon b.\, (b \Rightarrow v{=}\mathsf{H} \mid v{=}\mathsf{L})$$

From this definition, it follows that f(H) = T and f(L) = F. The use of the ε-operator makes the value of f(Z) be 'εb.F', i.e. a fixed but unknown value of type '*bool*'.

Using f, we can formulate the correctness of Imp as follows:

$$\mathsf{Def}(i) \wedge \mathsf{Imp}(i, o) \supset \mathsf{Inv}(\mathsf{f} \circ i,\ \mathsf{f} \circ o)$$

This correctness statement asserts that if the condition Def holds of the input i then Imp is correct with respect to the specification Inv. The condition 'Def(i)' is a ***validity condition*** on the correctness of the inverter, defined by:

$$\mathsf{Def}(i) \equiv \forall t.\, i(t){=}\mathsf{H} \vee i(t){=}\mathsf{L}$$

The condition Def(i) ensures that the input i is never floating—which must be the case for the predicate Inv to represent a valid abstraction of the inverter's behaviour. If the inverter is used in an environment in which Def(i) is satisfied, we know that the abstraction is valid and we can use the simplified behaviour represented by Inv in place of the more detailed behaviour given by Imp. This can help to simplify the descriptions of designs containing inverters that are always strongly driven.

7 Formalising Temporal Abstraction

Temporal abstraction involves relating formal specifications that describe hardware behaviour using different notions of discrete time. This type of abstraction is used when a design description gives more detail about how a device behaves over time than is desired for its abstract specification. With the mechanism of temporal abstraction, information about a device's behaviour at moments of time that are not of interest can be hidden from the abstract specification, allowing the specification to concentrate on how the device behaves at significant or 'interesting' points of time.

In the simplest case of temporal abstraction, each single unit of discrete time at the abstract level of description corresponds to several units of time at the more detailed level of description. Here, the abstract specification of a device defines its behaviour at fewer points of time than the design description—i.e. the grain of discrete time is 'coarser' at the abstract level than at the detailed level.

To express this in logic, we must specify for each single *unit* or step of 'high level' time a corresponding *interval* or sequence of steps of 'low level' time. Suppose, for example, the unit of time from t' to $t'+1$ at the abstract level of description corresponds to the interval of time from t to $t+n$ at the more detailed level of description. In this case, high level time t' corresponds with low level time t and high level time $t'+1$ corresponds with low level time $t+n$. This correspondence can be specified formally by defining a mapping f from points of high level time to points of low level time such that:

$$f(t') = t \quad \text{and} \quad f(t'+1) = t+n$$

We can use such a mapping $f{:}num{\rightarrow}num$ to specify any correspondence between successive units of high level time and contiguous intervals of low level time:

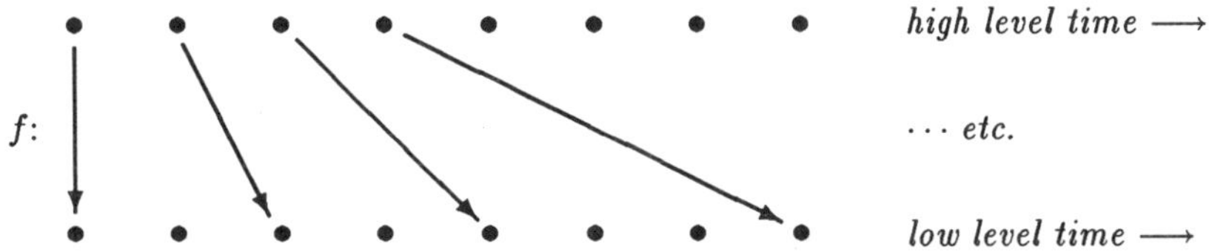

Not every function of type $num{\rightarrow}num$, however, specifies a valid temporal abstraction; we require that the time mapping function be *increasing*. This ensures that if time t_2 comes after time t_1 at the abstract level, then this relationship also holds between the corresponding points of time at the detailed level. The requirement that a time mapping, f, be increasing can be expressed using the predicate Incr, defined as follows:

$$\mathsf{Incr}(f) \equiv \forall n\, m.\ (n > m) \supset (f(n) > f(m))$$

We now consider how an increasing temporal abstraction function, such as f above, can be used to formulate correctness statements that involve specifications at different grains of time. Suppose that $sig{:}num{\rightarrow}bool$ is a signal involved in the detailed level description of a design. The function sig gives detailed information about the sequence of values present on a port of the device. The device's abstract specification will involve a corresponding signal, $sig'{:}num{\rightarrow}bool$ say, that gives the port's value at just those points of time that are significant at the abstract level. If f is an increasing function that specifies the correspondence between high and low level time, then the relationship that must hold between sig' and sig is:

$$\forall t.\ sig'(t) = sig(f(t)) \qquad (\text{i.e. } sig' = sig \circ f)$$

That is, at each instant of high level time the value of the abstract signal sig' must be the same as the value of the detailed signal sig at the corresponding instant of low level time.

Given that this correspondence must hold between the signals of an implementation description and those of an abstract specification, the required correctness relation is formulated as follows. Suppose that $\mathsf{Imp}(a, b)$ is the implementation description of a device, $\mathsf{Spec}(a, b)$ is its abstract specification, and f is a term that denotes a mapping of type $num \rightarrow num$. The specification Spec is related to the design description Imp by the temporal abstraction given by f if we can prove a correctness statement of the following form:

$$\mathsf{Incr}(\mathsf{f}) \wedge \forall a\, b.\ \mathsf{Imp}(a, b) \supset \mathsf{Spec}(a \circ \mathsf{f},\ b \circ \mathsf{f})$$

This states that whenever the signals a and b satisfy the temporally detailed specification Imp, then the abstracted signals $a \circ \mathsf{f}$ and $b \circ \mathsf{f}$ will satisfy the temporally abstract specification Spec.

The predicate Imp specifies the values allowed on the ports of the device at each instant of fine-grained time; the predicate Spec specifies the desired behaviour in terms of the values allowed on the ports at only some of these points of time. The function denoted by f defines the sequence of low level time points that lie between each unit of high level time. Proving this correctness statement involves showing that if a and b have the intermediate values allowed by Imp then the values of a and b at the abstracted time points will be allowed by Spec.

Notice that this correctness relation is formulated as an implication, rather than an equivalence. This reflects the idea that there may be several non-equivalent methods of implementing the device specified by Spec; the implementation in which the ports a and b take on the intermediate values defined by Imp is only *one* such method.

It is significant that this formulation of the temporal abstraction relation between Imp and Spec does not state that

- Spec is *a* temporal abstraction of Imp

but rather states that

- Spec is *the* temporal abstraction of Imp *given by* f

That is, the exact correspondence between time scales, given by the value of f, is an integral part of the statement of the correctness relation between Imp and Spec. There are several reasons why this information must be retained in the formulation of the correctness relation. One of these is to make it possible to 'compose' the correctness results that link several levels of structural and temporal abstraction, yielding one correctness statement relating the lowest level design description to the highest level abstract specification.

In the example given above, both signal a and signal b were abstracted using the same temporal abstraction function f. In general, however, it is not necessary that the same abstraction function be used for every signal of a device.

For example, some signals may be abstracted at points of time corresponding to rising edges of a clock, while others are abstracted at points of time corresponding to falling edges. If, however, different abstraction functions are used for different signals of a device in the same correctness statement, there are certain reasonablity conditions that must be imposed on the combination of functions used, beyond simply requiring that each of them be increasing. These constraints are imposed order to prevent apparent 'computation' being done by the abstraction step itself—which is only supposed to remove or restructure information.

7.1 Constructing Temporal Abstraction Functions

The first step in the formulation and proof of a correctness relation involving temporal abstraction is to construct the appropriate abstraction function mapping high level time to low level time. This can often be done most easily by simply identifying those points of low level time that will correspond to points of high level time—i.e. by identifying those points of low level time that will be 'mapped to' by the temporal abstraction function.

The advantage of defining abstraction functions this way is that some of the signals of the implementation description can themselves be used to identify the points of low level time that will correspond to points of high level time. In synchronous systems, the appropriate points of low level time can often be identified by the value of the clock; in asynchronous systems, handshaking signals can be used for the same purpose.

To construct an abstraction function f from high level time to low level time, it is sufficient to define a function $p{:}num{\rightarrow}bool$ that is true of just those points of low level time that are to correspond to points of high level time:

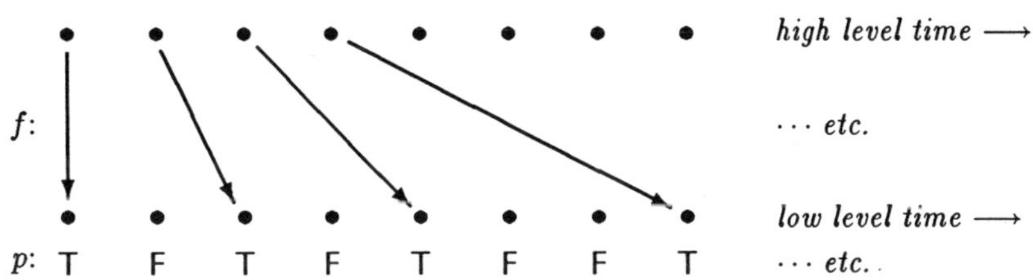

From such a function p, it is possible to derive the abstraction function f as follows. If the term 'Timeof $p\ n$' denotes the point of low level time at which p is true for the nth time, then f can be defined by:

$$\forall t.\ f(t) = \mathsf{Timeof}\ p\ t \qquad \text{(i.e. } f = \mathsf{Timeof}\ p\text{)}$$

It remains to define formally the function Timeof.

7.1.1 Defining Timeof

We first observe that the value of the term 'Timeof p n', as informally described above, may be undefined for some values of p and n. If the signal p is not true at an infinite number of time points, then there will exist some N such that p is true only N times. For all $n > N$, there will not exist points of time at which p is true for the nth time and 'Timeof p n' will be undefined. Thus, the value of 'Timeof p' will be a partial function. Such partial functions, however, are not directly definable in the logic; all functions defined must be total. The definition of Timeof will therefore be based on a *relation*, Istimeof, which can be defined directly.

The term 'Istimeof p n t', will have the meaning 'p is true for the nth time at time t'. The formal definition of the relation Istimeof is done by primitive recursion on the variable n. In the case where n is zero, the definition is:

$$\mathsf{Istimeof}\, p\, 0\, t \equiv p(t) \wedge \forall t'.\, t'<t \supset \neg p(t')$$

That is, time t when p is true for the first (i.e. the 0th) time if p is true at time t and is false at all times before t.

For the case when when p is true for the $(n+1)$th time, the definition is:

$$\begin{array}{l}\mathsf{Istimeof}\, p\, (n+1)\, t \equiv \\ \quad p(t) \wedge \exists t'.\, t'<t \wedge \mathsf{Istimeof}\, p\, n\, t' \wedge \forall t''.\, (t'<t'' \wedge t''<t) \supset \neg p(t'')\end{array}$$

meaning that t is the time when p is true for the $(n+1)$th time if p is true at time t, and there exists a time t' before t when p is true for the nth time, and p is false at all times between t' and t.

This primitive recursive definition captures the idea that 'p is true for the nth time at time t'. There is no guarantee, however, that such times t exist for all values of p and n. In order to use this definition, it is necessary to show that if the signal p is true 'infinitely often' then for all n there is a unique time t at which p is true for the nth time.

The condition that p must be true at an infinite number of points can be formalised by the predicate Inf defined by:

$$\mathsf{Inf}(p) \equiv \forall t.\, \exists t'.\, t'>t \wedge p(t')$$

The predicate Inf can be used to express the fact that if p is true infinitely often then the time at which p is true for the nth time exists for all n:

$$\mathsf{Inf}(p) \supset \forall n.\, \exists t.\, \mathsf{Istimeof}\, p\, n\, t \qquad (1)$$

The proof of this statement proceeds by induction on n and use of the well ordering property of natural numbers:

$$\forall P.\, (\exists n.\, P\, n) \supset (\exists n.\, P\, n \wedge \forall n'.\, n'<n \supset \neg P\, n)$$

It is also the case that Istimeof defines a *unique* time t for each value of n. Formally:

$$\forall t_1\, t_2.\ \mathsf{Istimeof}\ p\ n\ t_1 \wedge \mathsf{Istimeof}\ p\ n\ t_2 \supset (t_1 = t_2) \qquad (2)$$

Given these theorems, the relation Istimeof can now be used to define the function Timeof. Using the ε-operator the definition of Timeof is given by:

$$\forall p\, n.\ \mathsf{Timeof}\ p\ n = \varepsilon t.\ (\mathsf{Istimeof}\ p\ n\ t)$$

That is, 'Timeof $p\ n$' denotes some time, say t, such that 'Istimeof $p\ n\ t$' is true or, if no such time exists, 'Timeof $p\ n$' denotes an arbitrary natural number. With the use of the ε-operator, this definition makes the term 'Timeof p' always denote a total function; 'Timeof $p\,n$' is defined for all values of n, even when the signal p is true at only a finite number of points.

If, however, the signal p is true infinitely often, then for all n there will exist a time t such that 'Istimeof $p\,n\,t$' is true, and this time will be unique. Thus, if Inf(p) holds, 'Timeof $p\,n$' will in fact denote the unique time at which p is true for the nth time, as desired. More formally, an immediate consequence of (1) is:

$$\mathsf{Inf}(p) \supset \mathsf{Istimeof}\ p\ n\ (\mathsf{Timeof}\ p\ n)$$

which gives the following easily-proved lemma:

$$\mathsf{Inf}(p) \supset p(\mathsf{Timeof}\ p\ n)$$

Using (2) the following two lemmas can also be proven:

$$\mathsf{Inf}(p) \supset (\mathsf{Timeof}\ p\ n) < (\mathsf{Timeof}\ p\ (n+1))$$

$$\mathsf{Inf}(p) \supset \forall t.\ (\mathsf{Timeof}\ p\ n) < t \wedge t < (\mathsf{Timeof}\ p\ (n+1)) \supset \neg p(t)$$

These lemmas can be conveniently collected together using the predicate Next defined by:

$$\mathsf{Next}\ t_1\ t_2\ sig \equiv t_1 < t_2 \wedge sig(t_2) \wedge \forall t.\ (t_1 < t \wedge t < t_2) \supset \neg sig(t)$$

giving the following theorem

$$\mathsf{Inf}(p) \supset \forall n.\ \mathsf{Next}\ (\mathsf{Timeof}\ p\ n)\ (\mathsf{Timeof}\ p\ (n+1))\ p \qquad (3)$$

This theorem states that if p is true infinitely often then 'Timeof p' is well defined and denotes the desired function from high level time points to low level time points.

7.1.2 Using Timeof to Define Abstractions

Having formally defined the function Timeof and shown that it is well defined for functions p that are true at an infinite number of points, it is possible to use Timeof to define temporal abstraction functions for statements of correctness.

A temporal abstraction function that maps high level time to low level time is just an increasing function $f{:}num{\rightarrow}num$. Any such function can be defined using Timeof and an appropriate function p that indicates the points of low level time that are abstracted to points of high level time. Formally:

$$\forall f.\ \mathsf{Incr}(f) \equiv \exists p.\ \mathsf{Inf}(p) \wedge f{=}\mathsf{Timeof}\ p$$

This theorem follows from the definition of Incr and the properties of Timeof given by theorem (3).

With this theorem in mind, we can use Timeof to construct an alternative formulation of temporal abstraction as follows. If p is a term that denotes a function indicating the points of low level time that are to be abstracted to points of high level time, then the correctness relation that must hold between a specification Spec and an implementation Imp is:

$$\mathsf{Inf(p)} \wedge \forall a\, b.\ \mathsf{Imp}(a,b) \supset \mathsf{Spec}(a \circ (\mathsf{Timeof\ p}),\ b \circ (\mathsf{Timeof\ p})) \qquad (4)$$

This states that Imp is correct with respect to the temporally abstract specification Spec if whenever a and b satisfy Imp then the abstract signals constructed by sampling a and b when p is true will satisfy Spec. If 'when' is an infix operator defined as follows:[2]

$$\forall s\, t.\ s \mathbin{\mathsf{when}} t = s \circ (\mathsf{Timeof}\ t)$$

then this correctness statement can be written:

$$\mathsf{Inf(p)} \wedge \forall a\, b.\ \mathsf{Imp}(a,b) \supset \mathsf{Spec}(a \mathbin{\mathsf{when}} \mathsf{p},\ b \mathbin{\mathsf{when}} \mathsf{p})$$

Since $\forall p.\ \mathsf{Inf}(p) \supset \mathsf{Incr}(\mathsf{Timeof}\ p)$, the correctness statement (4) implies a correctness statement of the old form:

$$\mathsf{Incr(f)} \wedge \forall a\, b.\ \mathsf{Imp}(a,b) \supset \mathsf{Spec}(a \circ \mathsf{f},\ b \circ \mathsf{f})$$

with f=Timeof p. Furthermore, since every increasing function f can be expressed using 'Timeof p' with an appropriate function p, this alternative form of correctness relation for temporal abstraction is essentially equivalent to the original one.

The following example shows how Timeof can be used in the correctness proof of a unit delay register, and describes some ways in which the ideal form of correctness given above must be modified in practice.

[2] In previous work, I have called this operator 'Abs'. The mnemonicly superior name 'when' was suggested by the work of Halbwachs, Lonchampt and Pilaud [5].

7.2 An Example

A commonly used register-transfer level device is the unit delay:

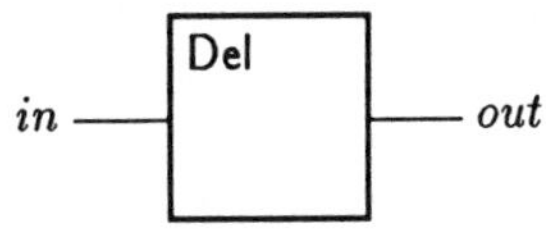

which can by specified in higher order logic by:

$$\mathsf{Del}(in, out) \equiv \forall t.\, out(t{+}1) = in(t)$$

The unit delay device is a temporal abstraction; its implementation is specified at a more detailed grain of time. One possible implementation is the rising edge triggered D-type flip flop:

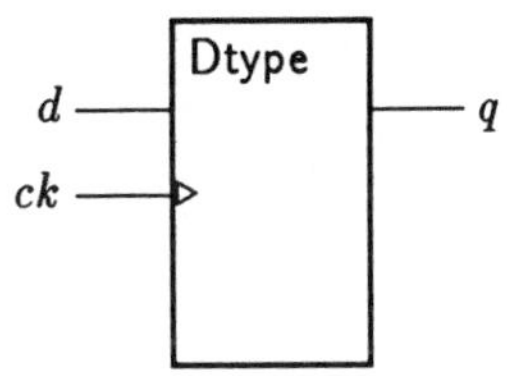

which, ignoring detailed timing information,[3] can be specified by the predicate Dtype defined as follows:

$$\mathsf{Dtype}(ck, d, q) \equiv \forall t.\, q(t{+}1) = (\mathsf{Rise}\; ck\; t \Rightarrow d(t) \mid q(t))$$

where the term '$\mathsf{Rise}\; ck\; t$' has the meaning 'the clock, ck, rises at time t':

$$\mathsf{Rise}\; ck\; t \equiv \neg ck(t) \wedge ck(t{+}1)$$

We want to show that the D-type flip flop is a correct implementation of the register-transfer level unit delay. Using the formulation of correctness for temporal abstraction described above, we must prove a correctness statement of the following form:

$$\mathsf{Inf}(\mathsf{p}) \wedge \forall\, ck\; d\; q.\, \mathsf{Dtype}(ck, d, q) \supset \mathsf{Del}(d\; \mathsf{when}\; \mathsf{p},\; q\; \mathsf{when}\; \mathsf{p})$$

where p is a term that denotes a function of type '$num{\to}bool$'. It is not possible, however, to prove a correctness statement of this form; for we can show that:

$$\neg\exists\, p.\, \mathsf{Inf}(p) \wedge \forall\, ck\; d\; q.\, \mathsf{Dtype}(ck, d, q) \supset \mathsf{Del}(d\; \mathsf{when}\; p,\; q\; \mathsf{when}\; p)$$

[3] For clarity, a greatly simplified specification of Dtype is used in this example. For D-type flip flop specifications that include information about detailed timing behaviour see [6] or [8].

The trouble is that the abstract time scale for Del must be defined in terms of the value of the clock, ck. Informally, Dtype implements a unit delay by sampling its input when the clock rises and holding this value on the output until the next rise of the clock. In this way the D-type delays by one clock period the *sequence* consisting of the values of the signal d at the times of successive clock rises.

This suggests that the temporal abstraction function for this proof should map successive points of high level time to the successive points of low level time at which the clock rises. Using 'Rise ck', the required temporal abstraction function is given, for each value of ck, by the term 'Timeof(Rise ck)'. Using this, we can reformulate the correctness statement for the Dtype implementation of Del as:

$$\begin{array}{l} \forall ck.\ \mathsf{Inf}(\mathsf{Rise}\ ck) \wedge \\ \qquad \forall d\, q.\ \mathsf{Dtype}(ck, d, q) \supset \mathsf{Del}(d\ \mathsf{when}\ (\mathsf{Rise}\ ck),\ q\ \mathsf{when}\ (\mathsf{Rise}\ ck)) \end{array}$$

This states that the clock rises infinitely often, and whenever the signals ck, d and q are in the Dtype relationship, the abstracted signals—i.e. the signals d and q sampled at successive rises of the clock—will be in the Del relationship. Thus there is a 'family' of sampling functions defined by 'Rise ck', giving a temporal abstraction from Dtype to Del for each value of the clock ck.

Because we are not interested in constructing temporal abstraction functions for signals that do not satisfy Dtype(ck, d, q), this correctness statement can be further modified to have the following form:

$$\begin{array}{l} \forall ck\, d\, q.\ \mathsf{Dtype}(ck, d, q) \supset \\ \qquad \mathsf{Inf}(\mathsf{Rise}\ ck) \wedge \mathsf{Del}(d\ \mathsf{when}\ (\mathsf{Rise}\ ck),\ q\ \mathsf{when}\ (\mathsf{Rise}\ ck)) \end{array}$$

As it stands, however, this correctness statement is false; for Dtype does not constrain the value of ck, and it is not the case that Inf(Rise ck) holds for all possible values of ck. The correctness statement will therefore be modified to have Inf(Rise ck) as an assumption, giving:

$$\begin{array}{l} \forall ck.\ \mathsf{Inf}(\mathsf{Rise}\ ck) \supset \\ \qquad \forall d\, q.\ \mathsf{Dtype}(ck, d, q) \supset \mathsf{Del}(d\ \mathsf{when}\ (\mathsf{Rise}\ ck),\ q\ \mathsf{when}\ (\mathsf{Rise}\ ck)) \end{array} \tag{5}$$

which is the final form of the correctness statement for this device.

The proof of (5) is straightforward. The main step is an induction on the number of time steps between adjacent rises of the clock, showing that the value of d that is sampled at a rise of ck is held on q until the next rise. The correctness statement follows easily from this result and the properties of Timeof given by (3).

The assumption that the clock rises infinitely often is a *validity condition* on the temporal abstraction; the correctness statement asserts that Del presents a valid abstract view of Dtype provided the condition 'Inf(Rise ck)' is satisfied. To use the Del abstraction, this condition on the clock must be met by the environment in which the Dtype is placed. The validity condition 'Inf(Rise ck)' is as unrestrictive as possible, given the simple D-type model used. The clock ck is not required to be regular or to have a minimum clock period; the assumed liveness condition is sufficient for the proof to go through.

This proof gives a very simple example of the most common type of temporal abstraction, where contiguous intervals of low level time correspond to successive units of high level time. Examples involving detailed timing information or several different temporal abstractions in the same proof are more complex but involve the same general approach as this simple example.

There are other types of temporal abstraction that can be used to formulate correctness statements, but that can not be constructed using Timeof. In one form of temporal abstraction, for instance, the occurrence of an event at any time during an interval of low level time corresponds to a signal being true at a point of high level time. This type of abstraction can be used to deal with asynchronous inputs to synchronous systems, hiding from the system's abstract specification irrelevant information about the exact timing of the asynchronous input signals.

8 Conclusion

In hardware verification, designs that are proven correct ought to have formal descriptions that model reality reasonably well. The more accurate the model of device behaviour, the less likely it is for design errors to escape discovery by formal verification. Formal specifications, on the other hand, must be clear and concise, in order to be seen to reflect the designer's intent. This means that most of the details of a device's behaviour must be left out of its formal specification; only the essentials can be included. The idea of abstraction is therefore fundamental to a formalisation of hardware design correctness.

This paper has shown how four basic abstraction mechanisms for hardware verification can be formalised in higher order logic and used to control the size and complexity of device specifications. For clarity, the examples given were kept very simple; each example involved only one simple abstraction step. More substantial examples have been done, in which a series of nested abstractions is used to relate the bottom level design description to the top level abstract specification.

The role of abstraction in hardware verification is not restricted to that of complexity control. Structuring a correctness proof to span several levels of abstraction allows it to serve as structured *documentation* for a design. The top level specification gives a concise description of the essential features of the device; if more information about its operation is required, this is provided by the specifications at the next lower level of abstraction. If still more information is desired, it can be obtained from the next level of abstraction, and so on, right down to the level of primitive components. The abstraction mechanisms that are used in the proof serve to explain how each level of abstraction is related to the next lower level.

Abstractions often involve *validity conditions* that state when a specification represents a valid abstract view of a more detailed design description. Formal correctness proofs make such abstraction validity conditions explicit; for the proofs to go through, they must either be satisfied or included as assumptions. This means that a formal proof clearly documents validity assumptions that might otherwise lead to misunderstandings by remaining implicit. The conditions under which a device's abstract specification reflects the actual behaviour of the device will be clear from the assumptions present in its proof of correctness.

Validity conditions that arise in the correctness proofs of some of the components of a device can sometimes help to simplify the task of writing specifications for other components of the device. For example, the validity condition 'Inf(Rise ck)' generated in the Dtype proof simplifies the task of specifying the desired behaviour of clock generator circuits for use with the Dtype device. We know that the clock signal must satisfy this condition; we can therefore take the abstract specification of the clock generator to be just the validity condition itself.

Some abstractions can be seen as specifications of the 'protocol' for communication at the external ports of a device. Consider, for instance, the temporal abstraction given for the D-type flip flop implementation of the unit delay. From the correctness statement, it is clear that values are input to the delay device only at rising edges of the clock. The abstract specification of this device states that the value input to the device is delayed by one unit of time; the temporal abstraction used in the correctness statement explains what it *means* for a value to be 'input to the device'.

Abstraction mechanisms similar to the ones discussed in this paper are likely to play an important role in the automatic *synthesis* of designs. In this paper, abstraction mechanisms were described as *removing* information from design descriptions to yield abstract specifications. Synthesis of designs involves the opposite process—adding detail to formal specifications until an implementable design description is obtained. The abstraction mechanisms commonly used in post-design verification may also suggest heuristics for design synthesis from formal specifications.

Acknowledgements

Thanks are due to the members of the Cambridge Hardware Verification Group, especially Albert Camilleri, Francisco Corella, Inder Dhingra, Mike Gordon and Jeff Joyce, who made valuable comments on drafts of this paper. I would also like to thank Graham Birtwistle of the University of Calgary for his suggestions. This research is funded by the Royal Commission for the Exhibition of 1851.

References

[1] Camilleri, A.J., Gordon, M.J.C. and Melham, T.F., 'Hardware Verification Using Higher Order Logic,' to appear in: *Proceedings of the IFIP International Conference: From H.D.L. Descriptions to Guaranteed Correct Circuit Designs*, Grenoble, September 9-11, 1986.

[2] Gordon, M.J.C., 'HOL: A Machine Oriented Formulation of Higher Order Logic,' Technical Report No. 68, Computer Laboratory, University of Cambridge, 1985.

[3] Gordon, M.J.C., 'HOL: A Proof Generating System for Higher Order Logic,' in: *VLSI Specification, Verification and Synthesis*, G.M. Birtwistle and P.A. Subrahmanyam, eds., (this volume).

[4] Gordon, M.J.C., 'Why Higher Order Logic is a Good Formalism for Specifying and Verifying Hardware,' in: *Formal Aspects of VLSI Design: Proceedings of the 1985 Edinburgh Conference on VLSI*, G.J. Milne and P.A. Subrahmanyam, eds., North-Holland, Amsterdam, 1986.

[5] Halbwachs, N., Lonchampt, A. and Pilaud, D., 'Describing and Designing Circuits by means of a Synchronous Declarative Language,' to appear in: *Proceedings of the IFIP International Conference: From H.D.L. Descriptions to Guaranteed Correct Circuit Designs*, Grenoble, September 9-11, 1986.

[6] Hanna, F.K. and Daeche, N., 'Specification and Verification using Higher-Order Logic: A Case Study,' in: *Formal Aspects of VLSI Design: Proceedings of the 1985 Edinburgh Conference on VLSI*, G.J. Milne and P.A. Subrahmanyam, eds., North-Holland, Amsterdam, 1986.

[7] Hanna, F.K. and Daeche, N., 'Specification and verification of digital systems using higher-order predicate logic,' *IEE Proceedings-E*, Vol. 133, Part E, No. 5, September 1986, pp. 242-254.

[8] Herbert, J.M.J., Ph.D. Thesis, The University of Cambridge, to appear in 1987.

[9] Hoare, C.A.R., 'A Calculus of Total Correctness for Communicating Processes,' *Science of Computer Programming*, Vol. 1, No. 1, 1981.

[10] Leisenring, A., *Mathematical Logic and Hilbert's ε-Symbol*, Macdonald and Co. Ltd., London, 1969.

[11] Winskel, G., 'Lectures on models and logic of MOS circuits,' Proceedings of the Marktoberdorf International Summer School on Logic Programming and Calculi of Discrete Design, July 1986.

10

Formal Validation of an Integrated Circuit Design Style

I. S. Dhingra[1]

University of Cambridge
Computer Laboratory
Corn Exchange Street
Cambridge CB2 3QG

December, 1986

1 Introduction

A specific design style is only ever used if it meets the required needs of the task in hand. The task in hand now is that of generating large, complex, application specific systems on silicon in a fairly short space of time with the confidence that they will perform to the required specification. In the past the development of a large circuit might have been done using a team of engineers over a period of few years, *e.g.* the development of the 68000 microprocessor. This method of circuit development is not acceptable in the present day due to the time and manpower spent in iterating to get the design correct. What is needed is a design technique which is easy to follow and gives very high degree of confidence in the first time correct implementation of the circuit.

Before such a design technique can be developed we need to look at the sort of mistakes that are made which slow down the development of the circuit. The sort of errors that are generally made can be classified into two categories—*timing errors* and *logical errors*. Designers use simulators to model the behaviour of circuits before their implementation. The critical point to note here is that the degree of confidence one has in the design after it has been simulated is no more than the confidence one has in the model of the primitives used in the simulation. The sort of errors that are not easy to catch at the simulation stage are the complex timing errors. This is because the process of manufacturing is not ideal and so there are slight differences from one batch of devices to the next.

So, the sort of design technique needed is a synchronous design scheme. Such a design scheme dictates that all storage elements be updated by the use of a

[1] The author is jointly funded by the British Science and Engineering Research Council, and Racal Research based in Reading, England.

global clock. This reduces the timing problems to the areas between the clocked elements and the problems of identifying difficult timing paths and race hazards is simplified. It also frees the designer to think more about the algorithm and the logical design of the system rather than the detailed timing requirements of the logic.

The price paid for this simplification of the design style is an increase in the size of the chip. This price is probably worth paying since the adoption of a difficult asynchronous design would probably lead to the introduction of obscure timing errors which may be difficult to model and may not be discovered until the chip has been fabricated.

The technology we are interested in is CMOS, and associated with it are hazards not encountered with the simple NMOS technology. For example in CMOS all latches are clocked with a pair of clock lines which are inverses of each other. The constraints on the clock lines are that the overlap between the clocks and the rise and fall times of the clock edges should be less than the delay across a gate (*i.e.* less than a few nano-seconds). This is a stringent requirement for a VLSI technology and may be very difficult to meet as the device geometries get smaller and the capacitive and resistive effects due to the clock lines become more significant. This not only introduces clock stagger across the chip but also degrades the clock rise and fall times—both of which are hazardous and may cause the latches to become transparent for a short time and hence corrupt data. To over come the problem of clock stagger on a large chip requires fine tuning of the clock delays across the chip which would be difficult to model at the design stage and still provide no guarantee as to the probability of it working. To overcome the poor clock rise and fall times requires one to use excessively large clock drivers and clock buffers at regular intervals which leads to rather complex clock structures. Unfortunately it is also desirable not to have very fast clock rise and fall times, since this generates electromagnetic interference and hence causes neighbouring lines to get corrupted, a highly undesirable feature!

The solution to these problems is to do with the design style, not the deployment of clever distributed clock drivers. Presented here is a technique for formally analysing such design styles. Our work is based on a design style known as CLIC [8] which is tolerant to considerable clock overlap and slow clock rise and fall times. It is an extension of the work done by Goncalves and De Man [4] where they present a design style they call NORA that generalises yet another design style known as DOMINO [7].

Common to all of these design styles is the use of dynamic logic rather than static logic. The reason for this comes from the fact that in static CMOS, all logic functions are duplicated—once using the p-type transistors and then again using the n-type transistors. This has the advantage that the static power consumption of the circuit is very low, but it also means that almost twice the amount of silicon area is being used than necessary. However in the context of a synchronous system we have a continuously running global clock which could be used for other purposes: *i.e.* instead of using static logic blocks we use dynamic logic blocks which are clocked by the global clock. This does increase the power consumption a little but it is still less than the power that may be used by a similar NMOS circuit.

This dynamic design feature was used in the DOMINO design style which only used the n-type gates and provided only non-inverting logic. The NORA design style on the other hand used both n-type and p-type devices and so provided the full freedom of inverting and non inverting logic. However a requirement with the NORA design technique was that the clock rise and fall times should be kept fairly small to avoid data corruption by the latchs becoming transparent for a short time. This problem begins to dominate as one approaches VLSI as described above. The CLIC design style overcomes this by use of a two phase clocking scheme.

After a short overview of the CLIC design style and its design rules, we give formal means of developing *correct* CLIC gates. This is based on a simple transistor model with charge storage capability and a simple four value model of the signals on wires. The CLIC design style is fully described in [8]. The notation used there is quite different from ours and the clock labeling scheme is completely different which does not easily lend itself to formal analysis. So a summary of this design technique is first given.

2 Overview of the CLIC Design Style

2.1 Introduction

In this section a brief overview is given of synchronous design methodology and dynamic logic, together with how these two ideas are married together to form the basis of the CLIC design style.

2.1.1 Synchronous Design Methodology

The basic principle behind any synchronous design philosophy is that the system is separated into blocks of purely combinatorial logic with no data storage facility, interleaved by register latches which hold the data between clock pulses. There is a global system clock which is used to clock the register latches, and, the period of this is such as to allow all combinatorial logic blocks to finish evaluation of results. So, on the tick of the clock, new data appears on the inputs of the combinatorial logic blocks, and the old results are passed as inputs to the next stage by use of the register latches. By definition there is no feedback within the purely combinatorial blocks. This principle is illustrated in figure 1.

```
         .-------.        .-------.        .-------.        .-------.
-->--+--| Logic |--->---| Latch |--->---| Logic |--->---| Latch |---+--->
     |  `-------'        `-------'        `-------'        `-------'    |
     |                                                                  |
     `------------------------------------------------------------------'
```

Figure 1: Synchronous Logic Concept

2.1.2 Dynamic Logic

A dynamic logic gate has two phases of operation, precharge and evaluate, which is controlled by a system clock. For example a simple pseudo-NMOS dynamic gate is shown in figure 2. The **output** is precharged high while the **clk** input is low. Then as **clk** goes high the path to **Vdd** is turned off and the path to **Gnd** is turned on. Now the output will either remain floating high or will be discharged low depending on the inputs.

```
                  ------+------ Vdd
                        |
                       -'
            .--------0||
            |          -.
            |           |
            |          .-+-+------- output
            |          |   |
            |         -'   `-
            |     ---||     ||---
            | ip1   -.     .-    ip2
            |        |     |
            |        `--+--'
            |           |
            |          -'
    clk ----+---------||
                       -.
                        |
                  ------+------ Gnd
```

Figure 2: A simple dynamic logic gate

Many design styles use such dynamic gates. In particular the DOMINO logic design style [7] uses this sort of structures with each output terminated by a static inverter. The advantage of this is that the output of the inverter is low while the gate is in its precharged state. So this can be fed as input to other such gates resulting in a chain of dynamic gates separated by static inverters.

Now when the clock goes high, all the gates change to the evaluation phase. Since all inputs are initially low no output of a gate will change state unless the gate at the start of the chain changes. Effectively what we have here is a chain of evaluation going from the start of the chain to the end, much as the fall of one domino causes the next to fall which in turn causes the next to fall and so on.

2.1.3 The CLIC Design Style

This design style uses both dynamic logic features and the principles of synchronous design. It comes with a set of rules which guarantee that there will be no timing hazards. The basic principle behind this is very similar to that of the NORA design style as described in [4]. The major difference is that instead of using the simple clock and its inverse, a two phase non-overlapping clocking scheme is used which results in having four clock lines— ϕ_1, $\overline{\phi}_1$, ϕ_2 and $\overline{\phi}_2$. The remainder of this section tries to give a brief overview of this design style. A more detailed electrical analysis of this is given in [8].

2.2 Clock Definition and Generation for CLIC

The two phase clock for the CLIC design style is generated from a single square wave clock input. On every rising edge of the external clock input an internal narrow pulse is generated on the ϕ_1 clock line. Similarly on the falling edge of the external clock another pulse is generated on the ϕ_2 clock line. These two internal clock lines are then inverted to form the remaining two internal clock lines, namely $\overline{\phi}_1$ and $\overline{\phi}_2$ respectively. These are shown in figure 3 together with the external clock.

For the purposes of analysing the clock scheme we can divide the clock cycle

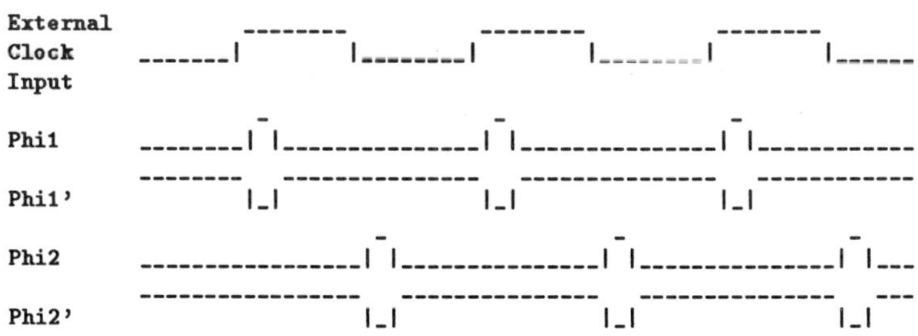

Figure 3: The external and the internal clock relationships

into eight distinct intervals as shown in figure 4. The shaded regions represent uncertainty in the value on the clock lines, *i.e.* the value could be Hi, Lo or something in between. The essential requirements of the clock scheme are that the duration of the clock pulses, *i.e.* the intervals t_3 and t_7, should be long enough so that the internal gates of the chip have enough time to precharge their outputs.

```
          /------ One Cycle -----\
          |                      |
           t1 t2 t3 t4 t5 t6 t7 t8
           |  |  |  |  |  |  |  |
           ^  ^  ^  ^  ^  ^  ^  ^

             ---------               ---------
Phi1         |||   |||               |||   |||
          ------   ---------------------   -----------------

          ------   ---------------------   -----------------
Phi1'        |||   |||               |||   |||
             ---------               ---------

                         ---------               ---------
phi2                     |||   |||               |||   |||
          ------------------   ---------------------   ------

          ------------------   ---------------------   ------
phi2'                    |||   |||               |||   |||
                         ---------               ---------
```

Figure 4: Full definition of the internal clock

2.3 CLIC Primitive Gates

All the logic gates used in the CLIC environment, except for the static inverter, are dynamic logic gates which are driven by one of the four clock lines. The primitives used in realising a logic function are a combination of these gates. There are four basic building blocks used in CLIC circuits—the N_Shell, the P_Shell, the Latch, and the Stat_Inv. These are illustrated in figure 5. Both the N_Shell and the P_Shell devices need extra components, namely transistors, to be "wired" into the them to make n-type and p-type gates respectively. It is these n-type and p-type gates which form the primitive gates as used in the CLIC design scheme.

In the remainder of this section we give a brief account of how each of these devices work. The working of the static inverter however are trivial and will not be discussed here.

```
      --------+------                 --------+------
              |                               |
             -'                              -'
         .--0||                          .--0||
         |   -.                          |   -.
         |    |                          |    |
         |    +----- op                  | /-----\
         |    |                 ip ======>|  P  |
         | /-----\                       | \-----/
 ip ======>|  N  |                       |    |
         | \-----/                       |    +----- op
         |    |                          |    |
         |   -'                          |   -'
phi  ----+---||                 phi' ----+---||
             -.                              -.
              |                               |
      --------+------ (a) N_Shell     --------+------ (b) P_Shell
```

```
          --------+------
                  |
                 -'
      .--------0||
      |          -.
      |           |
      |          -'                       phi'
      |   ---0||                            |
      | phi'     -.                        |\|
      |           |                        | \
ip ---+           +---- op        ip -----|L >O---- op
      |           |                        | /
      |          -'                        |/|
      |   ----||                            |
      | phi      -.                        phi
      |           |
      |          -'
      '--------||
                 -.
                  |
          --------+------    (c) Latch
```

```
          --------+------
                  |
                 -'
             .--0||
             |   -.
             |    |                   |\
     ip -----+    +---- op    ip -----| >O---- op
             |    |                   |/
             |   -'
             '---||
                 -.
                  |
          --------+------    (d) Static Inverter
```

Figure 5: The primitive building blocks of CLIC

2.3.1 The Latch

The Latch as used in the CLIC environment is a dynamic device. It needs to be updated at regular intervals otherwise the value held on its output may deteriorate. There are two types of latches in the CLIC system, one driven by ϕ_1 and $\overline{\phi}_1$, and the other driven by ϕ_2 and $\overline{\phi}_2$. The differences between the two are only in the clocks they are driven by and none other.

A typical latch is illustrated in figure 6 together with the clocks which drive it. The input is required to be stable during the times when there is a positive pulse on the ϕ line and a negative pulse on the $\overline{\phi}$ line. By stable is meant that the the input should either remain Hi or Lo, but should not be in the process of changing. This constraint on the input is necessary to stop the latch from latching onto a false value.

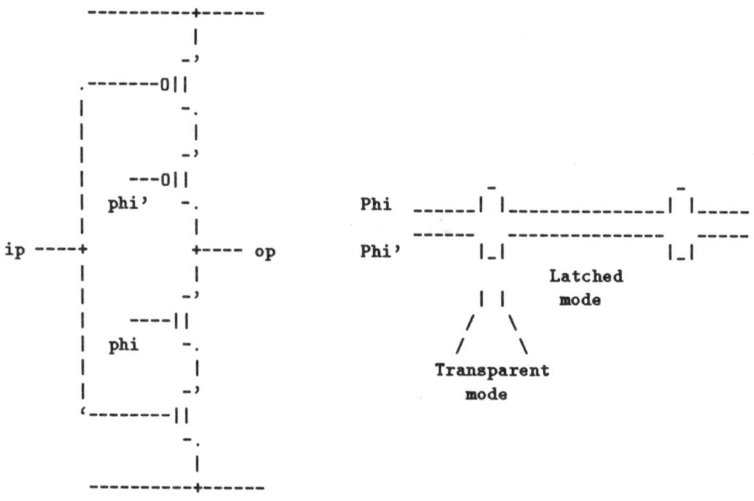

Figure 6: The Latch

If a latch is correctly clocked then it can be in one of two modes—Transparent mode or Latched mode.

Transparent mode $\phi =$ Hi, and $\overline{\phi} =$ Lo

The two transistors driven by the clock lines get switched on, so the value on the output of the latch is charged to the inverse of the input. It is also important to note that during this time the latch behaves as a static inverter, *i.e.* if there were any changes on the input then they would be reflected by a change on the output.

Latched mode $\phi = \mathsf{Lo}$, and $\overline{\phi} = \mathsf{Hi}$
: The two transistors driven by the clock lines get switched off, so the output node of the latch gets isolated from the power rails and so retains the previously charged value. During this time any changes on the input have no effect on the output. The output node is designed to hold the value long enough until the latch is refreshed again by going into the transparent mode for a short time.

The latch only needs to be in the transparent mode for a short time, long enough for the output node to be charged to the correct value. Even if there is considerable clock overlap and clock edges are poor, the latch will still lock onto the correct value provided the input is stable during the interval when there is a positive pulse on the ϕ line and a negative pulse on the $\overline{\phi}$ line. This restriction ensures that the latch behaves correctly as regards locking onto the input value.

2.3.2 P-type and N-type Logic Gates

The name of p-type and n-type logic gates arises from the fact that these gates use only p-type or n-type transistors respectively except for the precharging and enabling transistors. All p-type gates are driven by either ϕ_1 or ϕ_2 and all n-type gates are driven by either $\overline{\phi}_1$ or $\overline{\phi}_2$. Perhaps the best way to understand the working of these gates is by example.

Consider a simple two input n-type Nand gate as shown in figure 7. The working of this gate has two distinct phases—the Precharge period and the Evaluation period.

Precharge $\overline{\phi} = \mathsf{Lo}$
: During this period the output of the gate is pulled Hi by the enabled p-transistor. Any changes on the inputs during this time have no effect on the output since the bottom n-transistor is off and so the path to Gnd is effectively cut, *i.e.* the output cannot be pulled Lo.

Evaluation $\overline{\phi} = \mathsf{Hi}$
: When $\overline{\phi}$ goes Hi the top p-transistor goes off and the bottom n-transistor comes on. If both the inputs now go Hi then the output will be pulled down to logic level Lo, otherwise it will remain floating at Hi.

The correct answer is generated on the output of the gate at the end of the evaluation period and is held static until the next precharge period.

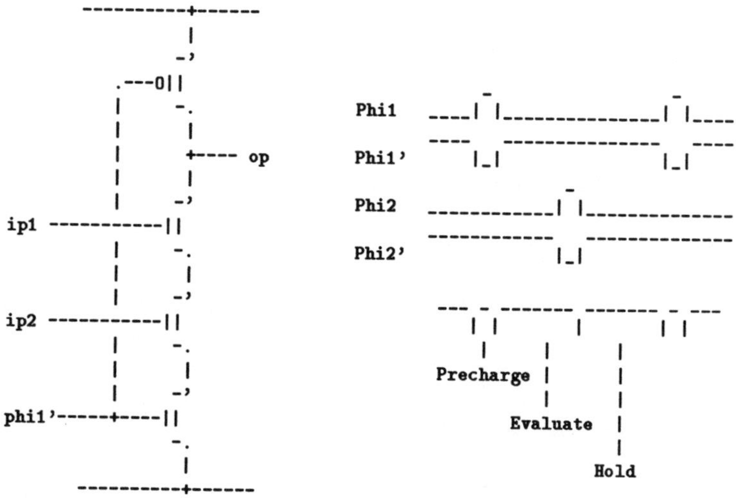

Figure 7: Two input N-type Nand gate

Note that the output may hold the wrong answer if the inputs are allowed to go Hi and then go Lo during the evaluation period. Since there is no pull-up during the evaluation period, if the output goes Lo then it will remain so until the next precharge period. So for example if the inputs are initially Hi and then go Lo, then at the end of the evaluation period the output and the inputs will all be Lo which is the wrong answer for a Nand gate. To overcome this problem a restriction is imposed which states that "there should be no Hi to Lo transitions on the inputs of n-type gates during the evaluation period."

Similarly the working of p-type gates can be understood by considering a two input p-type Nor gate. The structure of this is shown in figure 8. As in the case of n-type gates, this too has the precharge period and the evaluation period.

Precharge $\phi =$ Hi
: During this period the output of the gate is pulled Lo by the enabled n-transistor. Any changes on the inputs during this time have no effect on the output since the top p-transistor is off and so the path to Vdd is effectively cut, *i.e.* the output cannot be pulled Hi.

Evaluation $\phi =$ Lo
: When ϕ goes Lo the bottom n-transistor goes off and the top p-transistor comes on. If both the inputs now go Lo then the output will be pulled up to logic level Hi, otherwise it will remain floating at Lo.

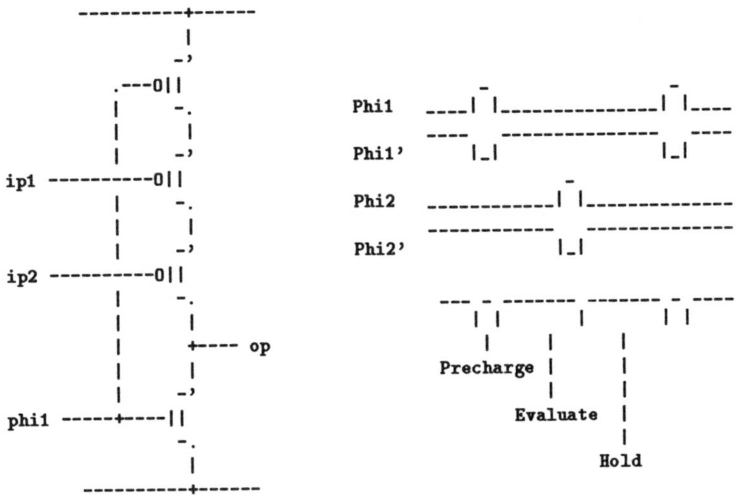

Figure 8: Two input P-type Nor gate

Just as before the correct answer is generated on the output of the gate at the end of the evaluation period and is held static until the next precharge period.

For similar reasons to those of the n-type gates, we need to impose the restriction that "there should be no Lo to Hi transitions on the inputs of p-type gates during the evaluation period."

2.4 Composition Rules for CLIC

From the previous section's work we note that the outputs of the n-type gates cannot have Lo to Hi transitions during the evaluation period which is exactly the requirements on the inputs of the p-type gates. Similarly the outputs of the p-type gates cannot have Hi to Lo transitions during the evaluation period which also meets the exact requirements for the inputs of the n-type gates. This is not by accident but is designed into the style so that we can compose these gates together. Rules like this and others which are necessary to make sure that the circuit designed with CLIC dynamic gates does not contain any timing hazards are presented in this section.

Perhaps the best method of presenting the CLIC composition rules is simply by listing them. Before giving these rules we introduce a few shorthands for the various types of gates and the clocks by which they may be driven.

P_Gate(ϕ) A p-type gate driven by ϕ where ϕ is either ϕ_1 or ϕ_2.

N_Gate($\overline{\phi}$) A n-type gate driven by $\overline{\phi}$ where $\overline{\phi}$ is either $\overline{\phi}_1$ or $\overline{\phi}_2$.

Latch(ϕ,$\overline{\phi}$) A latch driven by ϕ and $\overline{\phi}$ where these clock pairs are either ϕ_1 and $\overline{\phi}_1$ or ϕ_2 and $\overline{\phi}_2$.

If a clock phase ϕ for example is used in the statement of a rule then it is intended that the rule be interpreted as being in two parts—part one with all instances of ϕ and $\overline{\phi}$ replaced by ϕ_1 and $\overline{\phi}_1$ and part two with all instances of ϕ and $\overline{\phi}$ replaced by ϕ_2 and $\overline{\phi}_2$. Having got these we can now give the CLIC composition rules as follows:

Rule 1. An N_Gate($\overline{\phi}$) may be driven by:

(a) P_Gate(ϕ)
(b) N_Gate($\overline{\phi}$) buffered by a static inverter
(c) Latch(ϕ,$\overline{\phi}$)
(d) Latch(ϕ,$\overline{\phi}$) buffered by a static inverter

Rule 2. A P_Gate(ϕ) may be driven by:

(a) N_Gate($\overline{\phi}$)
(b) P_Gate(ϕ) buffered by a static inverter
(c) Latch(ϕ,$\overline{\phi}$)
(d) Latch(ϕ,$\overline{\phi}$) buffered by a static inverter

Rule 3. A Latch(ϕ_1,$\overline{\phi}_1$) may be driven by:

(a) N_Gate($\overline{\phi}_2$)
(b) N_Gate($\overline{\phi}_2$) buffered by a static inverter
(c) P_Gate(ϕ_2)
(d) P_Gate(ϕ_2) buffered by a static inverter
(e) Latch(ϕ_2,$\overline{\phi}_2$)
(f) Latch(ϕ_2,$\overline{\phi}_2$) buffered by a static inverter

Rule 4. A Latch(ϕ_2,$\overline{\phi}_2$) may be driven by:

(a) N_Gate($\overline{\phi}_1$)
(b) N_Gate($\overline{\phi}_1$) buffered by a static inverter
(c) P_Gate(ϕ_1)
(d) P_Gate(ϕ_1) buffered by a static inverter
(e) Latch(ϕ_1,$\overline{\phi}_1$)
(f) Latch(ϕ_1,$\overline{\phi}_1$) buffered by a static inverter

3 Formalising the Clic Design Style

The objective here is to use the various formal techniques available to capture the major concepts which go to make a useful design style. There are a number of level at which we could view the development of the circuit design, from the physics of the semiconductor devices to the top level specifications of a system given in vague terms using English like specification languages. A designer cannot hope to view all his circuit at all of these levels at once. He neither has the capacity nor the expertise in all of these areas, so the best he can do is to work with simple models of the lower level implementations of the various devices. The formalisation of the CLIC design style presented here will thus reflect the sort of devices a logic designer might reasonably be expected to work from. So we begin this section with a very brief introduction of the formalism used.

3.1 Introduction to Higher-Order Logic

The formalism used is that of typed higher-order logic developed by Mike Gordon at the University of Cambridge [5]. This logic uses standard predicate calculus notation. So for example "$\mathsf{P}(x)$" is interpreted as "x has property P." It has the usual logical operators $\sim$, $\wedge$, $\vee$, $\supset$ and $=$ denoting negation, conjunction, disjunction, implication and equivalence respectively. Also provided are the two quantifiers $\forall$ and $\exists$ which express the concepts of *all* and *some*, *e.g.* "$\forall x.\ \mathsf{P}(x)$" means that P holds for every value of x and "$\exists x.\ \mathsf{P}(x)$" means that P holds for some value of x. Finally conditionals of the form "if b then t_1 else t_2" are expressed as "$(b \rightarrow t_1 \mid t_2)$."

What makes this logic higher-order is that quantification is allowed over functions and predicates. So for example the induction axiom for natural number can be expressed as follows:

$$\forall \mathsf{P}.\ \mathsf{P}(0) \wedge (\forall n.\ \mathsf{P}(n) \supset \mathsf{P}(n+1)) \supset \forall n.\ \mathsf{P}(n)$$

The use of higher-order logic (HOL) as a hardware description language is explained in [6]. Essentially devices can be expressed as predicates with the labels of external ports of the device being synonymous with the arguments to the predicate. So for example a simple inverter with the input node labeled *ip* and the output node labeled *op* and delay δ would be defined as follows:

$$\mathsf{Invert}(ip, op) \quad =_{def} \quad \forall t.\ op(t+\delta) = \sim ip(t)$$

Where $=_{def}$ is simply a means of saying that the l.h.s. is defined to be equal to the r.h.s.. Joining together of devices in HOL is done by the conjunction of the predicates with the common ports having the same labels. Hiding of internal nodes is done be means of existentially quantifying them. This is explained in more detail in [2].

3.2 Overview

To see how to formalise the CLIC design style we first need to look at the forms of the correctness statements at the top level. For any given device, the correctness statement simply states that "the implementation together with some input-conditions on the inputs implies the specification together with some output-conditions on the outputs." This can be formally stated as follows:

$$\begin{pmatrix} \mathsf{Dev\_Imp}(ip_1, \ldots ip_n, op_1, \ldots op_m) \wedge \\ \mathsf{Ip\_Cond}\ ip_1 \ \wedge \cdots \wedge \ \mathsf{Ip\_Cond}\ ip_n \end{pmatrix} \supset \begin{pmatrix} \mathsf{Dev\_Spec}(ip_1, \ldots ip_n, op_1, \ldots op_m) \wedge \\ \mathsf{Op\_Cond}\ op_1 \ \wedge \cdots \wedge \ \mathsf{Op\_Cond}\ op_m \end{pmatrix} \tag{1}$$

So the derivation of the correctness statement of a device which is composed of two lower level devices simply requires that the input-conditions and the output-conditions match for all those lines which are to be connected between them. Then by simple logical manipulation we can show that the top level correctness statement can be derived purely from the lower level correctness statements.

This then is the overview of our methodology for the formalisation of a design style. Naturally the Ip_Cond and Op_Cond predicates will have to be defined to reflect the design rules for the particular design style. Also other parameters may need to be added to these predicates to formalise any peculiarities specific to the design style.

However before going on to giving the correctness statements for the various CLIC primitive gates we need a few preliminary axioms and definitions to get us off the ground. Namely we need to clarify the model of transistors we plan to use, the sort of values we plan to propagate around the circuit and reason about and the definitions of the clock.

3.3 Formal Definitions of CMOS Primitives

In formalising the CMOS primitives we need to declare the sort of values nodes in a circuit can have. These values will depend on the models we use for the primitive devices, namely the transistors. After studying the CLIC design style it seems that a unidirectional model for the transistor is adequate for describing the various gates *etc.* So the values we use reflect this choice of the transistor model.

Any node in the circuit can only have one of the following four values:

```
      Er
    /    \
  Hi      Lo
    \    /
      Zz
```

These four values are derived by using Bryants work [1] with only two strengths. Note that they form a complete lattice.

With this we introduce a simple operator $\sqcup$ which states what happens at a node when two signals meet. $\sqcup$ is simply defined to be the least upper bound on the above lattice of values. Now we are ready to define the basic primitives of the CMOS technology within the constraints of the above four valued algebra.

$$\mathsf{Vdd}(x) \quad =_{def} \quad \forall t.\ x(t) = \mathsf{Hi} \tag{2}$$

$$\mathsf{Gnd}(x) \quad =_{def} \quad \forall t.\ x(t) = \mathsf{Lo} \tag{3}$$

$$\mathsf{N\_Tran}(g, i, o) \quad =_{def} \quad \forall t.\ o(t) = ((g(t) = \mathsf{Lo}) \rightarrow \mathsf{Zz} \mid i(t)) \tag{4}$$

$$\mathsf{P\_Tran}(g, i, o) \quad =_{def} \quad \forall t.\ o(t) = ((g(t) = \mathsf{Hi}) \rightarrow \mathsf{Zz} \mid i(t)) \tag{5}$$

$$\mathsf{Join}(i_1, i_2, o) \quad =_{def} \quad \forall t.\ o(t) = i_1(t) \sqcup i_2(t) \tag{6}$$

$$\mathsf{Cap}_1(i, o) \quad =_{def} \quad \forall t.\ o(t) = (\sim (i(t) = \mathsf{Zz}) \rightarrow i(t) \mid i(t-1)) \tag{7}$$

Note that all further devices described here will be built out of these six primitive building blocks.

3.4 Formal Definition of Clock

The Clock as described earlier is derived from a single square wave input. The formal definition used is not derived from such a single input but it is simply defined to be that which such a circuit might generate. This is because we do not model the various gates to have delay. It would not be too difficult to derive the given definition as an abstraction of what might be generated by the circuit if we were to use a different model for the gates.

However we are now going to define a predicate Clock with four arguments which are the four lines which might be generated by the clock generator circuit. The way we do this is to define how one of the four clock lines behaves, and then relate the behaviour of the other three lines to it. Before giving the formal definition, here is a graphical representation with an arbitrary starting point.

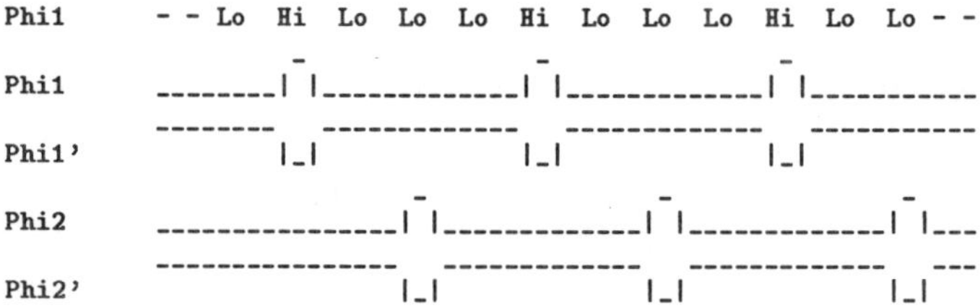

Figure 9: Graphical representation of the four clock lines

Note that the clock is cyclic over four units of time and the uncertainty states have been eliminated. This is done to help convey the basic principle of how the bulk of the work regarding the formalisation was done, rather than present even more unnecessary detail than already present. A full treatment to this is given in [3] where the correctness of the design style is shown at the finer grain of time and then related to the coarser grain of time at which this paper deals. Here then is the formal definition of the above graphical representation:

$$\mathsf{Clock}(\phi_1, \overline{\phi}_1, \phi_2, \overline{\phi}_2) \;=_{def}\; \begin{array}{l} \mathsf{Cycle}\ \phi_1 \quad \wedge \\ \mathsf{Shift}\ \ \phi_1\ \phi_2 \ \wedge \\ \mathsf{Invert}\ \phi_1\ \overline{\phi}_1 \ \wedge \\ \mathsf{Invert}\ \phi_2\ \overline{\phi}_2 \end{array} \tag{8}$$

Now definitions for the various predicates used to define Clock can be given. The simplest two, namely Shift and Invert are as follows:

$$\mathsf{Shift}\ \phi_1\ \phi_2 \;=_{def}\; \forall t.\ \phi_2(t) = \phi_1(t{+}2) \tag{9}$$

$$\mathsf{Invert}\ \phi\ \overline{\phi} \;=_{def}\; \forall t.\ \overline{\phi}\ t = \mathsf{NOT}(\phi\ t) \tag{10}$$

Where NOT is the negation function over the values Hi, Lo and Er.

So far we have defined the way in which all the clock lines are related to ϕ_1 but have not given a formal definition for Cycle. Before doing this we will state informally the behaviour of ϕ_1.

- ϕ_1 is cyclic over four units of time
- During *any* four consecutive units of time the value on ϕ_1 is Hi exactly once and Lo for the other three units.
- ϕ_1 can start in any of its four possible states.

Now we can formalise each of these three informal statements into logic to give the following definition for clock:

$$\begin{aligned}\mathsf{Cycle}\ \phi \ =_{def}\ & (\forall t.\ \phi(t) = \phi(t{+}4)) \wedge \\ & (\ \mathsf{Cycle\_1}\ \phi\ 0 \vee \\ & \ \ \mathsf{Cycle\_1}\ \phi\ 1 \vee \\ & \ \ \mathsf{Cycle\_1}\ \phi\ 2 \vee \\ & \ \ \mathsf{Cycle\_1}\ \phi\ 3\ )\end{aligned} \tag{11}$$

$$\begin{aligned}\mathsf{Cycle\_1}\ \phi\ t_0 \ =_{def}\ & (\phi(t_0) \quad\ \ = \mathsf{Hi}) \wedge \\ & (\phi(t_0{+}1) = \mathsf{Lo}) \wedge \\ & (\phi(t_0{+}2) = \mathsf{Lo}) \wedge \\ & (\phi(t_0{+}3) = \mathsf{Lo})\end{aligned} \tag{12}$$

This seems like a lengthy definition for Clock and it could have been shorter but for two reasons. Firstly that it closely mimics the way the designer has informally described the signals on the clock lines, and secondly it is of the form which allows some of the latter lemmas to be more easily derived.

However here are a few of the more elegant definitions I came up with at the time of thinking how to define the cyclic property of clock. These are all provably equivalent to the definition given above.

$$\begin{aligned}\mathsf{Cycle}\ \phi \ =_{def}\ & (\exists t.\ \ \phi(t) \quad\ = \mathsf{Hi}\ \wedge \\ & \qquad\ \ \phi(t{+}1) = \mathsf{Lo}\ \wedge \\ & \qquad\ \ \phi(t{+}2) = \mathsf{Lo}\ \wedge \\ & \qquad\ \ \phi(t{+}3) = \mathsf{Lo}\quad) \wedge \\ & (\forall t.\ \ \phi(t{+}4) = \phi(t)\ )\end{aligned} \tag{13}$$

$$\begin{aligned}\mathsf{Cycle}\ \phi \ =_{def}\ & \exists n.\ (0 \le n \le 3)\ \wedge \\ & \qquad \forall t.\ \phi(t) = ((\mathsf{MOD4}\ t\ =\ n) \rightarrow \mathsf{Hi} \mid \mathsf{Lo})\end{aligned} \tag{14}$$

Where MOD4 is the remainder of dividing its argument by 4.

Note how the various properties of Clock are separated into different predicates. This is done deliberately so that we can follow the formalisation more easily and also that it simply reads better.

3.5 Formal Composition Rules for CLIC

The rules governing the interconnection of CLIC gates have been described earlier in a rather lengthy and informal way. If we are to be able to formalise them then understanding the reason behind them is necessary. Consider for example an n-type gate driving a p-type gate. During the precharge period the output of the n-type gate is precharged Hi, which means that the inputs of the p-type gate are at the correct level, namely that the transistors are off and the output node of the p-type gate would be isolated if it went into the evaluation phase now. So when the clock changes and puts both of these gates into the evaluation phase, the output of the p-type gate does not change unless and until its inputs change. However if we had the other situation where the input of the p-type gate was held Lo then as soon as the gate went into its evaluation phase the output would change to Hi. Now no matter what happens to the inputs, the output cannot be changed to Lo until the next precharge period. So effectively the gate has erroneously changed its output value.

In summarising this we can say that the inputs of a p-type gate *must* not have Lo to Hi transitions during its evaluation phase and also the output of an n-type gate *does* not have a Lo to Hi transition during its evaluation phase. So to capture this sort of behaviour we need a single predicate which captures the output behaviour of the n-type gate and the constraints on the inputs of the p-type gate.

Here is the formal definition of such a predicate WB as used in our system. It relies on the fact that the gates are clocked and that the clock is correctly behaved.

$$\mathsf{WB}\ x\ \phi \ =_{def}\ \forall t.\ \begin{pmatrix} \phi(t{+}1) = \phi(t)\ \wedge \\ x(t) \quad = \phi(t) \end{pmatrix} \supset x(t{+}1) = x(t) \tag{15}$$

This says that the node x is defined to be "Well Behaved" with respect to ϕ where ϕ is one of the four clock line as defined by Clock.

So for example the output of an n-type gate driven by $\overline{\phi}_1$ satisfies "WB *op* ϕ_1," and this is exactly the required input conditions for a p-type gate driven by ϕ_1. In this context what "WB *op* ϕ_1" means is that while ϕ_1 is Lo, *i.e.* the n-type gate is in its evaluation phase, then the *op* cannot have a rising edge on it. This is exactly right since once the output of an n-type gate has been discharged then it cannot go to Hi again until the clock rises and precharges the gate as discussed above.

With this one predicate we have now condensed all of the rules which were listed in a rather informal way. However this predicate relies on the formal definition of Clock and must always be used in conjunction with it. This is not a restriction since CLIC is a dynamic design style and so all CLIC gates will require the existence of Clock for their correct behaviour.

3.6 Formal Derivations of CLIC Primitive Gates

Having got the preliminaries out of the way we can now begin the derivations of the correctness statements for the various CLIC gates. There are four types of gates in the CLIC design methodology, namely the n-type gate, p-type gate, the latch and the static inverter. Statements of correctness can be individually derived for the latch and the static inverter, but it would be foolish to simply derive a statement of correctness for each of the various n-type and p-type gates separately. Rather than doing this it is far better to derive some general theorems which will then be useful for generating the statement of correctness for the individual n-type and p-type gates.

3.6.1 N-type and P-type Logic Gates

For any general theorems to be proved of n-type or p-type gates we first need to extract out what is common to the various gates. A simple split would be to separate the set of components which perform the logic specific function into one bag and the remainder into an other. We call the remainder of an n-type gate the N_Shell, since it has a hole in it into which other components need to be inserted before it can function as an n-type CLIC gate.

This is illustrated in figure 10 and can be formally stated as follow:

$$
\begin{array}{lll}
\mathsf{N\_Shell}(\phi, o_1, ip, op) & =_{def} & \exists p_0\ p_1\ p_2\ p_3. \\
 & & \quad \mathsf{Gnd}(p_0) \quad\quad\quad\ \wedge \\
 & & \quad \mathsf{Vdd}(p_1) \quad\quad\quad\ \wedge \\
 & & \quad \mathsf{N\_Tran}(\phi, p_0, o_1) \wedge \\
 & & \quad \mathsf{P\_Tran}(\phi, p_1, p_2) \wedge \\
 & & \quad \mathsf{Join}(p_2, ip, p_3) \quad \wedge \\
 & & \quad \mathsf{Cap}_3(p_3, op)
\end{array} \tag{16}
$$

Cap_3 in the above definition is simply a capacitor with a "memory" of three units of time just as Cap_1 has a "memory" of one unit of time. Note that Cap_3 is derived by composing three Cap_1 devices together.

Before we can progress further we need to define the property which is held true of all those cluster of devices which may be inserted into this N_Shell. By studying the mechanism of an n-type gate we note that the cluster of devices which get inserted in the N_Shell perform one of two functions—they either maintain a link between the *ip* and the o_1 nodes of the N_Shell, or they don't. We call this property Opt_Link and it can be formally stated as follows:

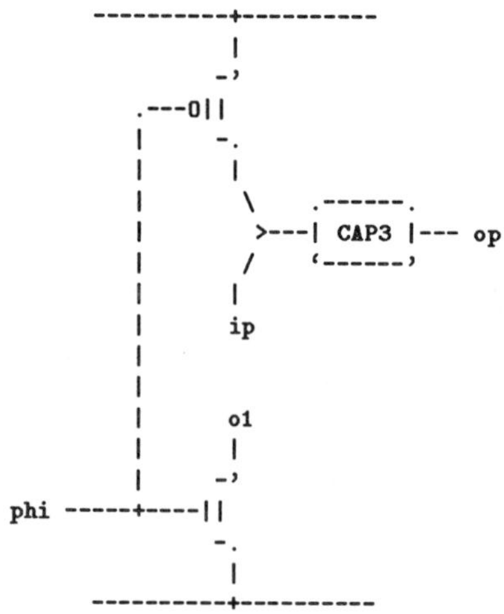

Figure 10: N_Shell as used in CLIC

$$\mathsf{Opt\_Link}(ip, op) \quad =_{def} \quad \forall t.\ (op\ t = ip\ t) \vee (op\ t = \mathsf{Zz}) \tag{17}$$

Here it is worth noting that the true property of two nodes being linked or not linked is not actually captured because we are using a directional flow of information model. The best we can do under the circumstances as stated, is say that the values on the two node are equal or that the input node to the N_Shell has a floating value on it. This is still not quite enough but we'll leave the extra conditions until later when they are needed. However there is enough for the following two properties of the N_Shell to be derived.

$$\begin{pmatrix} \mathsf{Clock}(\phi_1, \overline{\phi}_1, \phi_2, \overline{\phi}_2) \wedge \\ \mathsf{N\_Shell}(\overline{\phi}_1, a, b, op) \wedge \\ \mathsf{Opt\_Link}(a, b) \end{pmatrix} \supset \mathsf{WB}\ op\ \phi_1 \tag{18}$$

$$\begin{pmatrix} \mathsf{Clock}(\phi_1, \overline{\phi}_1, \phi_2, \overline{\phi}_2) \wedge \\ \mathsf{N\_Shell}(\overline{\phi}_1, a, b, op) \wedge \\ \mathsf{Opt\_Link}(a, b) \wedge \\ \phi_1(t) = \mathsf{Hi} \end{pmatrix} \supset \begin{pmatrix} \mathsf{Def}\ op\ t \wedge \\ \mathsf{Def}\ op\ (t{+}1) \wedge \\ \mathsf{Def}\ op\ (t{+}2) \wedge \\ \mathsf{Def}\ op\ (t{+}3) \end{pmatrix} \tag{19}$$

Where $\mathsf{Def}\ a\ t \quad =_{def} \quad (a(t) = \mathsf{Hi}) \vee (a(t) = \mathsf{Lo})$

The first theorem can be interpreted as saying that *if* the N_Shell is implemented correctly *and* it is correctly driven by clock *and* the cluster of devices placed in it are correctly behaved in that they have the property of Opt_Link *then* the output will be "Well Behaved," *i.e.* the output will not have Lo to Hi transitions during the evaluation phase. The second theorem simply says that given the same assumptions and assuming that at some time t the clock phase ϕ_1 goes Hi then the output will be "Well Defined" for the times t to $t+4$, *i.e.* the output will be either Hi or Lo.

Now that we have these general theorems we must ensure that this Opt_Link property is derivable for all the various sorts of cluster of elements that may be inserted in the N_Shell. For this to be truly general it will require us to talk of the structure of an arbitrary cluster of devices.

Any logic function which is implementable can be simplified into a network of transistors which only includes transistors in series and/or transistors in parallel with the outputs of the transistors joined together by the Join device. On the basis of this the following three theorems together allow us to show that any cluster containing only parallel and/or series transistor networks, can be shown to maintain the Opt_Link property across the two terminals by which the cluster is connected to the N_Shell.

$$\mathsf{N\_Tran}(g,i,o) \supset \mathsf{Opt\_Link}(i,o) \tag{20}$$

$$\begin{pmatrix} \mathsf{Opt\_Link}(a,b) \wedge \\ \mathsf{Opt\_Link}(b,c) \end{pmatrix} \supset \mathsf{Opt\_Link}(a,c) \tag{21}$$

$$\begin{pmatrix} \mathsf{Opt\_Link}(a,b) \wedge \\ \mathsf{Opt\_Link}(a,c) \wedge \\ \mathsf{Join}(b,c,d) \end{pmatrix} \supset \mathsf{Opt\_Link}(a,d) \tag{22}$$

To illustrate this let's look at a simple example namely the two input Nand gate of figure 7. The structure of this can be formally defined as follows:

$$\begin{aligned} \mathsf{N\_Nand\_Imp}(\phi, ip_1, ip_2, op) \quad =_{def} \quad & \exists p_1\ p_2\ p_3. \\ & \mathsf{N\_Shell}(\phi, p_1, p_3, op) \wedge \\ & \mathsf{N\_Tran}(ip_1, p_1, p_2) \quad \wedge \\ & \mathsf{N\_Tran}(ip_2, p_2, p_3) \end{aligned} \tag{23}$$

Now by using theorems 18, 19, 20, 21 and 22 we can derive the following two properties. This states that the output of a two input Nand gate is "well behaved" and that the output is also "well defined" over a certain interval of time with reference to the point when ϕ_1 is Hi.

$$\begin{pmatrix} \mathsf{Clock}(\phi_1, \overline{\phi}_1, \phi_2, \overline{\phi}_2) \quad \wedge \\ \mathsf{N\_Nand\_Imp}(\overline{\phi}_1, ip_1, ip_2, op) \end{pmatrix} \supset \mathsf{WB}\ op\ \phi_1 \tag{24}$$

$$\left(\begin{array}{ll}\mathsf{Clock}(\phi_1,\overline{\phi}_1,\phi_2,\overline{\phi}_2) & \wedge\\ \mathsf{N\_Nand\_Imp}(\overline{\phi}_1,ip_1,ip_2,op) & \wedge\\ \phi_1(t)=\mathsf{Hi} & \end{array}\right) \supset \left(\begin{array}{ll}\mathsf{Def}\ op\ t & \wedge\\ \mathsf{Def}\ op\ (t{+}1) & \wedge\\ \mathsf{Def}\ op\ (t{+}2) & \wedge\\ \mathsf{Def}\ op\ (t{+}3) & \end{array}\right) \tag{25}$$

So far we have only demonstrated that the output of n-type gates are "well behaved" and "well defined," but nothing has been said about the derivation of the logical behaviour of these gates. For this we need theorems considerably more complex than those given for Opt_Link. These properties include Link, No_Link and WB_Link which state under what circumstances a "link" exists across the two nodes of the cluster of devices inserted in the N_Shell. The line of thought regarding work on these is very similar to that followed for Opt_Link, so we shall not deal with them here. However the other two theorems for the two input nand gate giving its logical behaviour are presented below. They use an abstraction function Val_Abs which maps the values Hi and Lo to true and false respectively.

$$\begin{array}{l}\left(\begin{array}{ll}\mathsf{Clock}(\phi_1,\overline{\phi}_1,\phi_2,\overline{\phi}_2) & \wedge\\ \mathsf{N\_Nand\_Imp}(\overline{\phi}_1,ip_1,ip_2,op) & \wedge\\ \phi_1(t)=\mathsf{Hi} & \wedge\\ \mathsf{Def}\ ip_1\ (t{+}1) & \wedge\\ \mathsf{Def}\ ip_2\ (t{+}1) & \end{array}\right) \supset\\ (\ \mathsf{Val\_Abs}\ op\ (t{+}1) = \ \sim(\mathsf{Val\_Abs}\ ip_1\ (t{+}1)\ \wedge\ \mathsf{Val\_Abs}\ ip_2\ (t{+}1)))\end{array} \tag{26}$$

$$\begin{array}{l}\left(\begin{array}{ll}\mathsf{Clock}(\phi_1,\overline{\phi}_1,\phi_2,\overline{\phi}_2) & \wedge\\ \mathsf{N\_Nand\_Imp}(\overline{\phi}_1,ip_1,ip_2,op) & \wedge\\ \mathsf{WB}\ ip_1\ \overline{\phi}_1 & \wedge\\ \mathsf{WB}\ ip_2\ \overline{\phi}_1 & \wedge\\ \phi_1(t)=\mathsf{Hi} & \wedge\\ \mathsf{Def}\ ip_1\ (t{+}1) & \wedge\\ \mathsf{Def}\ ip_2\ (t{+}1) & \wedge\\ \mathsf{Def}\ ip_1\ (t{+}2) & \wedge\\ \mathsf{Def}\ ip_2\ (t{+}2) & \end{array}\right) \supset\\ (\ \mathsf{Val\_Abs}\ op\ (t{+}2) = \ \sim(\mathsf{Val\_Abs}\ ip_1\ (t{+}2)\ \wedge\ \mathsf{Val\_Abs}\ ip_2\ (t{+}2)))\end{array} \tag{27}$$

The treatment for p-type logic gates follows exactly the same line of argument, even to the point where considerable number of the intermediate results are common to both.

3.6.2 The Latch

This is also known as the $\mathrm{C^2MOS}$ latch and its structure is shown in figure 11. This is Formally captured as follows:

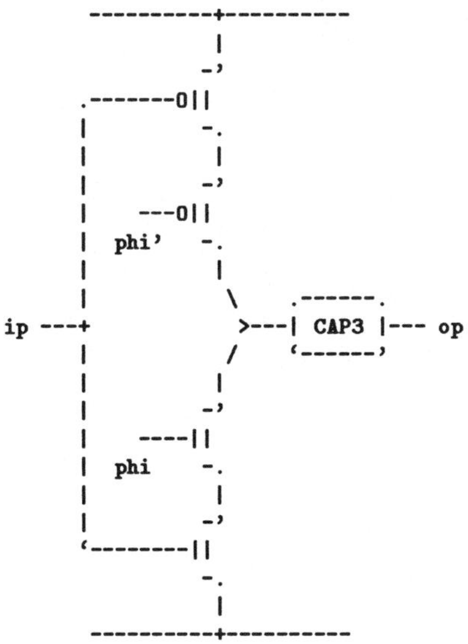

Figure 11: The Latch as used in CLIC

$$\begin{aligned}\mathsf{Latch\_Imp}(\phi, \overline{\phi}, ip, op) \;=_{def}\; & \exists p_0\ p_1\ p_2\ p_3\ p_4\ p_5\ p_6 . \\ & \mathsf{Gnd}(p_0) \wedge \\ & \mathsf{Vdd}(p_1) \wedge \\ & \mathsf{N\_Tran}(ip, p_0, p_2) \wedge \\ & \mathsf{N\_Tran}(\phi, p_2, p_4) \wedge \\ & \mathsf{P\_Tran}(\overline{\phi}, p_3, p_5) \wedge \\ & \mathsf{P\_Tran}(ip, p_1, p_3) \wedge \\ & \mathsf{Join}(p_4, p_5, p_6) \wedge \\ & \mathsf{Cap_3}(p_6, op)\end{aligned} \tag{28}$$

Since this is simply a one off result, the derivation is not important but what is important is the result so only that is presented. The full behaviour of the Latch device is summarised in the following three theorems.

$$\begin{pmatrix} \mathsf{Clock}(\phi_1, \overline{\phi}_1, \phi_2, \overline{\phi}_2) \wedge \\ \mathsf{Latch\_Imp}(\phi_1, \overline{\phi}_1, ip, op) \end{pmatrix} \supset \begin{pmatrix} \mathsf{WB}\ op\ \phi_1 \wedge \\ \mathsf{WB}\ op\ \overline{\phi}_1 \end{pmatrix} \tag{29}$$

$$\begin{pmatrix} \mathsf{Clock}(\phi_1, \overline{\phi}_1, \phi_2, \overline{\phi}_2) \wedge \\ \mathsf{Latch\_Imp}(\phi_1, \overline{\phi}_1, ip, op) \wedge \\ \phi_1(t) = \mathsf{Hi} \wedge \\ \mathsf{Def}\ ip\ t \end{pmatrix} \supset \begin{pmatrix} \mathsf{Def}\ op\ t \wedge \\ \mathsf{Def}\ op\ (t{+}1) \wedge \\ \mathsf{Def}\ op\ (t{+}2) \wedge \\ \mathsf{Def}\ op\ (t{+}3) \end{pmatrix} \tag{30}$$

$$\begin{pmatrix} \mathsf{Clock}(\phi_1, \overline{\phi}_1, \phi_2, \overline{\phi}_2) & \wedge \\ \mathsf{Latch\_Imp}(\phi_1, \overline{\phi}_1, ip, op) & \wedge \\ \phi_1(t) = \mathsf{Hi} & \wedge \\ \mathsf{Def}\ ip\ t & \end{pmatrix} \supset \begin{pmatrix} \mathsf{Val\_Abs}\ op\ t & = \sim\mathsf{Val\_Abs}\ ip\ t & \wedge \\ \mathsf{Val\_Abs}\ op\ (t{+}1) & = \sim\mathsf{Val\_Abs}\ ip\ t & \wedge \\ \mathsf{Val\_Abs}\ op\ (t{+}2) & = \sim\mathsf{Val\_Abs}\ ip\ t & \wedge \\ \mathsf{Val\_Abs}\ op\ (t{+}3) & = \sim\mathsf{Val\_Abs}\ ip\ t & \end{pmatrix} \tag{31}$$

The first of these captures the fact that the output may drive any of p-type or n-type gates, even both at the same time. The second theorem states that the output is "well defined" so the results can be abstracted into the boolean domain by use of the Val_Abs abstraction function. Finally the third gives the logical behaviour between the input and the output at the abstract level, *i.e.* on the clock tick the input is inverted and passed to the output where it is held static until the next clock tick.

3.6.3 The Static Inverter

This is the only device in the entire CLIC design style which does not need one of the clock lines for it to function correctly. It has only two external ports namely the input (*ip*) and output (*op*) ports and its behaviour could perfectly be defined without the use of the Clock predicate. However, to enable it to be used in conjunction with other dynamic CLIC devices, its correctness statement has to be given in the same *form*. So we begin by giving the formal definition for the structure of the gate as follows:

$$\begin{array}{lll} \mathsf{Stat\_Inv\_Imp}(ip, op) & =_{def} & \exists p_0\ p_1\ p_2\ p_3. \\ & & \quad \mathsf{Gnd}(p_0) \qquad \wedge \\ & & \quad \mathsf{Vdd}(p_1) \qquad \wedge \\ & & \quad \mathsf{N\_Tran}(ip, p_0, p_2)\ \wedge \\ & & \quad \mathsf{P\_Tran}(ip, p_1, p_3)\ \wedge \\ & & \quad \mathsf{Join}(p_2, p_3, op) \end{array} \tag{32}$$

Now the usual three properties can be derived for this gate. The first one being that it's output is "well behaved."

$$\begin{pmatrix} \mathsf{Clock}(\phi_1, \overline{\phi}_1, \phi_2, \overline{\phi}_2) \wedge \\ \mathsf{Stat\_Inv\_Imp}(ip, op) \end{pmatrix} \supset \begin{pmatrix} (\mathsf{WB}\ ip\ \overline{\phi}_1 \supset \mathsf{WB}\ op\ \phi_1) \wedge \\ (\mathsf{WB}\ ip\ \phi_1 \supset \mathsf{WB}\ op\ \overline{\phi}_1) \wedge \\ (\mathsf{WB}\ ip\ \overline{\phi}_2 \supset \mathsf{WB}\ op\ \phi_2) \wedge \\ (\mathsf{WB}\ ip\ \phi_2 \supset \mathsf{WB}\ op\ \overline{\phi}_2) \end{pmatrix} \tag{33}$$

Inspecting this theorem reveals that it is in a different *form* than the others so far. In fact it is not so different as to not allow logical inferences to be made using the same techniques. However if it were to be put in the same form as the ones so far, we would have four different theorems giving rise to the four different clauses. Remember that the inverter is used to invert the *polarity* of a gate so that a gate may drive its own sort, *e.g.* a p-type gate may drive another p-type gate only if it is buffered by an inverter. Since there are two different sorts of gates, n-type and p-type, and two clock phases, the need arises for four very similar theorems, or one containing all four clauses.

The remaining two theorems for this device are fairly standard. In fact they are even simplified a little to take advantage of the fact that this device is not clocked. The next theorem for instance simply states that "if the input is defined then so is the output." The last one gives the logical behaviour of the gate appropriately abstracted to the boolean level using the Val_Abs predicate.

$$\begin{pmatrix} \mathsf{Stat\_Inv\_Imp}(ip, op) \wedge \\ \mathsf{Def}\ ip\ t \end{pmatrix} \supset \mathsf{Def}\ op\ t \tag{34}$$

$$\begin{pmatrix} \mathsf{Stat\_Inv\_Imp}(ip, op) \wedge \\ \mathsf{Def}\ ip\ t \end{pmatrix} \supset (\mathsf{Val\_Abs}\ op\ t\ =\ \sim\mathsf{Val\_Abs}\ ip\ t) \tag{35}$$

3.7 Formalising the CLIC Circuit Design Methodology

So far we have outlined a method for deriving the correctness statement of any logic gate designed in the CLIC design style. If we are to design real circuits with these correctness statements, rather than just admire their elegance and still use the old rules of thumb, then we must provide formal means of doing so. *i.e.* a formal method of combining the correctness statements of an arbitrary number of gates resulting in a new correctness statement for the new circuit.

What the designer is interested in is simply obtaining the logical behaviour of the system so that he may satisfy himself that the system does what he intended it to do. So a technique is needed which allows the designer to compose the logical behaviour component of the specifications and leave the rest of the "checking" to the system. In our system simple logical inferences would be used to check the validity of connecting together the output of one gate to the input of an other. In fact the rule involved is the resolution of the predicates and their arguments governing the constraints on the inputs of devices, against the predicates and their arguments governing the properties of the outputs. Work is in hand at present to automate this process.

To illustrate the use of this consider the the implementation of a simple logical AND function with delay. Firstly, in order to get delay we will have to use a latch since this is the only device in the CLIC design style which allows behaviours between input and output to be mapped across time giving a controlled unit delay with respect to the clock. So the solution we shall choose to implement is to drive the output of an n-type nand gate, such as the one already used in an earlier example, by a latch. This is illustrated in figure 12.

```
                              phi2'
       |-----\                |\
       |      \               | \
ip ----| N     >O-------------|L >O----- op
       |      /               | /
       |-----/                |/
        phi1'                 phi2
```

Figure 12: A simple CLIC Circuit

The theorems needed for these two devices have already been derived earlier in this paper, namely theorems 27 and 31, for the nand gate and the latch respectively. The correctness statement for the nand gate is exactly as needed but that of the latch needs to be messaged into a form which allows these two to be combined. Given below are the two theorems for these two devices in their correct form just before they are to be combined. There are a number of important steps involved before we get to this state involving a lemma about clock but these are not covered here. Further details regarding these intermediate steps can be found in [3].

$$\left(\begin{array}{ll} \mathsf{Clock}(\phi_1, \overline{\phi}_1, \phi_2, \overline{\phi}_2) & \wedge \\ \mathsf{N\_Nand\_Imp}(\overline{\phi}_1, ip_1, ip_2, op) & \wedge \\ \mathsf{WB}\ ip_1\ \overline{\phi}_1 & \wedge \\ \mathsf{WB}\ ip_2\ \overline{\phi}_1 & \wedge \\ \phi_1(t) = \mathsf{Hi} & \wedge \\ \mathsf{Def}\ ip_1\ (t+1) & \wedge \\ \mathsf{Def}\ ip_2\ (t+1) & \wedge \\ \mathsf{Def}\ ip_1\ (t+2) & \wedge \\ \mathsf{Def}\ ip_2\ (t+2) & \end{array}\right) \supset \tag{36}$$
$$(\ \mathsf{Val\_Abs}\ op\ (t+2) = \ \sim(\mathsf{Val\_Abs}\ ip_1\ (t+2)\ \wedge\ \mathsf{Val\_Abs}\ ip_2\ (t+2)))$$

$$\begin{pmatrix} \mathsf{Clock}(\phi_1, \overline{\phi}_1, \phi_2, \overline{\phi}_2) & \wedge \\ \mathsf{Latch\_Imp}(\phi_2, \overline{\phi}_2, ip, op) & \wedge \\ \phi_1(t) = \mathsf{Hi} & \wedge \\ \mathsf{Def}\ ip\ t{+}2 & \end{pmatrix} \supset \begin{pmatrix} \mathsf{Val\_Abs}\ op\ (t{+}2) = & \sim\mathsf{Val\_Abs}\ ip\ (t{+}2) & \wedge \\ \mathsf{Val\_Abs}\ op\ (t{+}3) = & \sim\mathsf{Val\_Abs}\ ip\ (t{+}2) & \wedge \\ \mathsf{Val\_Abs}\ op\ (t{+}4) = & \sim\mathsf{Val\_Abs}\ ip\ (t{+}2) & \wedge \\ \mathsf{Val\_Abs}\ op\ (t{+}5) = & \sim\mathsf{Val\_Abs}\ ip\ (t{+}2) & \end{pmatrix} \tag{37}$$

Now we can combine these two by using Modus Ponens and Conjunction rules together with theorem 25 which satisfies the input constraint for the latch. The final result together with hiding the internal line using the existential quantifier looks like the following:

$$\begin{pmatrix} \mathsf{Clock}(\phi_1, \overline{\phi}_1, \phi_2, \overline{\phi}_2) & \wedge \\ \begin{pmatrix} \exists x.\ \mathsf{N\_Nand\_Imp}(\overline{\phi}_1, ip_1, ip_2, x) \wedge \\ \quad \mathsf{Latch\_Imp}(\phi_2, \overline{\phi}_2, x, op) \end{pmatrix} & \wedge \\ \mathsf{WB}\ ip_1\ \overline{\phi}_1 & \wedge \\ \mathsf{WB}\ ip_2\ \overline{\phi}_1 & \wedge \\ \phi_1(t) = \mathsf{Hi} & \wedge \\ \mathsf{Def}\ ip_1\ (t{+}1) & \wedge \\ \mathsf{Def}\ ip_2\ (t{+}1) & \wedge \\ \mathsf{Def}\ ip_1\ (t{+}2) & \wedge \\ \mathsf{Def}\ ip_2\ (t{+}2) & \end{pmatrix} \supset \begin{pmatrix} \mathsf{Val\_Abs}\ op\ (t{+}2) = & (\mathsf{Val\_Abs}\ ip_1\ (t{+}2)\ \wedge\ \mathsf{Val\_Abs}\ ip_2\ (t{+}2)) & \wedge \\ \mathsf{Val\_Abs}\ op\ (t{+}3) = & (\mathsf{Val\_Abs}\ ip_1\ (t{+}2)\ \wedge\ \mathsf{Val\_Abs}\ ip_2\ (t{+}2)) & \wedge \\ \mathsf{Val\_Abs}\ op\ (t{+}4) = & (\mathsf{Val\_Abs}\ ip_1\ (t{+}2)\ \wedge\ \mathsf{Val\_Abs}\ ip_2\ (t{+}2)) & \wedge \\ \mathsf{Val\_Abs}\ op\ (t{+}5) = & (\mathsf{Val\_Abs}\ ip_1\ (t{+}2)\ \wedge\ \mathsf{Val\_Abs}\ ip_2\ (t{+}2)) & \wedge \end{pmatrix} \tag{38}$$

Note that this theorem which gives the combined behaviour of the n-type nand gate and the latch is in the same form as the correctness statements for the two devices from which it is built. This particular feature helps our formal approach of combining CLIC gates together to be expanded to cover circuits of arbitrary complexity. The correctness statements will increase in size in combining large and complex circuits but will not change in their inherent structure.

4 Discussion and Future Work

A full formal presentation has been given for the CLIC design style which we believe to be suitable for VLSI. In particular a *form* for the correctness statement of CLIC gates has been developed which maintains uniformity of specifications across the many levels of hierarchy of circuit design—from the very simple logic gates to fairly complex structures using many macro blocks. The use of these formal techniques have been demonstrated on a simple example which uses many CLIC gates, both simple and complex. A major case study based on the design of a digital phase-locked loop is in progress which demonstrates the use of these techniques on large systems.

Naturally the work presented is only as good as the axioms on which it is based. Current models used for the primitive devices of integrated circuits are simplistic, with the view to making the proofs of correctness easier. These models are not inaccurate, but merely incomplete. They have only the features which are relevant to the design style; other properties are not modelled. Too simplistic a model of these devices, however, may allow a failure mode to pass unobserved. So proofs based on such models become void. More realistic models are needed for these primitives, together with means of showing that the simple models suffice in controlled environments. It is hoped that research in this area will support most of the work done to date using simpler models, by formally showing that the simpler models are adequate in the environment in which they are used. Some results in this area are already available [9], where the simulation model used in [1] is embedded in logic. However this is not yet developed to the point where it is usable for the dynamic behaviour of circuits.

Acknowledgements

I would like to thank all members of the Cambridge Hardware Verification Group. In particular Tom Melham and Jeff Joyce who made valuable comments on earlier drafts of this paper. Special thanks are due to Mike Gordon who provided the necessary stimulus for this work and who was a continuous source of encouragement. I would also like to thank Dave Orton of Racal Research and Glynn Winskel of Cambridge University for providing many stimulating discussions in hardware and formal methods respectively.

References

[1] Randal Everitt Bryant
"A Switch-Level Simulation Model for Integrated Logic Circuits," PhD. Thesis, also available as a Technical Report MIT/LCS/TR-259, Laboratory for Computer Science, MIT, Massachusetts, March 1981.

[2] A. Camilleri, M. Gordon and T. Melham
"Hardware Verification using Higher-Order Logic," In: *Proceedings of the IFIP International Conference: From H.D.L. Descriptions to Guaranteed Correct Circuit Designs*, Grenoble, September 9–11, 1986.

[3] I. S. Dhingra
Ph.D. Thesis. Computer Laboratory, University of Cambridge. 1987

[4] Nelson F. Goncalves and Hugo J. De Man
"NORA: A Racefree Dynamic CMOS Technique for Pipelined Logic Structures," *IEEE Journal of Solid-State Circuits*, SC–18 (**3**), June 1983 pp. 261–266.

[5] M. J. C. Gordon
"HOL: A Machine Oriented Formulation of Higher-Order Logic," Technical Report 68, Computer Laboratory, University of Cambridge, 1985.

[6] M. J. C. Gordon
"Why Higher-Order Logic is a Good Formalism for Specifying and Verifying Hardware," In: *Formal Aspects of VLSI Design*, edited by G. Milne and P. A. Subrahmanyam, North-Holland, 1986.

[7] R. H. Krambeck, Charles M. Lee and Hung-Fai Stephen Law
"High-Speed Compact Circuits with CMOS," *IEEE Journal of Solid-State Circuits*, SC–17 (**3**), June 1982 pp. 614–619.

[8] D. W. R. Orton
"Clocked Dynamic Logic for CMOS," Racal Research Internal Memo of 10*th* January 1984, Worton Drive, Worton Grange Industrial Est., Reading RG2 OSB, England.

[9] Glynn Winskel
"Models and Logic of MOS Circuits," in *VLSI Specification, Verification and Synthesis*, Ed. G. Birtwistle and P.A. Subrahmanyam, Kluwer 1987. (These proceedings)

11

A Compositional Model of MOS Circuits

by
Glynn Winskel
University of Cambridge,
Computer Laboratory,
Corn Exchange Street,
Cambridge CB2 3QG
England.

1. Introduction.

This paper describes a compositional model for MOS circuits. Following Bryant in [1] it covers some effects of capacitance and resistance used frequently in designs. Bryant's model has formed the basis of several hardware simulators. From the point of view of verification, however, it suffers the inadequacy that it is not compositional, an expression which will be explained further. This makes it hard to use the model to reason in a structured way.

In this paper we shall restrict our attention to the static behaviour of circuits. Roughly, a circuit's behaviour will be represented as the set of possible steady (or stable) states the circuit can settle into, ignoring oscillatory behaviour. Certainly a good understanding of such static behaviour is necessary in order to treat sequential circuits where the clock is taken to be slow enough so that one circuit settles into a steady state between changes. Because sequential circuits often use capacitance to precharge isolated parts of a circuit we shall include a treatment of charge sources due to capacitance in our analysis of static behaviour. Unfortunately while the model appears to generalise well to sequential circuits when all the memory is dynamic, there are difficulties describing static memory correctly. This problem is discussed in the concluding section. Note that Bryant can avoid this problem because a simulator need only produce one possible sequence of steady states of a circuit. The difficulty comes in accounting for *all* possible behaviours. While the model introduced here owes a great deal to Bryant's work, the reader is warned that we have not stuck closely to Bryant's development or terminology.

We take the view that it is useful to have a language to describe the construction of circuits. A term in the language describes a circuit and the term's structure can be used to direct reasoning about the circuit's behaviour. Naturally there should be atomic terms for basic components like transistors and then these are composed together using operations.

As a guide in choosing the operations, we borrow ideas from CCS and CSP [7, 5]. A circuit can be viewed as a process which communicates with its environment at its connection points. Assuming the connection points are named we can compose two circuits by connecting them at those points with names in common. In this way we can construct large circuits from smaller ones. Often once a connection point has been used to connect circuits we no longer want to connect to it, so remove its name, and in effect hide, or insulate, the point.

To summarise and make this more precise, a circuit is to be associated with a set of connection points which are named; the set of such points we shall call the *sort* of the circuit. There is an operation of *composition* on circuits. Two circuits are composed by connecting those points which share the same name. The sort of composition is thus the union of the sorts of its two constituent circuits. There is an operation of *hiding*; we can cease regarding a point of a circuit as a connecting point and remove its name. The sort of the resulting circuit is thus the sort of the original circuit with the name of the hidden point removed. These two operations, with atomic terms for basic components, will generate a language whose terms describe circuits.

To define the language, we assume a countably infinite set of *point names* Π, ranged over by $\alpha, \beta, \gamma, \cdots$, and a set of capacitance strengths K, with typical element k, and conductance strengths G, with typical element g. Strengths will give a qualitative measure of the relative sizes of basic devices in a way to be explained. The syntax of the language is given by:

$$\begin{aligned} c ::= & Pow\ (\alpha) \mid Gnd\ (\alpha) \mid \\ & cap_{kH}(\alpha) \mid cap_{kL}(\alpha) \mid \\ & pt(\alpha) \mid wre(\alpha,\beta) \mid res_g(\alpha,\beta) \mid \\ & ntran(\alpha,\beta,\gamma) \mid ptran(\alpha,\beta,\gamma) \mid \\ & c \bullet c \mid c \setminus \alpha. \end{aligned}$$

In the terms $wre(\alpha,\beta)$, $res_g(\alpha,\beta)$, $ntran(\alpha,\beta,\gamma)$, and $ptran(\alpha,\beta,\gamma)$ we insist that the points α, β, γ are distinct; so in these basic devices names are associated with unique connection points, a property that is inherited by any circuit described a compound term. Circuits are built up by composition; the term $c_0 \bullet c_1$ represents the composition of the circuits represented by c_0 and c_1. Hiding a point α in a circuit described by c is represented by the term $c \setminus \alpha$. The atomic term $Pow\ (\alpha)$ stands for a power source connected to the point α, while $Gnd\ (\alpha)$ stands a connection to ground at α. We shall draw these as

$Pow\ (\alpha)$: 

$Gnd\ (\alpha)$: 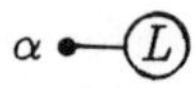

where the H signifies that power provides a high value of voltage and L that ground provides a low value. Another kind of source arises through charge storage. The term $cap_{kH}(\alpha)$ represents a capacitance of strength k, charged high, connected to point α at one end and to ground at the other. The term $cap_{kL}(\alpha)$ is a similar capacitance but charged low. They are drawn as:

$cap_{kH}(\alpha)$: 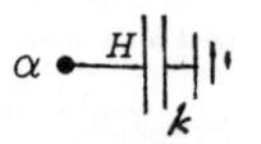

$cap_{kL}(\alpha)$: 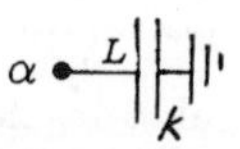

Note that all the capacitance terms we consider are associated with only one connection point because they are to be used simply as a sources of charge.

The terms $pt(\alpha)$, $wre(\alpha, \beta)$ and $res_g(\alpha, \beta)$ are various forms of connectors. The simplest is $pt(\alpha)$, a point named α. The term $wre(\alpha, \beta)$ represents a wire connecting points α and β while the term $res_g(\alpha, \beta)$ stands for a resistance with conductance strength $g \in G$ connecting points α and β—it is intended to model components like d-type transistors or be included in situations where resistance effects may count. We shall draw points, wires and resistances as:

$pt(\alpha)$: $\bullet\alpha$ $wre(\alpha, \beta)$:  $res_g(\alpha, \beta)$: 

The term $ntran(\alpha, \beta, \gamma)$ stands for an n-type transistor with a gate γ which when it is high connects points α and β with perfect conductance. The term $ptran(\alpha, \beta, \gamma)$ stands for a p-type transistor with a gate γ which connects points α and β, again with perfect conductance, when the gate γ is low. We draw transistors as:

$ntran(\alpha, \beta, \gamma)$: 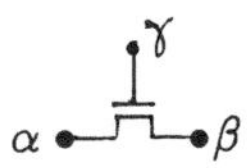 $ptran(\alpha, \beta, \gamma)$: 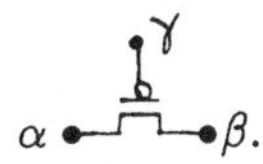

Should the effect of resistance be significant we can insert a suitable resistance between α and β, and, should the capacitance of the gate be significant, capacitances can be connected to model this. This is an idealised model of a transistor. It treats transistors as switches and ignores the fact that in reality transistors have thresholds (see section 4).

We explain the idea of qualitative strength on which Bryant's model and ours, which follows his, are based. The capacitance strengths K are assumed to form a finite totally ordered set, to which we adjoin a least element 0,

$$0 < k_1 < k_2 < \cdots < k_m.$$

(It seems most MOS designs can be described with $m = 3$.) The capacitances in a design are assigned a strength which says how they combine with other capacitances. The value 0 corresponds to negligible capacitance. The meaning of $k < k'$ in the strength order is that a capacitance assigned strength k is negligible with respect to one assigned strength k'. More exactly, the assumption made of the strength order can be stated: if a capacitance of strength k is combined with one of strength k' the result is equivalent to a capacitance of strength the maximum of k and k'. Accordingly, if a capacitance of strength k is charged low and connected to one of strength k' charged high, the result is, in effect, a capacitance of strength k' charged high. Accompanying this, if neither of two capacitances can be guaranteed to override the other, they are assigned the same strength.

Of course, a problem arises with this somewhat naive explanation. Two capacitances of the same strength k should combine to give a capacitance of strength k; the capacitance of one alone is not negligible relative to that of the combination. But if we combine many capacitances of the same strength k the result may not be a capacitance of strength k; it may not be negligible relative to one of strength k' even though $k < k'$. Still, for any particular circuit term, the number of such combinations is bounded. It follows that, with respect to a particular circuit term, we can interpret capacitance strengths so that when capacitance of strength k combines with one of strength k' the result is a capacitance of strength the maximum of k and k'. In more detail, there is a small number $\epsilon < 1$, dependent on the circuit term, so that we can interpret the capacitance strengths as intervals of positive real numbers (capacitance values)

$$^\bullet 0^\bullet, {}^\bullet k_1{}^\bullet, \cdots {}^\bullet k_i{}^\bullet, \cdots, {}^\bullet k_{i+i}{}^\bullet, \cdots, {}^\bullet k_m{}^\bullet,$$

where $^\bullet k^\bullet$ represents an interval $(^\bullet k, k^\bullet)$ with end points $^\bullet k < k^\bullet$, so

$$\frac{k_i{}^\bullet}{^\bullet k_{i+1}} < \epsilon$$

for $0 \leq i < m$. With respect to this interpretation a capacitance value C has strength k if $C \in {}^\bullet k^\bullet$. By taking the intervals $^\bullet k^\bullet$ wide enough and the number ϵ small enough there are choices of capacitances for the basic components $cap_k(\alpha)$ which ensure the assumptions of the strength order are never invalidated by compositions in the term.

To see how capacitances compose, assume a capacitance of size C, charged low to a voltage v with charge Q, is connected to a capacitance C', charged high to voltage v' with charge Q'. From the equation $Q = vC$, relating charge to voltage and capacitance, we see

$$v_{res} = \frac{Q + Q'}{C + C'}.$$

If now we assume C has strength k, and C' strength k' with $k < k'$ we know $C/C' < \epsilon$. By Taylor's theorem we derive

$$v_{res} = v' - (v' - v)(\frac{C}{C'} - (\frac{C}{C'})^2 + \textit{higher order terms}).$$

Hence v_{res} is to within $\epsilon(v' - v)$ of v'. We cannot conclude, in general, that if v' is high then v_{res} is high also. However, because the number of compositions of capacitances in a term is bounded, we can choose ϵ small enough so that whenever such a composition occurs in the circuit described by the term the resultant voltage is high.

There is a similar strength order on conductances G of resistances, again assumed finite, to which is adjoined a maximum element ∞,

$$g_1 < g_2 < \cdots < g_n < \infty.$$

(For most MOS designs $n = 3$ appears to be sufficient). Resistances are assigned a conductance strength so that a resistance of strength g', with $g < g'$, is negligible with respect to one of strength g. Wires are assumed to have perfect conductance and are given the special conductance strength ∞, which we take to dominate all other conductances.

Consider the situation

where a point β is connected both to ground at α via a resistance of conductance strength g and to power at γ via a resistance of strength g'. Then $g < g'$ means the value through the stronger conductance will dominate so the resulting value will be high. A simple application of Ohm's law explains how this can arise. Suppose the resistance of strength g has value R ohms and that of strength g' is relatively small, of value r ohms say. Let $v_\alpha, v_\beta, v_\gamma$ be the voltages at the corresponding points. By Ohm's law

$$\frac{v_\gamma - v_\beta}{v_\beta - v_\alpha} = \frac{Ir}{IR} = \frac{r}{R}.$$

Ensuring R is large relative to r ensures $v_\gamma - v_\beta$ is small enough so that v_β is close to the voltage v_γ and so high. Taking R large relative to r corresponds to R being associated with a conductance strength g, r with g', and $g < g'$. Similar arguments using Ohm's law give the conductance strengths of resistances in series and in parallel. If two resistances R and R' of strength g and g' are connected in series the resulting resistance $R + R'$ should be assigned conductance strength the *minimum* of g and g'. This is because, for example, if R'/R is very small then so is $R'/(R + R')$. Connected in parallel their resulting resistance, $RR'/(R + R')$, should be assigned strength the *maximum* of g and g'.

Such simplistic arguments suggest that if we connect arbitrarily many resistances of the same conductance strength g in parallel, or in series, the result will be equivalent to a resistance of strength g. This is clearly unrealistic in general, but with respect to any particular circuit term the number of such compositions is bounded so the assumptions needed for the strength order can be met. More precisely, with respect to the particular circuit term, the conductance strengths can be interpreted as intervals of resistance values so the properties required of the strength order hold in the circuit. The method is similar to that sketched for capacitances.

As we have explained, these qualitative orders of strengths are to be understood abstractly. The orders of capacitance and conductance strengths are only to be interpreted as referring to particular intervals of real numbers when the circuit description has been completed. The simple, seemingly naive, assumptions of the strength orders, on which the design is based, can be realised. Design, simulation and reasoning at the more abstract qualitative level postpones consideration of the exact sizes of capacitances and resistances, though, of course, they have to be dealt with at some stage before layout. It is a formidable research problem to bring these physical aspects within the domain of hardware verification. The difficulty is perhaps not so much that of not being able to solve a particular differential equation, as that of finding techniques which bridge between different levels of description. These are necessary to fulfil the ultimate goal and make the verification of real devices feasible, based only on the assumption that the layout is done correctly.

Our aim is to build a model for the behaviour of circuits described by our language so that the behaviour of a term $c_0 \bullet c_1$ is derived solely from the behaviours of c_0 and c_1, and similarly the behaviour of $c\backslash\alpha$ is derived solely from that of c. The advantage of a model like this is that proving a property of $c_0 \bullet c_1$, for instance, reduces to proving properties of c_0 and c_1; proof is directed by the algebraic structure of the terms. On such a compositional model it is possible to build a compositional proof system, a goal realised in [10]. Such an approach should be distinguished from that in [1]; the model and simulators of Bryant use a graphical representation of the whole circuit. Obtaining a compositional model is surprisingly involved. It rests on a detailed analysis of the idea of state necessary to support compositionality, and leads to the definition of a static configuration of a circuit.

2. Static Configurations.

A circuit computes by imposing a relation between the voltage values at its points, values which can often be assumed to be high (abbreviated to H) or low (abbreviated to L). A point takes the value H if it is only connected to sources of positive charge, like a power source or a positively charged capacitance, perhaps through resistances. It takes the value L if it is only connected to sources of negative charges. Sometimes, however, a point may take a voltage value which cannot be classified as H or L. For example, a point may be connected to both a power source and ground through resistances of roughly the same magnitude, or perhaps to two capacitances of roughly the same size but initially charged with opposite polarities. In either case the point cannot be assumed to have value H or L and so instead we take it to have the value X. It is useful to consider another case, that in which a point is not connected to any significant sources of charge. To such an isolated point we associate the value Z.

It may be helpful to think of the value X as the set $\{high, low\}$ because the value X results from a point being simultaneously connected to a source of high and low. Accordingly the value H can be identified with the singleton set $\{high\}$, similarly L with $\{low\}$, and Z with the empty set $\emptyset$ (Z is the value due to not being connected to any source at all). With this representation the inclusion order induces the following lattice $\mathbf{V}$ of values.

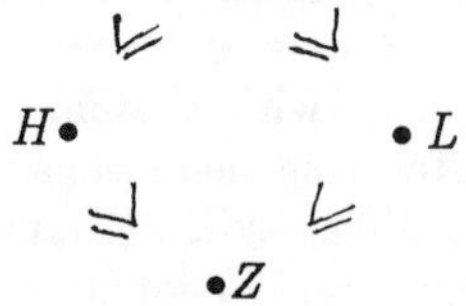

Thus X is the least upper bound of H and L with respect to the order $\leq$. We shall write $w_0 + w_1$ for the least upper bound of a pair of values w_0 and w_1 and, in general, we shall use ΣW for the least upper bound of a set of values W.

As has been said, the behaviour of a circuit will be represented as the set of steady states it can adopt. Because we wish to derive the behaviour of the composition $c_0 \bullet c_1$ of two circuits c_0, c_1 from the behaviour of c_0 and c_1 we cannot consider just those steady states which arise from the circuit in isolation, but must also take account of the steady states which can arise in composition with other circuits—clearly the behaviour of a transistor depends on what its gate is connected to. We shall use the expression "steady state" of a circuit to mean a stable state which the circuit can adopt in some environment. We are left with the task of formalising the notion of *static configuration*, where a static configuration consists of those attributes of a steady state which are relevant, in a sense we now make precise.

In its formalisation, the choice of those attributes which are relevant will be guided by the operations, composition and hiding, which we wish to perform on circuits. For composition, we require that the static configurations of a composition $c_0 \bullet c_1$ of two circuit should be determined by the static configurations of c_0 and c_1. For hiding, we require that the static configurations of the circuit $c \setminus \alpha$ formed by hiding the point α in the circuit c are determined by the static configurations of c.

At first sight it might appear that a static configuration of a circuit of sort Λ should be simply a *value function* $V : \Lambda \to \mathbf{V}$ giving the values of a steady state in some environment. However, not surprisingly, this fails to account for differences in behaviour due to effects of capacitance and conductance strength. A connection to a positively charged capacitance and another to a power source, have the same value functions and yet behave differently in the same context. Consider the two forms of sources

$$s_0 \equiv cap_{kL}(\gamma) \quad \text{and} \quad s_1 \equiv (Gnd\ (\delta) \bullet res_g(\delta, \gamma)) \setminus \delta,$$

where k is a capacitance strength and g a conductance strength. (We use the relation $\equiv$ to mean syntactic equality, so *e.g.* s_0 abbreviates the term to its right.) Both sources can be put in environments leading to values H, L, X at the point γ, and hence they are associated with the same value functions. But placing both in the same context,

$$C[-] \equiv (- \bullet ntran(\alpha, \beta, \gamma) \bullet res_g(\gamma, \epsilon) \bullet Pow\ (\epsilon)) \setminus \epsilon \setminus \gamma, \qquad (*)$$

we obtain circuits $C[s_0]$ and $C[s_1]$ which have different value functions.

$C[s_0]$:

$C[s_1]$:

The transistor in $C[s_1]$ may be disconnecting because it has the value X on its gate so there is a steady state of the circuit which has a value function V with $V(\alpha) = L$ and $V(\beta) = H$; the transistor in $C[s_0]$ must be connecting in any steady state because the power source will override the stored charge and this ensures $V(\alpha) = V(\beta)$, ruling out the possibility of $V(\alpha) = L$ and $V(\beta) = H$, for the value function of any steady state.

This example shows that static configurations should consist of further attributes in addition to just value functions; only then can it be possible to derive the value functions, and so the static configurations, of a compound circuit from those of its components. Furthermore, it shows there is a need for an attribute which expresses the degree to which values are secured at points. It motivates the introduction of a *strength function* $S : \Lambda \rightarrow \mathbf{S}$, where $\mathbf{S}$ is a set of strengths got by adjoining two elements 0 and ∞ to K and G. This is an attribute of the steady state of a circuit of sort Λ. A point α of a circuit in a steady state is not only associated with a value $V(\alpha)$ but is also associated with a strength $S(\alpha)$. If the point α has value $V(\alpha) = Z$ then it is assigned strength $S(\alpha) = 0$. Otherwise it either receives its value from capacitances of strength k or from power and ground via paths of conductance strength g (including ∞). It cannot receive contributions to its value $V(\alpha)$ from capacitances of different strengths or via paths of different conductance strengths by the assumptions of the strength orders. Nor can it receive contributions from both capacitances and power or ground, because then in a steady state the stored charge would drain away or be overriden. Hence the point α is associated with a unique strength in $K \cup G \cup \{0, \infty\}$, and this we take to be $S(\alpha)$.

It is useful to incorporate both the order on capacitances and that on conductances into a single order to reflect the fact that whenever a charged capacitance is connected to power or ground through even a fairly weak conductance the finite change stored on the capacitance is eventually overridden, and makes no contribution to the values in a steady state. We arrive at the concept of the *strength order*

$$0 < k_1 < \cdots < k_m < g_1 < \cdots < g_n < \infty.$$

which contains 0, the zero strength, to stand for the strength of a negligible capacitance or negligible conductance, and ∞, the strength of a perfect conductance. As notation for later, we shall use $s \cdot s'$ for the minimum and $s + s'$ for the maximum of two strengths s and s'. Now, for example, we can say that the strength of the point β in $cap_{kH}(\alpha) \bullet res_g(\alpha, \beta) \setminus \alpha$ will be $S(\beta) \geq k$ in any steady state. It is possible to have a steady state in which $S(\beta) = k$ if it is not connected to any stronger source in the environment, while it is possible that $S(\beta) > k$ for a steady state in an environment where β is connected to a stronger capacitance, or source of power or ground via a conductance. In all the steady states of $Pow\,(\alpha) \bullet res_g(\alpha, \beta) \setminus \alpha$, the point β will have strength $S(\beta) \geq g$. If it is connected to a source in the environment via a conductance of strength greater than g it will have strength $S(\beta) > g$.

Are the attributes of value function and strength function enough? Unfortunately not. To see this first define an "X-source"

$$x \equiv cap_{kH}(\epsilon) \bullet cap_{kL}(\epsilon)$$

which standing alone gives rise to value X at the point ϵ. If we compose x with a transistor with gate ϵ and then hide ϵ there are two possibilities; the transistor may be connecting or disconnecting. Hence in the term

$$c_0 \equiv (Gnd\ (\delta) \bullet ntran(\delta, \gamma, \epsilon) \bullet x) \setminus \delta \setminus \epsilon$$

the single unhidden point γ may or may not be connected to ground. Consequently the value and strength functions of steady states of the circuit c_0 in an environment coincide with those of the simple circuit

$$c_1 \equiv pt(\gamma)$$

consisting of a single point. However c_0 and c_1 behave differently in the context $C[-]$ given by $(*)$ above.

$C[c_0]$:

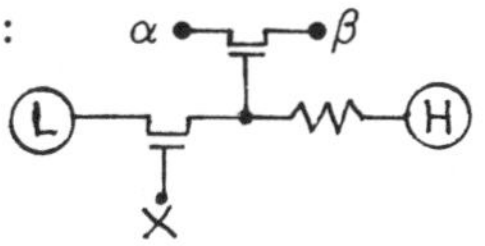

$C[c_1]$:

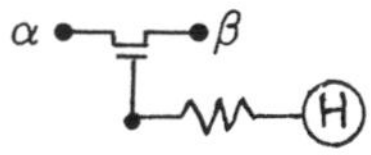

In the circuit $C[c_0]$ there is the possibility that the point γ will be pulled low, disconnecting α and β; it is therefore possible for their values to be different in the same steady state. In the circuit $C[c_1]$ there is the no such possibility and the points α and β must be connected and so share a common value.

This example highlights the need for a further attribute which distinguishes the circuits c_0 and c_1. Of course c_0 has an internal source while c_1 does not. An extra attribute called the *internal-value function* marks the contribution to the value function which is due to paths of conductance from internal sources. A steady state of a circuit of sort Λ has an internal-value function $I : \Lambda \rightarrow \mathbf{V}$ which to a point α in Λ associates the part $I(\alpha)$ of the value $V(\alpha)$ which is maintained by sources built into the circuit; so naturally $I(\alpha) \leq V(\alpha)$.

A further attribute is necessary. There are two circuits whose steady states are identical as far as their attributes of value, strength and internal-value are concerned, and yet behave differently in the same context. Let

$$d_0 \equiv (ntran(\delta, \gamma, \epsilon) \bullet x) \setminus \epsilon,$$

which connects the X-source x to the gate of a transistor. So d_0 may behave like a wire between δ and γ, or two disconnected points δ and γ. Let d_1 consist just of two disconnected points, so

$$d_1 \equiv pt(\delta) \bullet pt(\gamma).$$

Consider a steady state of d_0. There are no internal sources in d_0 so its internal-value function will satisfy $I(\delta) = I(\gamma) = Z$. If the transistor is connecting the strength and value function will have to be the same and otherwise not. Hence for any steady state of d_0 there will be a steady state of d_1 with the same attributes. Conversely, any such attributes of a steady state of d_1 will be shared by a steady state of d_0. With respect to the three attributes we have introduced so far the circuits d_0 and d_1 are the same. But they behave differently in the context $C[-]$ given by $(*)$ above.

$C[d_0]$:

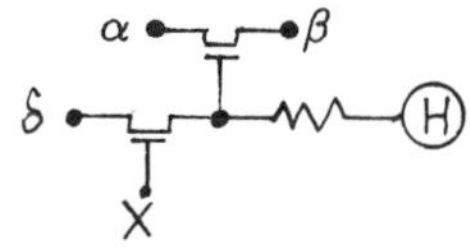

$C[d_1]$:

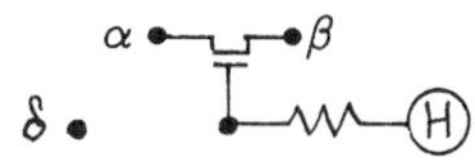

The argument is similar to that for introducing the internal-value function. In the circuit $C[d_0]$ there is the possibility that the point γ will be pulled low through the circuit being placed in an environment where δ is connected directly to ground. This would disconnect α and β so that their values can be different. In the circuit $C[d_1]$ there is the no such possibility and the points α and β must be connected and their values the same.

The final attribute we introduce is one to express a generalised form of connectivity. It is the fact that in the two circuits d_0 has the possibility that δ and γ are connected and d_1 not, which accounts for their different behaviour in the same context. However, because of strengths, values do not effect one another symmetrically across a resistance, so the relation $\rightsquigarrow$ we introduce will not be symmetric in general, although it will be reflexive and transitive.

To motivate the relation $\rightsquigarrow$, consider how values effect eachother across a resistance $res_g(\alpha, \beta)$. First note that in any steady state

$$S(\alpha) \cdot g \leq S(\beta).$$

If $S(\alpha) = k$, a capacitance strength, then α receives its value from capacitances of strength k and cannot be linked to power or ground. The point β is connected to α by the resistance and so receives its value from the same capacitances. Its strength, $S(\beta)$ is equal to k too. In this case $S(\alpha) \cdot g \leq S(\beta)$. If $S(\alpha) = g'$, a conductance strength, then α must receive its value from power or ground via paths of conductance strength g'. In series with the resistance from α to β these form paths of conductance strength $g \cdot g'$. Hence $S(\alpha) \cdot g = g' \cdot g \leq S(\beta)$ in this case. The relation $\alpha \rightsquigarrow_1 \beta$ is taken to hold for a steady state of $res_g(\alpha, \beta)$ iff

$$S(\alpha) \cdot g = S(\beta).$$

If $S(\alpha) \cdot g = S(\beta)$ the value at α effects the value at β so $V(\alpha) \leq V(\beta)$ and $I(\alpha) \leq I(\beta)$, whereas if $S(\alpha) \cdot g < S(\beta)$ it is overridden. More generally, we shall take the $\alpha \rightsquigarrow \beta$ to be true in a steady state of any circuit if the points α and β are equal or connected by any path

$$\alpha \rightsquigarrow_1 \gamma_1 \rightsquigarrow_1 \cdots \rightsquigarrow_1 \gamma_u \rightsquigarrow_1 \beta.$$

A steady state of a circuit of sort Λ has a *flow relation* $\rightsquigarrow: \Lambda \times \Lambda$ which, when it relates two points $\alpha \rightsquigarrow \beta$, ensures that values at α contribute to those at β so that $V(\alpha) \leq V(\beta)$, and also $I(\alpha) \leq I(\beta)$.

Example. Consider a steady state of $res_g(\alpha, \gamma) \bullet res_g(\gamma, \beta)$ in which $S(\alpha) = S(\beta) = g$ and $S(\gamma) = \infty$. Then we have $\gamma \rightsquigarrow \alpha$ and $\gamma \rightsquigarrow \beta$ but not $\alpha \rightsquigarrow \gamma$ or $\beta \rightsquigarrow \gamma$.

This shows that conducting paths, in the sense of paths conducting values, are not just chains of resistances according to this model; here there is a chain of conductances between α and β and yet no flow relation between them. In Bryant's words the path is "blocked" through the point γ having a higher strength than the endpoints α and β. Despite the chain of resistances linking α to β they are effectively disconnected, and the value at one point does not effect the value at the other. Connecting two resistances in series between two points, without hiding the connection point, is not equivalent to, nor does it imply, the effect of putting a resistance between them (a phenomenon analogous to that due to interruption points in parallel programs). It is partly for this reason that the model is based on the relation $\rightsquigarrow$ rather than on conductance matrices.

All these arguments suggest the form of a static configuration of a circuit of sort Λ; it should consist of a strength, value function, internal-value function and a flow relation. Of course it remains to detail the axioms which relate the different attributes, and later to demonstrate that the attributes are sufficient to give a compositional model for the static behaviour of circuits.

Definition. Let Λ be a set of points in Π. A *static configuration* of sort Λ is a structure

$$\langle S, V, I, \rightsquigarrow \rangle$$

where

$S : \Lambda \to \mathbf{S}$ (the *strength function*),
$V : \Lambda \to \mathbf{V}$ (the value function),
$I : \Lambda \to \mathbf{V}$ (the *internal* value function) and
$\rightsquigarrow$ is a reflexive, transitive relation on Λ (the *flow relation*),

which satisfy

(i) $S(\alpha) = 0 \leftrightarrow V(\alpha) = \emptyset$,
(ii) $I(\alpha) \leq V(\alpha)$,
(iii) $\alpha \rightsquigarrow \beta \to S(\alpha) \geq S(\beta)$,
(iv) $\alpha \rightsquigarrow \beta \to I(\alpha) \leq I(\beta) \wedge V(\alpha) \leq V\beta$,
(iii) $\alpha \rightsquigarrow \beta \wedge S(\beta) \in K \cup \{0\} \to S(\alpha) = S(\beta)$,
(vi) $\alpha \rightsquigarrow \beta \wedge S(\alpha) = S(\beta) \to \beta \rightsquigarrow \alpha$.

for any points α, β in Λ.

Write sort(σ) for the sort of a configuration σ. Write Sta$[\Lambda]$ for the set of static configurations of sort Λ, and Sta for the set of all static configurations.

Most of the properties have appeared in the earlier discussion, but it is worth commenting on why they hold again. Property (i) says a signal of no strength has value Z and *vice versa*; this fits the interpretations of 0 as mim-inmum strength and Z as negligible value. Naturally the contribution to the value from internal sources cannot exceed the resultant value—property (ii). Property (iii) says flow cannot go from a point at weaker strength to a point at greater strength. This is clear by considering how a relation $\alpha \rightsquigarrow \beta$ is built out of a chain of relations $\alpha \rightsquigarrow_1 \cdots \rightsquigarrow_1 \beta$ and noting (iii) holds across each single step $\rightsquigarrow_1$. Property (iv) formalises the idea that $\rightsquigarrow$ represents the direction in which values at points contribute to values at other points. The property (v) states that if $\alpha \rightsquigarrow \beta$ and $S(\beta)$ is a capacitance strength then $S(\alpha)$ is the same strength. This follows as capacitance strengths are always ranked below those of conductance in the strength order. Assume $\alpha \rightsquigarrow_1 \beta$ and $S(\beta) \in K$. Then $S(\alpha) \cdot g = S(\beta)$ for some conductance strength g for which $S(\beta) \leq g$. This can only occur with $S(\alpha) = S(\beta)$. The property follows when $\alpha \rightsquigarrow \beta$ because this relation is built out of steps $\rightsquigarrow_1$. Property (vi) follows in a similar way. It expresses the fact that if $\alpha \rightsquigarrow \beta$ and α and β are at the same strength then $\beta \rightsquigarrow \alpha$, so α and β are connected. To justify (vi) assume that $\alpha \rightsquigarrow_1 \beta$ and $S(\alpha) = S(\beta)$. Then $\alpha \rightsquigarrow_1 \beta$ arises iff there is a conductance, of strength g say, between α and β so that $S(\alpha) \cdot g = S(\beta)$. Hence $S(\beta) \cdot g = S(\alpha)$ too making $\beta \rightsquigarrow_1 \alpha$.

There are sufficiently many axioms, and the idea of static configuration sufficiently complicated, to raise the question of their completeness—do static configurations all share some property that does not follow from the axioms (i)–(vi)? We have argued for their soundness. To show completeness we need an argument showing that there are no properties shared by all static configurations of circuit terms which do not follow from those written down in the definition. Later, we shall show that every structure $\langle S, V, I, \rightsquigarrow \rangle$ on a finite set of points which satisfies all the axioms (i)–(vi) can be realised as the static configuration of a circuit built–up from resistances and sources connected to those points. Then any property of structures $\langle S, V, I, \rightsquigarrow \rangle$ which holds of all static configurations of circuits must also hold of all finite structures satisfying the definition above.

In a static configuration, two points α and β may be both in the relation $\alpha \rightsquigarrow \beta$ and $\beta \rightsquigarrow \alpha$. In this case the points α and β must have the same S, V and I values. Sometimes neither $\alpha \rightsquigarrow \beta$ nor $\beta \rightsquigarrow \alpha$, *i.e.* the two points are incomparable with respect to the flow relation $\rightsquigarrow$. This means that the points are effectively disconnected. We introduce notation to describe these circumstances.

Definition. Let $\sigma = \langle S, V, I, \rightsquigarrow \rangle$ be a static configuration of sort Λ. For $\alpha, \beta \in \Lambda$ define

$$\alpha \sim \beta \equiv \alpha \rightsquigarrow \beta \wedge \beta \rightsquigarrow \alpha$$

$$\alpha \parallel \beta \equiv \neg(\alpha \rightsquigarrow \beta) \wedge \neg(\beta \rightsquigarrow \alpha).$$

In the special case where the effects of resistance are ignored there is only only one perfect conductance strength ∞, the flow relation $\rightsquigarrow$ coincides with the equivalence relation $\sim$. In this situation any two points are either connected or disconnected.

Proposition. *Assume a static configuration σ has the property that for all points x in the sort of σ, either $S(x) = 0$ or $S(x) = \infty$. Then in σ the relation $\rightsquigarrow$ is symmetric and so an equivalence relation on points with $x \rightsquigarrow y$ iff $x \sim y$ for all points x, y of σ.*

Proof. The proof is a little exercise in using the axioms which define a static configuration. ∎

Now we present the obvious first-order syntax for describing static configurations. Static configurations are structures $\sigma = \langle S, V, I, \rightsquigarrow \rangle$. As such they can be described by predicates based on the functions S, V, I, relation $\rightsquigarrow$ together with relations of equality. The axioms for static configurations are examples of the kind of predicate we have in mind. Also, for example, the predicate

$$(V(\gamma) = H) \rightarrow (\alpha \rightsquigarrow \beta \wedge \beta \rightsquigarrow \alpha)$$

which is true of all those static configurations where if the point γ is high then the points α and β are connected. As this predicate suggests, such assertions will be useful for giving the meaning of circuit terms as sets of static configurations and for specifying their behaviour. The treatment is a little different from that usual for predicate calculus because we shall allow point names in assertions which apply to static configurations where the names do not appear in the sort. However variables bound by quantifiers only range over points in the sort. Actually we are presenting a simple example of a "free logic" where terms need not necessarily denote; it is of the kind discussed in [8] for instance.

Notation. Because we shall work with partial functions, mathematical expressions may sometimes be undefined. We indicate a mathematical expression e is defined by writing $e\downarrow$.

We first present the formal syntax of the language.

The syntax of assertions: *Variables:* We assume a set of variables Var which range over points Π, and have typical members $x, y, z, \cdots$.

Value terms: Terms, denoting values in $\{H, L, Z, X\}$ have the form

$$t ::= H \mid L \mid Z \mid X \mid V(\pi) \mid I(\pi),$$

where π is a point term, *i.e.* an element of Π or a variable in Var.

Strength terms: Terms, denoting values in $\mathbf{S} = K \cup G \cup \{0, \infty\}$ have the form

$$e ::= s \mid S(\pi) \mid e \cdot e \mid e + e$$

where $s \in \mathbf{S}$, and π is a point term.

Assertions for static configurations: The set of assertions is generated by

$$\begin{aligned} \phi ::= & \pi_0 = \pi_1 \mid \pi_0 \rightsquigarrow \pi_1 \mid \\ & t_0 = t_1 \mid t_0 \leq t_1 \mid \\ & e_0 = e_1 \mid e_0 \leq e_1 \mid \\ & \mathrm{tt} \mid \mathrm{ff} \mid \phi \wedge \phi \mid \phi \vee \phi \mid \neg\phi \mid \\ & \exists x.\phi \mid \forall x.\phi \end{aligned}$$

where $\pi_0.\pi_1$ are point terms, t_0, t_1 are value terms and e_0, e_1 are strength terms. We shall regard $\phi \rightarrow \theta$ as abbreviating $\neg\phi \vee \theta$ and $\phi \leftrightarrow \theta$ as abbreviating $\phi \rightarrow \theta \wedge \theta \rightarrow \phi$, and use previously introduced abbreviations like $x \sim y$ for $x \rightsquigarrow y \wedge y \rightsquigarrow x$.

The semantics of assertions: We give the semantics of closed assertions. For brevity, assume a static configuration σ has the form $\langle S_\sigma, V_\sigma, I_\sigma, \rightsquigarrow_\sigma \rangle$. First we treat value terms which are denoted by partial functions Sta $\rightharpoonup$ **V**, as follows:

$$[\![H]\!]\sigma = H \text{ for all } \sigma \in \text{Sta}$$
$$[\![L]\!]\sigma = L \text{ for all } \sigma \in \text{Sta}$$
$$[\![Z]\!]\sigma = Z \text{ for all } \sigma \in \text{Sta}$$
$$[\![X]\!]\sigma = X \text{ for all } \sigma \in \text{Sta}$$
$$[\![V(\alpha)]\!]\sigma = \begin{cases} V_\sigma(\alpha) & \text{if } \sigma \in \text{Sta and } \alpha \in \text{sort}(\sigma), \\ \text{undefined} & \text{otherwise,} \end{cases}$$
$$[\![I(\alpha)]\!]\sigma = \begin{cases} I_\sigma(\alpha) & \text{if } \sigma \in \text{Sta and } \alpha \in \text{sort}(\sigma), \\ \text{undefined} & \text{otherwise.} \end{cases}$$

Strength terms are denoted by partial functions Sta $\rightharpoonup$ **S**, as follows:

$$[\![s]\!]\sigma = s \text{ for all } \sigma \in \text{Sta}$$
$$[\![S(\alpha)]\!]\sigma = \begin{cases} S_\sigma(\alpha) & \text{if } \sigma \in \text{Sta and } \alpha \in \text{sort}(\sigma), \\ \text{undefined} & \text{otherwise,} \end{cases}$$
$$[\![e_0 \cdot e_1]\!]\sigma = \begin{cases} s_0 \cdot s_1 & \text{if } [\![e_0]\!]\sigma = s_0 \text{ and } [\![e_1]\!]\sigma = s_1 \\ \text{undefined} & \text{otherwise,} \end{cases}$$
$$[\![e_0 + e_1]\!]\sigma = \begin{cases} s_0 + s_1 & \text{if } [\![e_0]\!]\sigma = s_0 \text{ and } [\![e_1]\!]\sigma = s_1 \\ \text{undefined} & \text{otherwise.} \end{cases}$$

Each assertion is denoted by the subset of static configurations at which it is true, defined by the following induction on the length of assertions:

$$[\![\alpha_0 = \alpha_1]\!] = \{\sigma \in \text{Sta} \mid \alpha_0 \in \text{sort}(\sigma) \text{ and } \alpha_1 \in \text{sort}(\sigma) \text{ and } \alpha_0 = \alpha_1\}$$
$$[\![\alpha_0 \rightsquigarrow \alpha_1]\!] = \{\sigma \in \text{Sta} \mid \alpha_0 \in \text{sort}(\sigma) \text{ and } \alpha_1 \in \text{sort}(\sigma) \text{ and } \alpha_0 \rightsquigarrow_\sigma \alpha_1\}$$
$$[\![t_0 = t_1]\!] = \{\sigma \in \text{Sta} \mid [\![t_0]\!]\sigma \downarrow \text{ and } [\![t_1]\!]\sigma \downarrow \text{ and } [\![t_0]\!]\sigma = [\![t_1]\!]\sigma\}$$
$$[\![t_0 \leq t_1]\!] = \{\sigma \in \text{Sta} \mid [\![t_0]\!]\sigma \downarrow \text{ and } [\![t_1]\!]\sigma \downarrow \text{ and } [\![t_0]\!]\sigma \leq [\![t_1]\!]\sigma\}$$
$$[\![e_0 = e_1]\!] = \{\sigma \in \text{Sta} \mid [\![e_0]\!]\sigma \downarrow \text{ and } [\![e_1]\!]\sigma \downarrow \text{ and } [\![e_0]\!]\sigma = [\![e_1]\!]\sigma\}$$
$$[\![e_0 \leq e_1]\!] = \{\sigma \in \text{Sta} \mid [\![e_0]\!]\sigma \downarrow \text{ and } [\![e_1]\!]\sigma \downarrow \text{ and } [\![e_0]\!]\sigma \leq [\![e_1]\!]\sigma\}$$
$$[\![\mathrm{tt}]\!] = \text{Sta}, \qquad [\![\mathrm{ff}]\!] = \emptyset$$
$$[\![\phi_0 \wedge \phi_1]\!] = [\![\phi_0]\!] \cap [\![\phi_1]\!] \text{ and } [\![\phi_0 \vee \phi_1]\!] = [\![\phi_0]\!] \cup [\![\phi_1]\!]$$
$$[\![\neg\phi]\!] = (\text{Sta} \setminus [\![\phi]\!])$$
$$[\![\exists x.\phi]\!] = \{\sigma \mid \exists \alpha \in \text{sort}(\sigma).\ \sigma \in [\![\phi[\alpha/x]]\!]\}$$
$$[\![\forall x.\phi]\!] = \{\sigma \mid \forall \alpha \in \text{sort}(\sigma).\ \sigma \in [\![\phi[\alpha/x]]\!]\}$$

This completes our formal language for describing static configurations. We shall write $\sigma \models \phi$ iff $\sigma \in [\![\phi]\!]$. In fact, because of the completeness theorem for such (free) predicate calculi, there is a proof system very like that for ordinary predicate calculus so that $\phi_0 \vdash \phi_1$ (*i.e.* ϕ_1 can be proved from ϕ_0) iff $[\![\phi_0]\!] \subseteq [\![\phi_1]\!]$ (*i.e.* ϕ_0 entails ϕ_1)—see [10] for more details.

3. The Static Behaviour of Circuits.

It remains to give a semantics to circuit terms. Each circuit term is to be denoted by the set of static configurations it can be in. This depends on being able to characterise the static configurations of a circuit $c_0 \bullet c_1$ in terms of those of c_0 and c_1, and the static configurations of a circuit $c \setminus \alpha$ in terms of those of c. The static configurations of c are its steady states in any environment. Hence a static configuration σ of $c_0 \bullet c_1$ should be obtained as a composition, call it $\sigma_0 \bullet \sigma_1$, of static configurations σ_0 of c and σ_1 of c_1. We cannot expect all such compositions $\sigma_0 \bullet \sigma_1$ to make sense. Afterall, quite possibly σ_0 and σ_1 may disagree on the value $V(\alpha)$ or strength $S(\alpha)$ at a point in common to both their sorts. Similarly, a static configuration σ of $c \setminus \alpha$ should be obtained by restricting a static configuration ρ of c to sort $\mathrm{sort}(\sigma) \setminus \{\alpha\}$ but only in the case where in ρ there is no essential contribution to the value $V(\alpha)$ through the point α. This motivates the introduction of partial functions of type

$$\bullet : \mathrm{Sta} \times \mathrm{Sta} \rightharpoonup \mathrm{Sta},$$
$$\setminus \alpha : \mathrm{Sta} \rightharpoonup \mathrm{Sta},$$

for any $\alpha \in \Pi$. The denotation of $c_0 \bullet c_1$ and $c \setminus \alpha$ will be derived from them.

Notation. Let $\Lambda \subseteq \Pi$. Write $\Lambda \setminus \alpha$ for $\Lambda \setminus \{\alpha\}$. Let $f : \Lambda \to U$ be a function on Λ. Write $f \setminus \alpha$ for $f \lceil (\Lambda \setminus \alpha)$, the function restricted to $\Lambda \setminus \{\alpha\}$. Let $R \subseteq \Lambda \times \Lambda$ be a relation. Write $R \setminus \alpha$ for the restriction $R \cap (\Lambda \setminus \alpha)^2$.

Assume σ_0 is a static configuration of a circuit c_0 and σ_1 is a static configuration of a circuit c_1. When can σ_0 and σ_1 be composed to give a static configuration of c_0 and c_1? When their strengths and values agree at common points; only then do σ_0 and σ_1 make consistent assumptions about the environment. Then the resulting flow relation should be the transitive closure of the flow relations in the components and the internal-value function should be spread out accordingly.

Definition. Let $\sigma_0 = \langle S_0, V_0, I_0, \rightsquigarrow_0 \rangle$ be a static configuration of sort Λ_0 and $\sigma_1 = \langle S_1, V_1, I_1, \rightsquigarrow_1 \rangle$ be a static configuration of sort Λ_1. Define their *composition* to be

$$\sigma_0 \bullet \sigma_1 = \begin{cases} \langle S, V, I, \rightsquigarrow \rangle & \text{if } S_0 \lceil \Lambda_1 = S_1 \lceil \Lambda_0 \text{ and} \\ & \quad V_0 \lceil \Lambda_1 = V_1 \lceil \Lambda_0 \\ \text{undefined} & \text{otherwise} \end{cases}$$

where

$$\begin{aligned} S &= S_0 \cup S_1, \\ V &= V_0 \cup V_1, \\ \rightsquigarrow &= (\rightsquigarrow_0 \cup \rightsquigarrow_1)^* \quad \text{and} \\ I(\alpha) &= \Sigma\{I_0(\beta) \mid \beta \in \Lambda_0 \text{ and } \beta \rightsquigarrow \alpha\} + \\ &\quad \Sigma\{I_1(\beta) \mid \beta \in \Lambda_1 \text{ and } \beta \rightsquigarrow \alpha\} \end{aligned}$$

for any $\alpha \in \Lambda_0 \cup \Lambda_1$.

Suppose σ is a static configuration of a circuit c. When does σ restrict to a configuration of the circuit $c \setminus \alpha$ in which the point α has been hidden? When the value (and strength) at α result from the combined effect of internal sources and the contribution from the unhidden points $\Lambda \setminus \alpha$, more precisely when

$$V(\alpha) = I(\alpha) + \Sigma\{V(\beta) \mid \beta \in \Lambda \setminus \alpha \text{ and } \beta \rightsquigarrow \alpha\}.$$

Then the hiding of α makes no difference. Of course hiding a point α can make no difference to a circuit which does not have α in its sort and we take care of this case too in the following definition.

Definition. Let $\sigma = \langle S, V, I, \rightsquigarrow\rangle$ be a static configuration of sort Λ and α a point in Π. Define *hiding*

$$\sigma \setminus \alpha = \langle S \setminus \alpha,\ V \setminus \alpha,\ I \setminus \alpha,\ \rightsquigarrow \setminus \alpha\rangle$$

if $\alpha \notin \Lambda$ or $V(\alpha) = I(\alpha) + \Sigma\{V(\beta) \mid \beta \in \Lambda \setminus \alpha \text{ and } \beta \rightsquigarrow \alpha\}$, and to be undefined otherwise.

We take the behaviour of a circuit term to be the set of possible static configurations it can settle into in some static environment.

The semantics of circuit terms:

It is convenient to use $\{\sigma \mid \phi\}$ where ϕ is an assertion on static configurations to mean $\{\sigma \mid \sigma \models \phi\}$. Define the semantic function $[\![\ ]\!]$ from circuit terms to $P(\text{Sta})$ to be the map defined by the following structural induction.

$$[\![Pow\ (\alpha)]\!] = \{\sigma \in \mathrm{Sta}[\alpha] \mid S(\alpha) = \infty \wedge I(\alpha) = H\}$$

$$[\![Gnd\ (\alpha)]\!] = \{\sigma \in \mathrm{Sta}[\alpha] \mid S(\alpha) = \infty \wedge I(\alpha) = L\}$$

$$\begin{aligned}[\![cap_{kU}(\alpha)]\!] = \{\sigma \in \mathrm{Sta}[\alpha] \mid\ & S(\alpha) \geq k \wedge \\ & (S(\alpha) = k \rightarrow I(\alpha) = U) \wedge (S(\alpha) > k \rightarrow I(\alpha) = Z)\}\end{aligned}$$

$$[\![pt(\alpha)]\!] = \{\sigma \in \mathrm{Sta}[\alpha] \mid I(\alpha) = Z\}$$

$$[\![wre(\alpha, \beta)]\!] = \{\sigma \in \mathrm{Sta}[\alpha, \beta] \mid I(\alpha) = Z \wedge I(\beta) = Z \wedge \alpha \sim \beta\}$$

$$\begin{aligned}[\![res_g(\alpha, \beta)]\!] = \{\sigma \in \mathrm{Sta}[\alpha, \beta] \mid\ & I(\alpha) = Z \wedge I(\beta) = Z \wedge \\ & S(\alpha)\cdot g \leq S(\beta) \wedge S(\beta)\cdot g \leq S(\alpha) \wedge \\ & (S(\alpha)\cdot g = S(\beta) \leftrightarrow \alpha \rightsquigarrow \beta) \wedge (S(\beta)\cdot g = S(\alpha) \leftrightarrow \beta \rightsquigarrow \alpha)\}\end{aligned}$$

$$\begin{aligned}[\![ntran(\alpha, \beta, \gamma)]\!] = \{\sigma \in \mathrm{Sta}[\alpha, \beta, \gamma] \mid\ & I(\alpha) = Z \wedge I(\beta) = Z \wedge I(\gamma) = Z \wedge \\ & \gamma \parallel \alpha \wedge \gamma \parallel \beta \wedge (\alpha \parallel \beta \vee \alpha \sim \beta) \wedge \\ & (V(\gamma) = H \rightarrow \alpha \sim \beta) \wedge (V(\gamma) = L \rightarrow \alpha \parallel \beta)\}\end{aligned}$$

$$\begin{aligned}[\![ptran(\alpha, \beta, \gamma)]\!] = \{\sigma \in \mathrm{Sta}[\alpha, \beta, \gamma] \mid\ & I(\alpha) = Z \wedge I(\beta) = Z \wedge I(\gamma) = Z \wedge \\ & \gamma \parallel \alpha \wedge \gamma \parallel \beta \wedge (\alpha \parallel \beta \vee \alpha \sim \beta) \wedge \\ & (V(\gamma) = L \rightarrow \alpha \sim \beta) \wedge (V(\gamma) = H \rightarrow \alpha \parallel \beta)\}\end{aligned}$$

$$[\![c \bullet d]\!] = \{\sigma \bullet \rho \mid \sigma \in [\![c]\!] \text{ and } \rho \in [\![d]\!] \text{ and } \sigma \bullet \rho \downarrow\}$$

$$[\![c \setminus \alpha]\!] = \{\sigma \setminus \alpha \mid \sigma \in [\![c]\!] \text{ and } \sigma \setminus \alpha \downarrow\}.$$

As the reader will notice, even capacitance sources have fairly complicated behaviours involving conditional assertions, in order to cope with strengths and the fact that they can be overridden. Sources of power and ground have simpler definitions because they cannot be overridden. Resistances introduce the flow relation in a manner dependent on strengths, as we explained when introducing the flow relation. The behaviour of wires is got as a special case, by assuming perfect conductance. We stipulate that a transistor either connects or disconnects according to the voltage on the gate. Notice for both kinds of transistors a value of X on the gate can be associated with a static configuration where it is connecting or with one where it is disconnecting. The denotations of compositions and hiding of terms are obtained in a pointwise fashion.

Further terms can be defined. For instance, we can define a general source $sce_{sU}(\alpha)$, at α, of strength s and value $U \in \mathbf{V}$ where $s = 0$ iff $U = Z$, to have denotation

$$[\![sce_{sU}(\alpha)]\!] = \{\sigma \in \text{Sta}[\alpha] \mid S(\alpha) \geq s \wedge \\ (S(\alpha) = s \rightarrow I(\alpha) = U) \wedge (S(\alpha) > s \rightarrow I(\alpha) = \emptyset)\}$$

Such a source only makes a contribution at α if the strength of α is exactly s. If s is a capacitance strength k then such a weakened source with value $U = H$ or L can be realised by the charged capacitance $cap_{kU}(\alpha)$. Similarly, if s is a conductance strength g it can be realised by passing signals from ground or power through a resistance of strength g. Sources with value X can be got by connecting two sources with values H and L at a common point. Writing $c = c'$ iff $[\![c]\!] = [\![c']\!]$, for circuit terms c, c', we have

$$sce_{kH}(\alpha) = cap_{kH}(\alpha), \qquad sce_{kL}(\alpha) = cap_{kL}(\alpha),$$
$$sce_{gH}(\alpha) = (Pow\ (\beta) \bullet res_g(\beta, \alpha)) \setminus \beta, \ sce_{gL}(\alpha) = (Gnd\ (\beta) \bullet res_g(\beta, \alpha)) \setminus \beta,$$
$$sce_{gX}(\alpha) = sce_{gH}(\alpha) \bullet sce_{gL}(\alpha).$$

These are examples of some basic equivalences between circuit terms, which can be proved from the denotational semantics. Of course, such ideas were used informally in motivating the definition of static configuration. The one remaining case, $sce_{0Z}(\alpha)$ can be realised as a single connection point standing alone, *i.e.*

$$sce_{0Z}(\alpha) = pt(\alpha).$$

To explain the way sources combine, it is helpful to use an ordering between pairs sU, where $s \in \mathbf{S}$ and $U \in \mathbf{V}$ and $s = 0$ iff $U = Z$. On such pairs define

$$sU \leq s'U' \text{ iff } s < s' \text{ or } (s = s' \text{ and } U \leq U').$$

This forms a finite distributive lattice (meet $\cdot$ and join $+$), mentioned in Bryant's thesis and Hayes work [4], which may be drawn as:

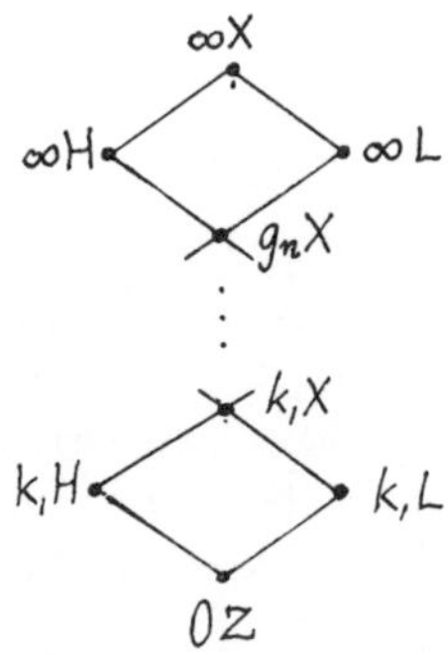

The way sources compose can now be summarised by:

$$sce_{sU}(\alpha) \bullet sce_{s'U'}(\alpha) = sce_{sU+s'U'}(\alpha).$$

As further examples of equivalences on terms which follow from the semantics we look at composing resistances in series and parallel:

$$(res_g(\alpha, \beta) \bullet res_{g'}(\beta, \gamma)) \setminus \beta = res_{g \cdot g'}(\alpha, \gamma).$$
$$(res_g(\alpha, \beta) \bullet res_{g'}(\alpha, \beta)) = res_{g+g'}(\alpha, \beta).$$

In other words combined in series two resistances yield a resistance of conductance strength the minimum of the strengths of the components, while combined in parallel they are equivalent to one of strength the join. Again this agrees with the informal explanation given earlier but this time it follows from the general definition of the semantics of circuit terms.

Some equivalences highlight problems with the model. For example, according to the model we could produce a wire by sealing in a source to the gate of a transistor:

$$\begin{aligned} wre(\alpha, \beta) &= (Pow\ (\alpha) \bullet ntran(\alpha, \beta, \gamma)) \setminus \gamma \\ &= (Gnd\ (\alpha) \bullet ptran(\alpha, \beta, \gamma)) \setminus \gamma. \end{aligned}$$

A first complaint is that this is unrealistic because transistors have significant resistance. But this objection can be catered for simply by building in resistances. More importantly it indicates how effects caused by transistors having thresholds are ignored by the model, a topic we shall return to in the final section.

Earlier we pointed out the problem of knowing whether or not we had written down sufficient axioms for static configurations. It is sufficient to show every finite static configuration can be realised as a static configuration of a circuit term. This is established in the following proposition.

Proposition. *Any static configuration σ of sort Λ a finite subset of Π is the static configuration of some circuit c i.e. $\sigma \in [\![c]\!]$.*

Proof. Given σ, the idea is to define the required circuit to be the composition of the following finite set of components:

$$\{sce_{S(\alpha)U}(\alpha) \mid \alpha \in \Lambda \text{ and } I(\alpha) = U \neq Z\} \cup$$
$$\{res_g(\alpha, \beta) \mid \alpha \rightsquigarrow \beta \text{and } g \in G \text{ is the least s.t. } S(\alpha) \cdot S(\beta) \leq g\}.$$

This uses our knowledge of how to build all sources $sce_{sU}(\alpha)$ as circuits. From the definition of sources, resistances and composition, it follows that $\sigma \in [\![c]\!]$. ∎

We would like a language of assertions with which to specify properties of circuits. A circuit denotes a subset of static configurations Sta. A circuit specification *Spec* should pick out the subsets of circuits which satisfy it, and so semantically should denote a subset $[\![Spec]\!] \subseteq P(\text{Sta})$ of the powerset of static configurations. We have already introduced assertions ϕ describing static configurations and, like circuits, they denote subsets $[\![\phi]\!] \subseteq \text{Sta}$, so as they stand they do not have the right type to be taken as circuit specifications. However, there is a standard way to to take assertions of type Sta and make them into assertions of type $P(\mathrm{Sta})$. For ϕ an assertion about static configurations, define $\diamond\,\phi$ and $\Box\,\phi$ to be circuit specifications with denotations

$$[\![\diamond\,\phi]\!] = \{C \in P(\mathrm{Sta}) \mid C \cap [\![\phi]\!] \neq \emptyset\},$$
$$[\![\Box\,A]\!] = \{C \in P(\mathrm{Sta}) \mid C \subseteq [\![\phi]\!]\}.$$

A circuit term c satisfies $\diamond\,\phi$ iff $[\![c]\!] \in [\![\diamond\,\phi]\!]$, *i.e.* $\exists \sigma \in [\![c]\!].\ \sigma \models \phi$, while c satisfies $\Box\,\phi$ iff $[\![c]\!] \subseteq [\![\phi]\!]$, *i.e.* $\forall \sigma \in [\![c]\!].\ \sigma \models \phi$. Thus a circuit satisfies $\diamond\,\phi$ means it *can* satisfy ϕ, whereas it satisfies $\Box\,\phi$ means it *must* satisfy ϕ, which is consistent with the way these two words were used informally in section 2. Compound specifications may be made by combining these basic specifications using the usual logical operations. The operators $\diamond$ and $\Box$ are then each definable from the other as $\diamond\,\phi$ is equivalent to $\neg\Box\,\neg\phi$ (*i.e.* they are both satisfied by the same things), and $\Box\,\phi$ is equivalent to $\neg\diamond\,\neg\phi$. However both $\diamond$ and $\Box$ specifications characterise circuit behaviour in the sense of the following proposition. A circuit specification *Spec* denotes a subset $[\![S]\!] \subseteq P(\mathrm{Sta})$, and, for a circuit term c, we define

$$c \models Spec \Leftrightarrow [\![c]\!] \in [\![Spec]\!],$$

when we say a circuit term c satisfies *Spec*.

Proposition. *Let c and c' be a circuit terms. Define*

$$c \preceq_{can} c' \Leftrightarrow \forall \text{ assertions } \phi.\ c \models \diamond\,\phi \Rightarrow c' \models \diamond\,\phi, \text{ and}$$
$$c \preceq_{must} c' \Leftrightarrow \forall \text{ assertions } \phi.\ c \models \Box\,\phi \Rightarrow c' \models \Box\,\phi.$$

Then

$$c \preceq_{can} c' \Leftrightarrow [\![c]\!] \subseteq [\![c']\!], \text{ and}$$
$$c \preceq_{must} c' \Leftrightarrow [\![c']\!] \subseteq [\![c]\!].$$

Hence if two circuit terms c, c' satisfy the same specifications $\diamond\,\phi$, or the same specifications $\Box\,\phi$, then $c = c'$ (i.e. they have the same denotation).

Proof. The proof use the fact that any finite configuration σ is a characterised by an assertion A_σ in the sense that

$$\rho \models A_\sigma \Leftrightarrow \rho = \sigma.$$

Suppose $c \preceq_{can} c'$. If $\sigma \in [\![c]\!]$ then $c \models \diamond\, A_\sigma$ so $c' \models \diamond\, A_\sigma$, whence $\sigma \in [\![c']\!]$. Hence $c \preceq_{can} c' \Rightarrow [\![c]\!] \subseteq [\![c]\!]'$, and the converse is obvious.

Suppose $c \preceq_{must} c'$. Clearly $c \models \Box \bigvee_{\sigma \in [\![c]\!]} A_\sigma$. It follows that $c' \models \Box \bigvee_{\sigma \in [\![c]\!]} A_\sigma$. If $\rho \in [\![c']\!]$ then $\rho \models \bigvee_{\sigma \in [\![c]\!]} A_\sigma$ which implies $\rho = \sigma$ for some $\sigma \in [\![c]\!]$. Hence $c \preceq_{must} c' \Rightarrow [\![c']\!] \subseteq [\![c]\!]$, and the converse is obvious. ∎

Now we see how to give a specification of the NMOS inverter. The specification takes account of the fact that the strengths of its outputs are not the same. Its circuit is constructed by the term

$$c \equiv (ntran(\alpha, \beta, \gamma) \bullet res_g(\beta, \delta) \bullet Pow\ \delta \bullet Gnd\ (\alpha)) \setminus \alpha \setminus \delta.$$

Informally, its output β behaves like a direct connection to ground when its input γ is high and like a weakened power source when γ is low. Formally, if we define

$$\begin{aligned} Sce_{sU}(x) \equiv\; & s \leq S(x) \wedge \\ & s = S(x) \rightarrow I(x) = U \wedge \\ & s < S(x) \rightarrow I(x) = Z, \end{aligned}$$

then we can say c satisfies the specification

$$\begin{aligned} \Box\, (& V(\gamma) = H \rightarrow Sce_{\infty L}(\beta) \wedge \\ & V(\gamma) = L \rightarrow Sce_{gH}(\beta)). \end{aligned}$$

We would like a proof system in which to demonstrate that a circuit meets a specification. In [10] a complete proof system is presented for proving a circuit term satisfies a specification of the form $\Box\, \phi$. A proof system incorporating specifications $\Diamond\, \phi$ will most likely follow similar lines, though it has not been done yet.

To conclude this section, we note an inadequacy which arises if we restrict to specifications of the form $\Box\, \phi$. A related difficulty appears in the model of [2, 3] where terms like *Pow* $(\alpha) \bullet$ *Gnd* (α) cause problems because they denote the empty relation and satisfy any specification. We have

$$[\![Pow\ (\alpha) \bullet Gnd\ (\alpha)]\!] \neq \emptyset,$$

and so, when we come to the logic, the circuit *Pow* $(\alpha) \bullet$ *Gnd* (α) will not satisfy $\Box$ ff. However, there are other circuit terms which do denote $\emptyset$ and so will satisfy $\Box$ ff, and indeed any specification $\Box\, \phi$. These are terms which represent circuits whose only possible behaviour is to oscillate. For example the term

$$c \equiv (ntran(\alpha, \beta, \gamma) \bullet res_g(\beta, \delta) \bullet Pow\ \delta \bullet Gnd\ \gamma \bullet wre(\gamma, \beta)) \setminus \alpha \setminus \beta \setminus \gamma \setminus \delta$$

with g a conductance strength strictly between 0 and ∞, "ties–back" the output of an NMOS inverter to its input, and then insulates all points from the environment. It can be drawn as:

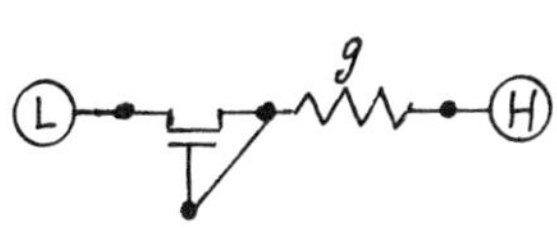

The circuit c denotes the emptyset, $[\![c]\!] = \emptyset$. The circuit is a little peculiar in that it has an empty sort. However composing it with any circuit with nonempty sort yields a circuit with nonempty sort which denotes $\emptyset$. For this reason the proof system for circuits in [10], where all specifications have the form $\Box\, \phi$, is akin to the logic of partial correctness assertions (Hoare logic); there a purely oscillating circuit satisfies $\Box\, \phi$ for any assertion ϕ just as a diverging program satisfies all partial correctness assertions. However, a purely oscillating circuit like that above will not satisfy $\Diamond$ *true*, so a proof system extended to include specifications $\Diamond\, \phi$ would not have this inadequacy.

4. Problems.

A compositional model for the static behaviour of MOS circuits has been presented. Certainly such a model is necessary for a satisfactory treatment of dynamically changing circuits. Their behaviour can often be viewed as going through a sequence of static configurations as data changes in synchrony with one or several clocks. The data and the clock pulses are held long enough for the circuit to settle into a static configuration. It seems obvious therefore to represent a possible history for such a circuit as a sequence of static configurations. Copying the approach in [2] we could try to extend our work to the dynamic case by simply including a time parameter t, a natural number, and associating devices with assertions expressing time dependencies between static configurations at different times. For example, then we could express facts like a capacitance would contribute a charge at time $t+1$ if it was precharged at time t. While this works reasonably well for circuits with purely dynamic memory there is a hard problem in modelling static memory accurately. The main difficulty, it seems, is that short-term capacitance effects influence the physical behaviour but are not captured easily in any extension of our model. The strength orders we use assume that the environment is stable long enough for sources of current to override charges stored by capacitance. This assumption breaks down in some stages needed to explain the behaviour of devices like the following register:

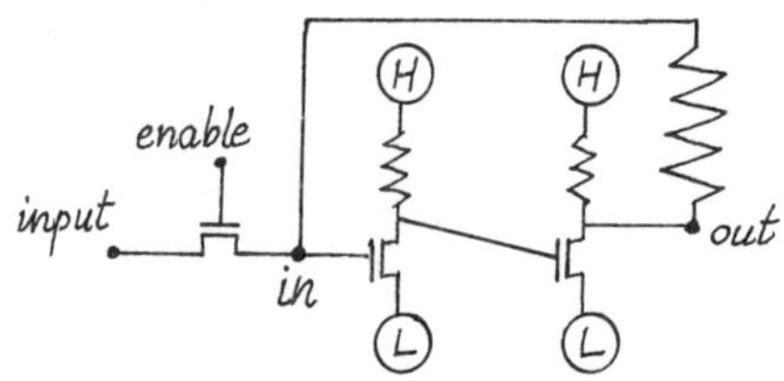

When *enable* is high, a strong signal at *in* overrides that already present, and its value is established, after two inversions, at *out*. When *enable* becomes low the value is preserved at *out* (the input value is stored). Speaking loosely, the latter stage relies on capacitance to maintain the value at *in* until the feedback loop takes over. This occurs over a very short time when the assumptions behind the strength order are violated. I cannot, at present, see how to extend the model to account for effects over such a short time. Through failing to cope adequately with such short–term effects, the obvious model I have sketched allows more possibilities than are physically possible. By the way, the work of Bryant does not address this problem; a simulator need only generate one possible course of behaviour and Bryant does this by making a unit–delay assumption. (In his simulators all transistors switch with the same delay after their gates change.)

Even for just static behaviour there are inadequacies in the model. Chief among them, it seems, is that it fails to cope with the fact that transistors do not behave purely as switches which connect or disconnect according to the voltage on the gate. An n-type transistor $ntran(\alpha, \beta, \gamma)$ only connects when the difference between the voltage on its gate γ and the minimum of that on α and

β exceeds a certain positive threshold. This is ignored in switch level models like those of [1] and the model here. Consequently the model we have described allows certain static configurations which are not physically feasible. Consider the term $Pow\ (\alpha) \bullet ntran(\alpha, \beta, \gamma) \bullet wre(\alpha, \beta)$.

It has a static configuration in which 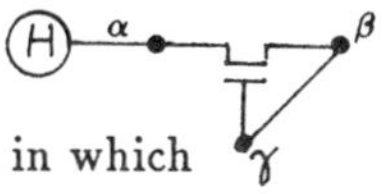

$$I(\alpha) = I(\beta) = V(\alpha) = V(\beta) = H \text{ and } \alpha \sim \beta.$$

This is not possible for a real n-type transistor. The gate γ and the source β have the same voltage so the difference in their voltages cannot exceed the positive threshold necessary for α and β to connect. Similarly the term $Pow\ (\alpha) \bullet ntran(\alpha, \beta, \gamma) \bullet wre(\alpha, \beta) \bullet Gnd\ (\beta)$ has a static configuration in which

$$I(\alpha) = I(\beta) = V(\alpha) = V(\beta) = X \text{ and } \alpha \sim \beta.$$

Again this is not realistic.

Such inaccuracies can lead to an incorrect circuit being proved correct and a correct circuit being proved incorrect. For example according to the model presented the circuit

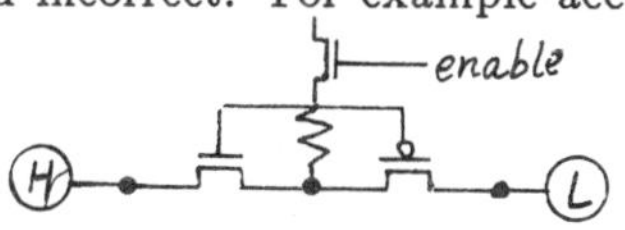

can store high and low, though showing it acts like a register encounters the same problems as for the real register above. It will not in reality because of thresholds. The following diagram describes a circuit which does compute exclusive-or (it is taken from [9], Figure 8.7, page 317).

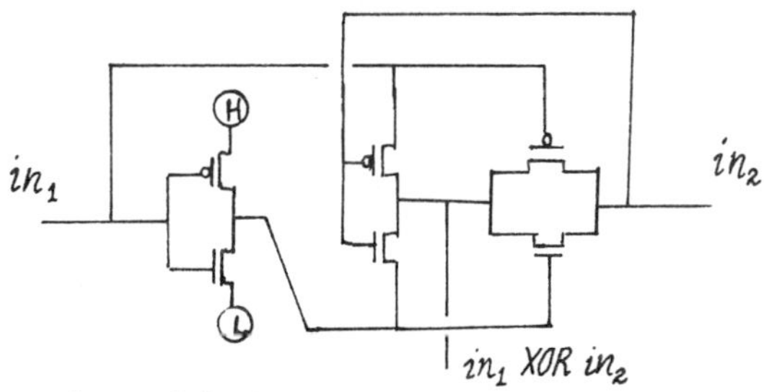

However, according to the model when both inputs in_1, and in_2 are high not only may the output be low, as required but also X–a possibility that is ruled out in reality because of thresholds.

Interestingly, although the model is inaccurate, we could not prove the design incorrect according to the model if we just use specifications of the form $\Box\ \phi$. Certainly if the design satisfies $\Box\ \phi$ according to the model then it will in reality, because the model allows more static configurations than are really possible. We could not prove it correct either. There is a $\Diamond$-assertion which holds of the circuit according to our model, *viz.* $\diamond A_\sigma$ where A_σ is the assertion characterising the unrealistic static configuration we have described, which does not hold of the real design.

It looks as if the model can be extended to cope with thresholds while keeping the same qualitative style. It appears to need another attribute to be attached to static configurations—a measure at each point of how strongly the value has been degraded. N-type transistors can transmit high values. But they do so at the cost of degrading the voltage to that of roughly the gate voltage minus the threshold. The measure of degradation would count the number of such degradations. The idea needs to be worked through more carefully however, and a decision taken as to how to manage different thresholds values and their relationship to each other.

Much needs to be done. Extending the model to sequential circuits with static memory seems hard. Coping with thresholds to give a more accurate treatment of static behaviours seems more tractable. Dealing with it will surely press home the need for a more careful analysis of the relationship between qualitative models of the kind here and explanations based more closely on the physics of devices. Some work, [6] and [11], has been done on the relationship between different models of hardware. But the development of a general framework in which to carry out the verification of hardware at all stages in its development, from logical design to layout, calls for a spectrum of models at different levels of abstraction with formal relations between them. We cannot claim to have anything like this at present.

Acknowledgements.

I would like to thank the members of the Hardware Verification Group here in Cambridge for help and encouragement and especially Mike Gordon for laying bare some of the issues in modelling and verifying hardware which led to this work. Mark Greenstreet helped explain the difficulties which arise from ignoring switching thresholds in transistors. Randy Bryant has been encouraging. Finally many thanks to Graham Birtwistle for his organisation and encouragement.

References.

[1] Bryant, R.E., "A switch-level model and simulator for MOS digital systems", IEEE Transactions on Computers C-33 (2) pp. 160–177, February 1984.

[2] Camilleri, A., Gordon, M., and Melham, T., "Hardware verification using higher order logic", To appear in the proceedings of the IFIP International working conference, Grenoble, France, September 1986. Also available as a report 91 of the Computer Laboratory, University of Cambridge, 1986.

[3] Gordon, M.J.C., "Why higher order logic is a good formalism for specifying and verifying hardware", in "Formal aspects of VLSI design: Proceedings of the 1985 Edinburgh Conference on VLSI", G.J.Milne and P.A.Subrahmanyam, eds., North-Holland, Amsterdam, 1986.

[4] Hayes, J., "A unified switching theory with applications to VLSI design", Proc. IEEE 70 (10) pp. 1140–1155 Oct. 1982.

[5] Hoare, C.A.R., "Communicating sequential processes", Prentice-Hall, 1985.

[6] Melham, T.F., "Abstraction Mechanisms for Hardware Verification" in: VLSI Specification, Verification and Synthesis G.M. Birtwistle and P.A. Subrahmanyam, eds, (this volume).

[7] Milner, R., A Calculus of Communicating Systems. Springer Lecture Notes in Comp. Sc. vol. 92, 1980.

[8] Scott, D.S., Identity and existence in intuitionistic logic. In proc. of Durham conference on Applications of Sheaves 1977, Lecture notes in Math., Springer–Verlag, 1979.

[9] Weste, N., and Eshraghian, K., "Principles of CMOS VLSI Design", Addison Wesley Publishing Company, 1985.

[10] Winskel, G., "Lectures on models and logic of MOS circuits", Proceedings for the Marktoberdorf Summer School, July 1986. Being published by Springer.

[11] Winskel, G., "Relating two models of hardware", Submitted to the conference "Category theory and computer science", Edinburgh, September 1987.

12

A Tactical Framework for Hardware Design

A Tactical Framework for Hardware Design*

Steven D. Johnson, Bhaskar Bose, and C. David Boyer
Computer Science Department
Indiana University
Bloomington, Indiana

Introduction

This work explores an algebraic approach to digital design. A collection of transformations has been implemented to derive circuits, but "hardware compilation" is not a primary goal. Paths to physical implementation are needed to explore the approach at practical levels of engineering. The potential benefits are more descriptive power and a unified foundation for design automation. However, these prospects mean little when the algebra is done manually. Hence, a basic, automated algebra is prerequisite to our experimentation with design methods.

The transformation system discussed in this paper supports a basic design algebra. At this early stage of its development, the system is essentially an editor for deriving digital system descriptions from certain applicative specifications and for interactively manipulating them. It is integrated with existing assemblers for programmable hardware packages, such as the PAL components used to implement the examples presented in Sections 3 and 5. The transformation system and back-end assemblers are implemented in Scheme [18], a Lisp dialect, by Bose [2].

The coherent manipulation of design descriptions is a significant problem. Digital engineering is governed by a variety of theories addressing orthogonal problem aspects. It is not enough to unify these treatments in some

*Research reported herein was supported, in part, by the National Science Foundation, under grants numbered DCR 84-05241 and DCR 85-21497, and by the Westinghouse Educational Foundation.

metalanguage, for they must also be integrated. In any design instance, independent aspects become dependent facets, and hence, designs simultaneously decompose in distinct hierarchies. In this paper, we look at the separation of algorithm and architecture, the logical hierarchy of functional units, and the organization of physical components. The direction of this work is to find ways to maintain orthogonal views of a description as it is manipulated toward an implementation.

The examples in this paper are implemented as externally clocked systems, often called "synchronous systems" or "clocked sequential systems." The underlying model of sequential behavior is simply realized by a global clock. There is nothing to preclude other realizations of timing, although we have not done so. The examples show a transformational approach to synchronous design. They follow a structured digital design strategy to obtain comparable implementations. They also reflect a more general design methodology.

Section 1 is an informal discussion of how the approach is related to standard methods. The key idea, and central subject of this paper, is a class of transformations called *system factorizations*. These are used to isolate a level of detail while maintaining the global coherence of a circuit description.

We use an algebra on applicative notation. Section 2 gives a brief look at this language and its use to describe sequential systems. Functional recursion equations are used as specifications. Part of the algebra is to correctly translate these specifications into sequential-system descriptions, and Section 2 concludes with a simple characterization relating the two forms of description.

Section 3 is a first example of design derivation, and it serves several purposes. Most important is the illustration of the transformation algebra, and in particular, the factorization of subsystems. At the same time, a generic treatment of synchronous communication is discussed. We then explain how the transformation system is integrated with PAL assemblers to obtain physical implementations. Integration with VLSI tools is in progress. Section 3.3 is a brief discussion promoting the free use of functional abstraction in describing sequential systems. This notion is not explicitly used later, so Section 3.3 may be skipped.

Section 4 explores system factorizations in more detail, laying out basic laws from which transformations are composed. The parameters in factorization are discussed.

Section 5 reviews the derivation and implementation of a garbage collector. This example shows how the basic algebra presented in this paper applies to a non-trivial design exercise. The accompanying figures (5–1 through 5–6) are developed directly from representations used by the transformation system. It is the form of these figures, not their content, that is of interest. They are included to convey the sense of engineering in this syntactic framework. Boyer gives details of the collector's design [4].

The garbage collector's derivation is targeted to a bit-slice implementation in PAL components. Repeated factorizations lead to an appropriate physical decomposition and bring a substantial portion of the design into programmable technology. This derivation strategy also bypasses the problem of incorporating representations in higher level descriptions. This issue is discussed in the conclusion.

1. The Relationship to Conventional Synchronous Design

Synchronous design isolates the algorithmic aspect of a problem—often specified as a finite-state machine, flowchart, or procedure—and refines a preliminary architecture once control is understood. We view this as a translation between dialects of recursive expression, a perspective that is projected in the ideogram

$$\mathcal{S}_{\Theta/\Gamma} \longrightarrow \mathcal{I}_{\Theta/\Gamma} \longrightarrow \mathcal{C}_{\Theta/\Gamma} \longrightarrow \mathcal{C}'_{\Gamma}\|\Theta_{\Gamma} \longrightarrow \cdots$$

This formula expresses nothing formal beyond the notion of transformation. $\mathcal{S}$ is a problem specification, which for us is a system of function definitions. $\mathcal{S}$ is expressed in terms of a vocabulary, Θ/Γ, of primitive constants, operations, predicates, and so forth, called the *basis*, or "ground type." The basis is an aggregate of complex objects (Θ) and concrete objects (Γ). Some examples are arrays of integers, queues of characters, sets of points. The complex/concrete distinction is subjective and hierarchical; Γ represents an intended level of description.

The first engineering task is transforming $\mathcal{S}$ to a form conducive to hardware implementation. Previous papers develop this phase of design derivation [12, 14, 13]. $\mathcal{I}$ stands for "*iterative*"—that class of recursion schemata that characterizes sequential control. In software, $\mathcal{S} \rightarrow \mathcal{I}$ is sometimes called "recursion removal," but implementing recursion in Θ is not precluded. An explicit treatment of control is one argument for the power of a functional algebra, but that is apart from the subject here. In this paper, all specifications are iterative to begin with.

$\mathcal{C}$ is a sequential-system description. We have shown that such descriptions are mechanically derivable from iterative specifications [14]. The transformation $\mathcal{I} \rightarrow \mathcal{C}$ is implemented and this phase of design derivation is automatic. $\mathcal{C}$ has the same ground type as $\mathcal{I}$, but its interpretation is "behavioral." Where before, variables ranged over values, in $\mathcal{C}$ they denote sequences of values. We are interested in other interpretations of $\mathcal{C}$ as well, such as its description of physical connectivity or graphical layout [16, 3].

$\mathcal{C}$ is an abstract description because its sequences hold complex objects. It is refined to a more concrete description by factoring Θ as an *abstract component*. The formula $\mathcal{C}'\|\Theta$ is meant to suggest a communicating system. $\mathcal{C}'$ is subject to further refinement and Γ to further decomposition. What makes this kind of transformation a factorization is that terms inside the *sub-system* Θ may be retained in $\mathcal{C}'$.

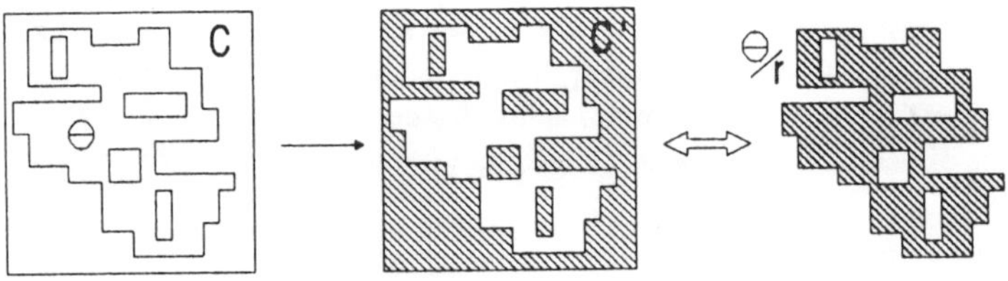

One purpose of a *system factorization* is to segregate problem dependent concrete terms from terms that are intrinsic to the type abstraction. The entailed analysis may depend both on the structure of the basis and also on prior assumptions of how Θ is implemented. Our present concern is exploring how to parameterize syntactic transformations in support of such analysis.

For the purpose of contrast, we would describe standard synchronous design methods as

$$\mathcal{I}_\Gamma \| \Theta_\Gamma \longrightarrow \mathcal{C}'_\Gamma \| \Theta_\Gamma \rightarrow \cdots$$

The engineer makes an estimate of the architecture required to solve a problem, then develops control algorithm $\mathcal{I}$ to govern that architecture. A hardware realization of control is then compiled or systematically derived; that is, $\mathcal{C}'$ is Θ's controller. Communication is implicit in the control specification language. For example, in finite-state [10] and ASM [22] notations, variables denote the presence of a value *now*; and this is also true in higher level hardware description languages.

The basis Θ/Γ of a functional specification corresponds to the estimate of architecture. The initial transformations secure the global design description as control is isolated. That is, a correct description of connectivity is maintained as the control is incorporated in architecture. However, complex (i.e., complex typed) behaviors arise in descriptions thus derived. Factorizations isolate concrete terms that are actually implemented from those that serve to preserve the global coherence of a description. It is a thesis of this work that many aspects of digital engineering can be addressed in this manner.

2. Notation and Background

Descriptions are built from applicative *terms*, such as $f(x, \mathsf{c})$, where operation symbols like f and constant symbols like c come from a given basis and x is a variable. Functional values are allowed; the lambda-expression $\lambda\mathcal{X}.\mathcal{E}$ denotes the function with rule $\mathcal{E}$ and parameter $\mathcal{X}$. In function definitions it is conventional to suppress the lambda, writing

$$AndNot(u, v) \stackrel{\mathrm{d}}{=} and(u, not(v))$$

instead of

$$AndNot = \lambda(u, v)\ .\ and(u, not(v)).$$

Function names are written in *slanted* font, variables in *italic* font and constants in sans serif font. The names of defined functions, such as *AndNot*, are capitalized. Product formation is expressed with brackets. For instance, we might describe a multiple output function

$$BitSum(a, b, c_{\text{in}}) \stackrel{\mathrm{d}}{=} [s, c_{\text{out}}] \quad \text{where} \quad \begin{cases} s = parity(a, b, c_{\text{in}}) \\ c_{\text{out}} = majority(a, b, c_{\text{in}}) \end{cases}$$

A *recursion equation* is a simultaneous system of function-defining equations. There are examples throughout this paper. Function definitions usually involve *conditional* expressions, sometimes written as

$$P \rightarrow E, E'$$

and sometimes as case-statements such as

$$\textit{per instruction viz } \begin{array}{l} \mathsf{Q} : E_1 \\ \mathsf{R} : E_3 \\ \mathsf{W} : E_2 \end{array}$$

This form expresses a simple selection; the *per*-expression is always a variable and the case labels are always constants.

An expression is *simple* if it contains no recursively defined function symbols; otherwise it is *recursive.* A simple function definition is called a *combination.* A recursive term is *tail-recursive* if its only defined symbol is in the outermost applied position. A recursion equation is *iterative* if each branch of every conditional is either simple or tail-recursive. All but one of the examples in this paper are iterative.

Circuits are also described with applicative expressions, but the interpretation is sequential. The *signal* $\boxed{f}(X, \boxed{c})$ denotes the sequence S in which $S_i = f(X_i, c)$. The boxes are a convention to distinguish the sequential interpretation and are omitted later. Signal variables are upper case.

Abstraction is extended so that, for example

$$\boxed{AndNot}(A, B) \;\equiv\; \boxed{and}(A, \boxed{not}(B))$$

and

$$\boxed{\lambda(u,v) \,.\, f(u, g(u,v))}(X, Y) \;\equiv\; \boxed{f}(X, \boxed{g}(X, Y))$$

The two-way selector $\boxed{if}$ is a sequential conditional.

$$\text{If} \quad X = \boxed{if}(P, S, S') \quad \text{then} \quad X_k = \begin{cases} S_k, & \text{if } P_k \text{ is true,} \\ S'_k, & \text{if } P_k \text{ is false.} \end{cases}$$

The operator '!' prefixes a value to a signal. It expresses storage or delay.

$$\text{If} \quad X = \check{v} \,!\, S \quad \text{then} \quad X_0 = \check{v} \text{ and } X_{k+1} = S_k.$$

A *system description* is a collection of simultaneous signal-defining equations. For example,

$$\begin{array}{l} Toggle(C, T) \stackrel{\mathrm{d}}{=} Q \\ \quad \textbf{where} \\ \qquad Q = \phi \,!\, \boxed{if}(C, \boxed{\mathsf{false}}, Z) \\ \qquad Z = \boxed{if}(T, \boxed{not}(Q), Q) \end{array}$$

describes a storage element that is cleared to false whenever signal C is true and otherwise inverts its content whenever signal T is true. The symbol ϕ stands for an unspecified value; it can mean "don't know" or "don't care" or "don't care to say," depending on the context. In *Toggle* $\phi \in \{\mathsf{true}, \mathsf{false}\}$. According to the sequential interpretation, the recursively defined signal Q is given by

$$Q_0 = \phi \quad \text{and} \quad \begin{array}{l} Q_{k+1} = \begin{cases} \mathsf{false} & \text{if } C_k = \mathsf{true} \\ Z_k & \text{if } C_k = \mathsf{false} \end{cases} \\ Z_k = \begin{cases} not(Q_k) & \text{if } T_k = \mathsf{true} \\ Q_k & \text{if } T_k = \mathsf{true} \end{cases} \end{array}$$

The left-hand sides of signal equations may be groupings when the defining expression involves multiple-output operations. For example, one might describe an adder using the *BitSum* combination:

$$\begin{aligned} & Adder([A_n \ldots A_0],\ [B_n \ldots B_0],\ C_0) \stackrel{\mathrm{d}}{=} [C_{n+1}\ S_n\ \cdots\ S_0] \\ & \text{where} \quad [C_1,\ S_0] = BitSum(A_0, B_0, C_0) \\ & \qquad\qquad \vdots \\ & \qquad\qquad [C_{n+1},\ S_n] = BitSum(A_n, B_n, C_n) \end{aligned}$$

The *Adder* combination is "combinational" and could stand in either the discrete or sequential interpretation (From here on, the boxes are omitted). Our system currently has no means to recognize ellipses, or to understand making n a parameter in *Adder*. Such constructs are desirable and are used throughout this paper.

At the core of this work is a proposition relating recursion equations to system descriptions. If $\mathcal{P}$, $\mathcal{T}$, and $\mathcal{R}_i$ are all simple expressions, the function definition

$$\mathbf{F}(x_1, \ldots, x_n) \;=\; \mathcal{P} \rightarrow \mathcal{T}, \mathbf{F}(\mathcal{R}_1, \ldots, \mathcal{R}_n)$$

is equivalent to the system description

$$\begin{aligned} & \mathbf{C}(\mathring{x}_1, \ldots, \mathring{x}_n) \stackrel{\mathrm{d}}{=} [P\ T] \\ & \quad \text{where} \quad X_1 = \mathring{x}_1 \ !\ \mathcal{R}_1 \\ & \qquad\qquad \vdots \\ & \qquad\qquad X_n = \mathring{x}_n \ !\ \mathcal{R}_n \\ & \qquad\qquad P = \mathcal{P} \\ & \qquad\qquad T = \mathcal{T} \end{aligned}$$

That is, in $\mathbf{C}(v_1, \ldots, v_n)$ the signal T contains $\mathbf{F}(v_1, \ldots, v_n)$ as soon as P contains true [14]. Thus, a function with $\mathbf{F}$'s form *transliterates* to a sequential-system description with the same terms.

The system $\mathbf{C}$ has the characteristics of externally clocked synchronous hardware: all combinational feedback passes through storage. Under minimal assumptions, an instance of $\mathbf{C}$ can be constructed from any iterative recursion equation. Thus, iterative schemata serve as a conservative specification language for synchronous-system designs.

3. Derivation of a Single Pulser

The *Single Pulser* is from a textbook by Winkel and Prosser on digital design [22]. This first example of design derivation is presented to give an intuitive idea of the algebra and to show how implementations are assembled from lower level system descriptions.

A second purpose is to develop a treatment of communication. In the general system description $\mathbf{C}$ at the end of Section 2, the "ready" signal P is scant acknowledgment that circuits coordinate with external behaviors. The Single Pulser, being almost purely communicative, sheds a harsh light on that characterization, making it plain that more expressiveness is needed. We attack the problem by introducing communication as an attribute of the ground type. This approach is self serving because it is on the abstraction of communication that we illustrate the notion of system factorization.

We begin with a specification of a two state process that outputs a unit pulse for every pulse on its synchronized input i.

$$\begin{aligned} ACK(i) = now?(i) &\rightarrow put(\mathsf{true}, NAK(next(i))), \\ &\quad\; put(\mathsf{false}, ACK(next(i))) \\ NAK(i) = now?(i) &\rightarrow put(\mathsf{false}, NAK(next(i))), \\ &\quad\; put(\mathsf{false}, ACK(next(i))) \end{aligned} \tag{1}$$

Let us call the basis of this specification PORT, whose intended model is boolean communication, and let P be its domain of values. Input involves two operations: the predicate $now?: P \rightarrow \{\mathsf{true}, \mathsf{false}\}$ gives the current value and $next: P \rightarrow P$ gives subsequent behavior. $put: \{\mathsf{true}, \mathsf{false}\} \times P \rightarrow P$ builds output. Two laws for PORT, used later, are $now?(put(b, p)) \equiv b$ and $put(now?(p), next(p)) = p$.

Each branch of ACK and NAK performs input and output exactly once. This is essential since (1) is to be considered a specification of synchronous control.

3.1. Construction of a System Description

In order to construct a system description, an iterative specification is is needed. Let us abuse the notation by moving the *puts* inside the calls to *ACK* and *NAK*. The assertion "$\{now?(o)\}$" stands as a reminder that this was done.

$$\begin{aligned} ACK(i,o) &= \{now?(o)\}\ now?(i) \rightarrow NAK(\mathit{next}(i), \mathit{put}(\mathsf{true}, o)), \\ &\qquad\qquad\qquad\qquad\quad ACK(\mathit{next}(i), \mathit{put}(\mathsf{false}, o)) \\ NAK(i,o) &= \{now?(o)\}\ now?(i) \rightarrow NAK(\mathit{next}(i), \mathit{put}(\mathsf{false}, o)), \\ &\qquad\qquad\qquad\qquad\quad ACK(\mathit{next}(i), \mathit{put}(\mathsf{false}, o)) \end{aligned} \tag{2}$$

A single function is built by first introducing a parameter, $c \in \{\mathsf{ack}, \mathsf{nak}\}$, to indicate whether *ACK* or *NAK* is in control; then extracting a combination for the conditional; and distributing it across recursive calls. This transformation is discussed in [12], [13], and [14], with many examples; it is an instance of Harel's compendium of folk theorems [9]. The result is

$$\begin{aligned} &Select(p, q, v_0, v_1, v_2, v_3) \stackrel{\mathrm{d}}{=} \mathit{per}\ p\ \mathit{viz}\ \mathsf{ack} : q \rightarrow v_0,\ v_1 \\ &\qquad\qquad\qquad\qquad\qquad\qquad\qquad\ \ \mathsf{nak} : q \rightarrow v_2,\ v_3 \\ &SP(c,i,o) = \{now?(o)\} \\ &\quad SP\big(Select(c, now?(i), \mathsf{nak}, \mathsf{ack}, \mathsf{nak}, \mathsf{ack}), \\ &\qquad\quad Select(c, now?(i), \mathit{next}(i), \mathit{next}(i), \mathit{next}(i), \mathit{next}(i)), \\ &\qquad\quad Select(c, now?(i), \mathit{put}(\mathsf{true}, o), \mathit{put}(\mathsf{false}, o), \\ &\qquad\qquad\qquad\qquad\qquad \mathit{put}(\mathsf{false}, o), \mathit{put}(\mathsf{false}, o))\big) \end{aligned} \tag{3}$$

For any i and o, $SP(\mathsf{ack}, i, o) \equiv ACK(i,o)$. Since *SP* is an instance of **C**, according to the characterization in Section 2 the corresponding sequential-system description is

$$\begin{aligned} &SP(\check{\imath}, \check{o}) = now?(O) \\ &\quad \text{where} \\ &\qquad C = \mathsf{ack}\ !\ Select(C, now?(I), \mathsf{nak}, \mathsf{ack}, \mathsf{nak}, \mathsf{ack}) \\ &\qquad I = \quad \check{\imath}\ !\ Select(C, now?(I), \mathit{next}(I), \mathit{next}(I), \mathit{next}(I), \mathit{next}(I)) \\ &\qquad O = \quad \check{o}\ !\ Select(C, now?(I), \mathit{put}(\mathsf{true}, O), \mathit{put}(\mathsf{false}, O), \\ &\qquad\qquad\qquad\qquad\qquad\qquad \mathit{put}(\mathsf{false}, O), \mathit{put}(\mathsf{false}, O)) \end{aligned} \tag{4}$$

The term $now?(O)$ replaces $[P,\ T]$ in the characterization. The signals clearly simplify, but it serves our purpose not to simplify them too much. I's selector isn't needed and O's is needed only for *put*'s first argument.

$$
\begin{aligned}
&SP(\check{\imath}, \check{o}) = now?(O) \\
&\quad \text{where} \\
&\qquad C = \mathsf{ack} \,!\, Select(C, now?(I), \mathsf{nak}, \mathsf{ack}, \mathsf{nak}, \mathsf{ack}) \\
&\qquad I = \quad \check{\imath} \,!\, next(I) \\
&\qquad O = \quad \check{o} \,!\, put(X, O) \\
&\qquad X = Select(C, now?(I), \mathsf{true}, \mathsf{false}, \mathsf{false}, \mathsf{false})
\end{aligned} \tag{5}
$$

I and O range over ports; they are complex signals that do not describe physical hardware. However, the surrounding system is concerned not with their content but only with the result of the operation *now?*. Therefore, let us encapsulate the defining expressions as combinations presenting the desired signals. In general, there are problem dependent subterms that should be retained, but the one instance in this case has already been identified as the signal X. Defining appropriate combinations *INPUT* and *OUTPUT* yields

$$
\begin{aligned}
&SP(\check{\imath}, \check{o}) = O' \\
&\quad \text{where} \\
&\qquad C = \mathsf{ack} \,!\, Select(C, I', \mathsf{nak}, \mathsf{ack}, \mathsf{nak}, \mathsf{ack}) \\
&\qquad I' = INPUT(\check{\imath}) \\
&\qquad O' = OUTPUT(\check{o}, X) \\
&\qquad X = Select(C, I', \mathsf{true}, \mathsf{false}, \mathsf{false}, \mathsf{false})
\end{aligned} \tag{6}
$$

and

$$INPUT(\check{p}) \stackrel{\mathrm{d}}{=} now?(P) \text{ where } P = \check{p} \,!\, next(P)$$

and

$$OUTPUT(\check{p}, B) \stackrel{\mathrm{d}}{=} now?(P) \text{ where } P = \check{p} \,!\, put(B, P)$$

A schematic can be drawn from these equations:

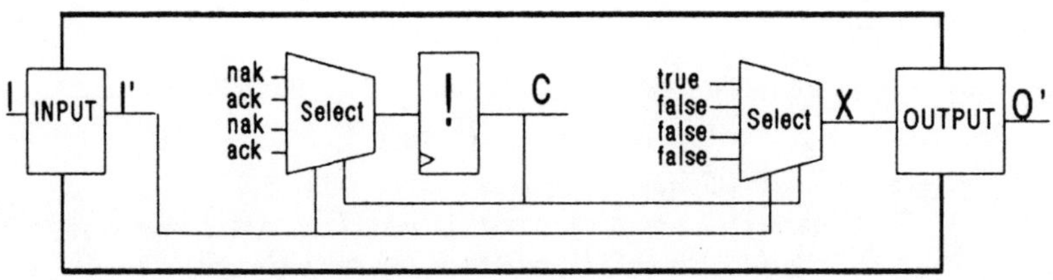

Component abstractions *INPUT* and *OUTPUT* hide the complex behaviors over ports from the concrete boolean signals in (6), leaving a description that could reasonably be implemented. This encapsulation is one kind of *system factorization.*

From the laws of PORT in the sequential interpretation, $INPUT(p) \equiv p$ and $OUTPUT(p, B) \equiv B$. Hence, we may rewrite (6) as

$$SP(I) = O \text{ where } \begin{cases} C = \mathsf{ack}\,!\; Select(C, I, \mathsf{nak}, \mathsf{ack}, \mathsf{nak}, \mathsf{ack}) \\ O = Select(C, I, \mathsf{true}, \mathsf{false}, \mathsf{false}, \mathsf{false}) \end{cases} \tag{7}$$

Description (7) is obtained more directly from the specification

$$\begin{aligned} ACK(i, o) = i \; &\rightarrow \; NAK(i, \mathsf{true}), \\ &\quad\; ACK(i, \mathsf{false}) \\ NAK(i, o) = i \; &\rightarrow \; NAK(i, \mathsf{false}), \\ &\quad\; ACK(i, \mathsf{false}) \end{aligned} \tag{8}$$

That is, the treatment of communication in PORT can be incorporated in the specification notation, as surely it should be. A refined semantics—at the very least a typing discipline—is needed to make sense of specifications that implicitly use PORT. However, this understanding is not needed in the transformation process; the same construction used to build (4) from (1) yields (7) from (8).

3.2. PAL Implementation of the Single Pulser

From (7), boolean equations are generated and targeted both to a PAL programmer and into VLSI layout facilities. Both (7) and the garbage collector in Section 5 are built with PALs. "PAL" stands for Programmable Array Logic, a general purpose logic device that also contains storage elements. Once programmed, a PAL circuit has the description

$$\lambda(Y_1, \ldots, Y_n) \,.\, [X_1^c, \ldots, X_m^c, X_1^d, \ldots, X_r^d]$$

where

$$\begin{aligned} X_1^c &= \quad f_1(Y_1, \ldots, Y_n, X_1^c, \ldots, X_m^c, X_1^d, \ldots, X_r^d) \\ &\vdots \\ X_m^c &= \quad f_m(Y_1, \ldots, Y_n, X_1^c, \ldots, X_m^c, X_1^d, \ldots, X_r^d) \\ X_1^d &= \phi \,!\, f_{m+1}(Y_1, \ldots, Y_n, X_1^c, \ldots, X_m^c, X_1^d, \ldots, X_r^d) \\ &\vdots \\ X_r^d &= \phi \,!\, f_{m+r}(Y_1, \ldots, Y_n, X_1^c, \ldots, X_m^c, X_1^d, \ldots, X_r^d) \end{aligned}$$

Each f_i is a boolean function subject to certain restrictions, such as the number of or's it can do. The values of n, m, and r vary in different packagings. Outputs X^c are combinational and signals X^d are stored in clocked, D-type flip-flops. A PAL "program" is a description of the functions f_i.

Recall that the selector combination for the single pulser is

$$\mathit{Select}(p, q, v_0, v_1, v_2, v_3) \stackrel{\mathrm{d}}{=} \mathit{per}\ p\ \mathit{viz}\ \begin{array}{l} \mathsf{ack} : q \rightarrow v_0,\ v_1 \\ \mathsf{nak} : q \rightarrow v_2,\ v_3 \end{array}$$

The method for implementing (7) is to generate a state assignment for each selection branch. This is a standard synthesis technique following [22]. The state associated with alternative v_i is the binary value of i; represented in signals $[S_1, S_0]$. Boolean equations are developed in conjunction with an assignment of truth values for the symbolic constants. For (7), with nak represented as true, the following is essentially the input generated for the A+PLUS simplifier [1].

$$\begin{aligned} C^d &= (\overline{S}_1 \wedge \overline{S}_0) \vee (S_1 \wedge \overline{S}_0) \\ O^d &= \overline{S}_1 \wedge \overline{S}_0 \\ S_1^c &= (C \wedge \overline{I}) \vee (C \wedge I) \\ S_0^c &= (\overline{C} \wedge \overline{I}) \vee (C \wedge \overline{I}) \end{aligned}$$

The simplifier reduced the system to

$$\begin{aligned} S_0^c &= \overline{I} \\ S_1^c &= C \\ O^d &= \overline{S}_1 \wedge \overline{S}_0 \\ C^d &= \overline{S}_0 \end{aligned}$$

This is directed to the Altera LE-2 fuse burner to obtain a working single pulser in an Altera EP310 PAL [1].

The simplifier is understandably reluctant to remove signals, but the equations could clearly reduce further to

$$SP(I) = O \quad \text{where} \quad \begin{cases} C = \mathsf{ack}\ !\ I \\ O = \mathit{and}(I, \mathit{not}(C)) \end{cases} \tag{9}$$

as was the outcome in Winkel's and Prosser's text. Version (9) might be developed by a direct symbolic expansion of (7). We have not implemented this kind of analysis and make do with a state encoding passed through available simplifiers.

3.3. A Comment about Modeling

Since the ideas discussed below are not explicitly used later, this section can be skipped or read lightly.

The Single Pulser presents an opportunity to discuss the use of functional abstraction in bridging levels of hardware description. The underlying principle is that describing something as a function is the purest form of specification, carrying no suggestion of representation. A more complicated mathematical foundation is needed if functions are allowed in a *basis*, and Boute makes credible claims that functions aren't needed in this way [3]. Further examination of the issue is warranted; it is too early to preclude the free use of functions in hardware modeling.

The version of *SP* below uses *AndNot*, from Section 2, to suppress details of *O*'s defining equation in (9).

$$SP(I) = O \text{ where } \begin{cases} C = \phi \,!\, I \\ O = AndNot(I, C) \end{cases} \tag{10}$$

Now, consider *AndNot*'s truth table with the idea of abstracting its second argument to a function on the first.

u	v	$AndNot(v, u)$	$u\colon v \to \{\text{true}, \text{false}\}$
true	true	false	$u(v) = \text{false}$
true	false	false	
false	true	true	$u(v) = v$
false	false	false	

The right-most column shows that the functional abstraction of u is

$$Abstr(u) \stackrel{\text{d}}{=} u \;\rightarrow\; (\lambda v.\text{false}),\; (\lambda v.v)$$

The version of *SP* below uses the abstraction, treating signal C as a component whose function is governed by I.

$$SP(I) = O \text{ where } \begin{cases} C = \phi \,!\, Abstr(I) \\ O = C(I) \end{cases} \tag{11}$$

For this system description to be meaningful, it must make sense to apply signals. Let us revise the sequential interpretation of application to allow it.

$$\text{If} \quad X = F(\cdots S^i \cdots) \quad \text{then} \quad X_k = F_k(\cdots S^i_k \cdots).$$

```
                                        |-----------------------
L = \[v].v                              | Voltage abstractions
H = \[u].\[v].v                         |-----------------------
                                        | Symbolic interface
repr = \[A]. (A:<\v."L">):<"H">         |  repr: value --> name
REPR = <repr *>                         |  REPR: values --> names
                                        |-----------------------
SP = \I. <H ! I>:<I>                    | SP(I) = (delay I)(I)
                                        |-----------------------

REPR:< SP:<L L L H H H H L L H L L H H L L H H H L L L> >
          [L L L H L L L L L H L L H L L L H L L L L L ■]
```

The language is Daisy, a lazy dialect of Pure Lisp, whose use for hardware description is discussed in [14,16]. Sequences are represented as stream-like lists. `<* ! *>`, a `CONS`-expression, evaluates to the list `[* ! *]`; ':' means application; '\' is a lambda. `REPR` maps `repr` over a list. `SP`, taken from equation (12), maps `I` delayed over `I`.

FIGURE 3–1. EXECUTABLE DESCRIPTION OF SP

The interpretation of a base operation is then a constant sequence

$$\boxed{f} = (f, f, \ldots)$$

The values $\{(\lambda v.v), (\lambda v.\mathsf{false})\}$ suggest a pass transistor with a pull-false resistor. If I were already abstracted, and taking

$$Delay_{\check{x}}(I) \stackrel{\mathrm{d}}{=} \check{x} \,!\, I$$

we get

$$SP(I) = (Delay_{\phi}\ I)(I) \tag{12}$$

Hence, we might consider implementing SP in NMOS as

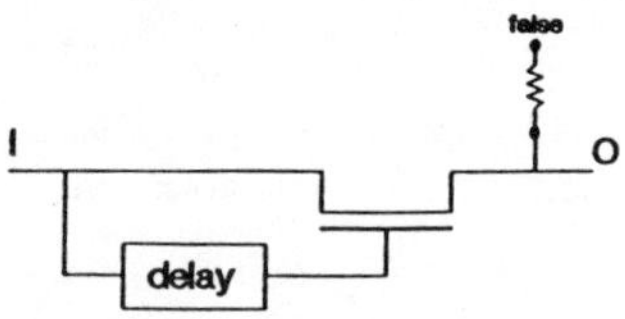

However, it remains to solve for independent basis of values, in order to eliminate the false in $\{(\lambda v.v), (\lambda v.\mathsf{false})\}$. A set that works is $\{(\lambda v.v), (\lambda u\,.\,\lambda v.v)\}$ as is demonstrated by the program in Figure 3–1. Whether this is a useful transistor model is a matter for study. Also of interest is the manner by which a logic description transforms to a gate description, and in general, the role of function abstractions in developing such transformations.

4. Factorization of System Descriptions

A system factorization encapsulates a subsystem in order to remove some collection of operations from the surrounding description. This is called "abstract component factorization" in [12] referring to the use of abstract data types in decomposing software systems into information hiding modules. After building a basic vocabulary of elementary transformations, we look at a small example to get a sense of what a factorization does.

Identification is giving a name to an expression by adding a signal equation for it. Identification of like terms by a single equation has the effect of eliminating redundant circuitry when the occurrence of a term corresponds to an instance of physical hardware.

Grouping is the collection of terms into a single equation. A set of equations $\{X_i = S_i\}$ may be rewritten as

$$[\cdots X_i \cdots] = [\cdots S_i \cdots]$$

Grouping is sometimes done to develop physical decompositions. The expression $[\cdots S_i \cdots]$ might describe a physical package; recall the n-bit *Adder* in Section 2, in which *BitSum* combinations stand for groupings.

As before, *combination* is the formation of a lambda-combinator, or "macro," or "repeated-pattern symbol" in drafting terminology. Examples are *AndNot*, *BitSum*, and *Adder* in Section 2; *Select*, *INPUT*, *OUTPUT*, and *SP* in Section 3.

Distribution is the symbolic application of a distributive law. A key instance is the law for selection that says

$$\mathit{if}[p, f(x), g(y)] \quad \equiv \quad (\mathit{if}[p, f, g])\,(\mathit{if}[p, x, y])$$

This rule extends to *Select* combinations—it was used in equation (5) of Section 3.1 to get the signal X— but we shall explore its use with the binary selector, as above.

Generalization is the introduction of extraneous don't-care arguments. It is typically done in order to distribute selection over non-uniform alternatives.

$$\mathit{if}[p,\ f(x,y),\ g(u,v,w)]$$

is rewritten as

$$(\mathit{if}[p,f',g])(\mathit{if}[p,x,u],\ \mathit{if}[p,y,v],\ \mathit{if}[p,\phi,w])$$

with f extended to

$$f'(a,b,c) \stackrel{\mathrm{d}}{=} f(a,b).$$

Collation is a merging of signals by instantiating ϕs. An example is to replace

$$\left\{\begin{matrix} X = \mathit{if}[p,s,\phi] \\ Y = \mathit{if}[p,\phi,t] \end{matrix}\right\} \quad \text{with} \quad \{XY = \mathit{if}[p,s,t]\}$$

We sketch this elementary algebra in a small example involving two signals with a common selection predicate.

$$\begin{aligned} U &= \mathit{if}[p,\ \mathit{add}(A,B),\ \mathit{inc}(C)] \\ X &= \mathit{if}[p,\ \mathit{dcr}(D),\ \mathit{add}(E,F)] \end{aligned} \tag{1}$$

The operations' names suggest arithmetic functions and it is assumed that all signals enjoy a uniform representation. The goal is to reduce the number of components by distributing selection in the system. With this in mind, (1) is expanded to

$$\begin{aligned} U &= \mathit{if}[p,V,W] \\ V &= \mathit{add}(\mathit{if}[p,A,\phi],\ \mathit{if}[p,B,\phi]) \\ W &= \mathit{inc}(\mathit{if}[p,\phi,C]) \\ X &= \mathit{if}[p,Y,Z] \\ Y &= \mathit{dcr}(\mathit{if}[p,D,\phi]) \\ Z &= \mathit{add}(\mathit{if}[p,\phi,E],\ \mathit{if}[p,\phi,F]) \end{aligned} \tag{2}$$

We get a single *add* by collating V and Z.

$$\begin{aligned} U &= \mathit{if}[p,\ VZ,\ W] \\ W &= \mathit{inc}(\mathit{if}[p,\phi,C]) \\ VZ &= \mathit{add}(\mathit{if}[p,A,E],\ \mathit{if}[p,B,F]) \\ X &= \mathit{if}[p,Y,\ VZ] \\ Y &= \mathit{dcr}(\mathit{if}[p,D,\phi]) \end{aligned} \tag{3}$$

The arguments to *inc* and *dcr* also merge and we can express W and Y as a single signal

$$WY = (\mathit{if}[p, \mathit{dcr}, \mathit{inc}])(\mathit{if}[p, D, C])$$

The component-expression $\mathit{if}[p, \mathit{dcr}, \mathit{inc}]$ is represented in concrete terms by encoding *dcr* and *inc* in an *instruction signal* ranging over symbols $\{\mathsf{up}, \mathsf{dn}\}$.

$$\begin{aligned} U &= \mathit{if}[p, VZ, WY] \\ X &= \mathit{if}[p, WY, VZ] \\ VZ &= \mathit{add}(\mathit{if}[p, A, E],\ \mathit{if}[p, B, F]) \\ WY &= \mathit{UPDN}(\mathit{if}[p, \mathsf{dn}, \mathsf{up}],\ \mathit{if}[p, D, C]) \end{aligned} \tag{4}$$

Instructions are interpreted by a synthesized abstraction of *WY's* behavior.

$$\mathit{UPDN}(I, S) \stackrel{\mathrm{d}}{=} \mathit{per}\ I\ \mathit{viz}\ \begin{array}{l} \mathsf{up}\text{: } \mathit{inc}(S) \\ \mathsf{dn}\text{: } \mathit{dcr}(S) \end{array}$$

Of course, there are alternative decompositions. For instance, one could introduce a component abstraction for each of U and X. Unary operations in (1) are generalized to get a uniform actual parameter:

$$\begin{aligned} U &= \mathit{if}[p,\ \mathit{add}(A, B),\ \mathit{inc}'(C, \phi)] \\ X &= \mathit{if}[p,\ \mathit{dcr}'(D, \phi),\ \mathit{add}(E, F)] \end{aligned} \tag{5}$$

and two combinations are introduced, with synthesized instruction signals.

$$\begin{aligned} U &= \mathit{ADDINC}(\mathit{if}[p, \mathsf{add}, \mathsf{up}],\ \mathit{if}[p, A, C],\ B) \\ X &= \mathit{DCRADD}(\mathit{if}[p, \mathsf{dn}, \mathsf{add}],\ \mathit{if}[p, D, E],\ F) \end{aligned}$$

where

$$\mathit{ADDINC}(I, S, T) \stackrel{\mathrm{d}}{=} \mathit{per}\ I\ \mathit{viz}\ \begin{array}{l} \mathsf{add}\text{: } \mathit{add}(S, T) \\ \mathsf{up}\text{: } \mathit{inc}'(S, T) \end{array} \tag{6}$$

and

$$\mathit{DCRADD}(I, S, T) \stackrel{\mathrm{d}}{=} \mathit{per}\ I\ \mathit{viz}\ \begin{array}{l} \mathsf{dn}\text{: } \mathit{dcr}'(S, T) \\ \mathsf{add}\text{: } \mathit{add}(S, T) \end{array}$$

Abstract components *UPDN*, *ADDINC*, and *DCRADD* are *system factorizations*. These kinds of transformations are motivated by a variety of global considerations, such as physical packaging, routing, and logical decomposition. One function of the transformation system is to execute the steps of a factorization correctly. The process has several implicit steps, including

(A) Determining a set of *subject terms.*

(B) Isolating these terms in the surrounding description.

(C) Executing some collation tactic.

(D) Generating a symbolic instruction signal.

(E) Synthesizing a correct component specification.

(F) Incorporating the component in the description.

The subject terms are those that apply an operation which is to be encapsulated. We have implemented two ways of doing (A). The first, called a *general factorization,* is to state the set of operations that are to be encapsulated. The subject terms are those in which one of the set is applied. The second, called a *signal factorization,* encapsulates a signal; in this case the subject terms are those in which that signal's name occurs as an argument.

Descriptions (3) and (4), above, are general factorizations. In (3), the set of subject terms are those using *add*; in (4), subject terms develop from the set $\{inc, dcr\}$. Description (6) results from two general factorizations of (1), each restricted to a single signal equation. Subject terms are determined by $\{add, inc\}$ for signal U and $\{add, dcr\}$ for signal X.

In a signal factorization, the encapsulated operations are those applied to a named signal. The combinations *INPUT* and *OUTPUT* in Section 3.1 are examples. The purpose is often to isolate (more) concrete signals from (more) complex signals. Subject terms $\{d_i(T,\ldots)\}$ are identified and grouped with the signal of interest, which typically has the form

$$T = \check{t}\,!\ \mathit{Select}(\cdots\ c_j(T,\ldots)\ \cdots)$$

If the basis is Θ/Γ, then T has type Θ; the operations $c_j\colon(\Theta\times\Gamma^r)\rightarrow\Theta$, inside T's equation, build new complex values; and those $d_i\colon(\Theta\times\Gamma^n)\rightarrow\Gamma^m$ in the surrounding system extract concrete values. The component abstraction encloses all references to and operations on T, taking an appropriate instruction signal as an input:

$$\begin{aligned}&\Theta(\check{t},I,\ldots) \stackrel{\mathrm{d}}{=} [\cdots d_i(T,\ldots)\cdots]\\ &\quad \text{where } T=\check{t}\,!\ \mathit{per}\ I\ \mathit{viz}\ \cdots\ \mathsf{instruction}_j\colon\ c_j(T,\ldots)\cdots\end{aligned}$$

Θ specifies how *any* implementation must behave in relation to the surrounding circuit. However, Θ is itself an abstract behavioral description. We do not derive implementations from such specifications and would like a compatible means of verification for establishing the correctness of implementations.

The factorizations now done by the transformation system are entirely syntactic and depend on convention. Factorization steps (A) through (F) raise theoretical and analytical issues. For example, generalization depends on assumptions of type-compatibility among the subject operations. It is not always possible to form reasonable combinations from a set of subject terms, due to clashes in collation. Factorizations are often motivated by unstated assumptions about the implementation. Despite these problems, we find this way of manipulating design descriptions to be natural and powerful. Fairly simple factorizations seem adequate for decomposing designs at levels of description that are typical in digital design.

5. Derivation of a Garbage Collector

This section sketches a non-trivial design exercise that has been carried to hardware. The accompanying figures are intended to be *impressionistic* and are included mainly to give a sense of the derivation process. Figures 5–1 through 5–5, at the end of this paper, are developed directly from the representations used in the transformations system, but they are highly condensed. They convey little of the logical design, whose details are explained by Boyer [4]. The transformation system is due to Bose [2].

The derivation strategy is governed by architectural considerations. The design is targeted for a bit-slice implementation in PALs. We have the means to prototype quickly in this technology, using a Logic Engine [17], which provides a large wire-wrap panel, secure power and clocking, and good digital display facilities. In addition, the Logic Engine houses a general purpose microprogramming environment, which is used to implement the peripheral system needed to exercise the design. The collector has been tested against heaps generated by a Scheme implementation [5].

PALs have a good switching capacity, and a bit-slice implementation of the collector eliminates the need for an internal register bus. On the other hand, bit-slicing in PALs precludes a direct implementation of arithmetic, due to intolerable propagation delays. Hence, conventional arithmetic components are used in the design.

Figure 5–1 is a specification for a *stop-and-copy* garbage collector [7]. This kind of collector divides memory into two semispaces, only one being

```
IDLE(M1,M2,H,D,C,U,A,GO,R) =
  GO → NEXT(M1,M2,*H*,D,C,0,0,GO,false),
        IDLE(M1,M2,H,D,C,U,A,true)

DRIVER(M1,M2,H,D,C,U,A,GO,R) =
  eq?(U,A) → IDLE(M2,M1,H,D,C,U,A,GO,true),
             NEXT(M1,M2,read(M1,U),D,C,U,A,GO,R)

NEXT(M1,M2,H,D,C,U,A,GO,R) =
  pointer?(H) → OBJ(M1,M2,H,read(M2,H),C,U,A,GO,R),
  bvec?(H) → DRIVER(M1,M2,H,D,C,addinc(U,btow(ptr(H))),A,GO,R),
             DRIVER(M1,M2,H,D,C,inc(U),A,GO,R)

OBJ(M1,M2,H,D,C,U,A,GO,R) =
  eq?(fwd,tag(D)) →
    DRIVER(write(M1,U,cell(H,D)),M2,H,D,C,inc(U),A,GO,R),
  eq?(pair,tag(H)) →
    PAIR1(write(M1,A,D),write(M2,H,cell(fwd,A)),H,D,C,U,A,GO,R),
  eq?(vec,tag(H)) →
    VEC(write(M1,U,cell(H,A)),M2,H,D,ptr(D),inc(U),A,GO,R),
  eq?(bvec,tag(H)) →
    BVEC( write(M1,A,D), M2, H, cell(D,btow(ptr(D))),
          btow(ptr(D)), U, A, GO, R)
  eq?(fbvec,tag(H)) →
    DRIVER(M1,M2,H,D,C,inc(U),A,GO,R),


PAIR1(M1,M2,H,D,C,U,A,GO,R) =
  PAIR2( write(M1,U,cell(H,A)), M2, H, read(M2,inc(ptr(H))),
         C, U, inc(A), GO, R)

PAIR2(M1,M2,H,D,C,U,A,GO,R) =
  DRIVER(write(M1,A,D),M2,H,D,C,inc(U),inc(A),GO,R)

VEC(M1,M2,H,D,C,U,A,GO,R) =
  VLOOP(write(M1,A,D),M2,H,read(M2,add(ptr(H),C)),dcr(C),U,A,GO,R)

VLOOP(M1,M2,H,D,C,U,A,GO,R) =
  eq?(C,-1) →
    DRIVER(M1,write(M2,H,cell(fwd,A)),H,D,C,U,incadd(A,ptr(D)),GO,R),
    VLOOP( write(M1,incadd(C,A),D), M2, H, read(M2,add(ptr(H),C)),
           dcr(C), U, A, GO, R)

BVEC(M1,M2,H,D,C,U,A,GO,R) =
  BLOOP( write(M1,U,cell(H,A)), M2, H, read(M2,add(ptr(H),ptr(D))),
         dcr(C), inc(U), A, GO, R)

BLOOP(M1,M2,H,D,C,U,A,GO,R) =
  eq?(C,-1) →
    DRIVER( M1, write(M2,H,cell(fwd,A)), H, D, C, U,
            incadd(A,btow(ptr(D))), GO, R)
    BLOOP( write(M1,incadd(C,A),D), M2, H, read(M2,add(ptr(H),C)),
           dcr(C), U, A, GO, R)
```

FIGURE 5–1. GARBAGE COLLECTOR SPECIFICATION

active at any time. When the allocator exhausts available space, the collector copies and compresses the heap image into the inactive space. The roles of the two semispaces are then exchanged. The specified collector compresses a heap that includes binary list cells, vectors, and relocatable segments of data; there is also a provision for bypassing non-relocatable segments. Details of the collection algorithm and representations of collected objects are not important to what follows. They are explained in [4]. Relevant details can be seen in the following portion of Figure 5–1.

$$\begin{array}{l} DRIVER(M_1, M_2, H, D, C, U, A, GO, R) = \\ \quad eq?(U, A) \rightarrow IDLE(M_2, M_1, H, D, C, U, A, GO, \text{true}), \\ \qquad\qquad\qquad NEXT(M_1, M_2, read(M_1, U), D, C, U, A, GO, R) \end{array}$$

The formal argument describes the "registers" at this level of description. Parameters M_1 and M_2 have type MEMORY and are subject to operations $read(\star,\star)$ and $write(\star,\star,\star)$. The design uses two distinct physical memories for the semispaces, and operates simultaneously on both during collection. Registers H and D each hold a CONTENT, with field manipulations $ptr(\star)$, $tag(\star)$, and $cell(\star,\star)$. C, U, and A are of type ADDRESS with arithmetic operations $inc(\star)$, $dcr(\star)$, $add(\star,\star)$, $btow(\star)$, $addinc(\star,\star)$, and $incadd(\star,\star)$. The *btow* operation rounds byte offsets to word boundaries; *addinc* and *incadd* are combinations of addition and incrementing. The remaining parameters, GO and R, are boolean communication ports. Ports are treated in the manner of Section 3.1.

Figure 5–1 is an unstructured register-transfer description. Each recursive function is a point of control, whose invocation could be thought of as a parallel assignment to the registers and a transfer of control. This suggests that the *NEXT* branch in *DRIVER* literally exchanges M_1 and M_2. Since this is unreasonable in standard technology, the alternative used in this design makes explicit the fact that memory paths are switched, and not memories themselves. A boolean register W records which memory is active. For instance, *DRIVER* becomes

$$\begin{array}{l} DRIVER(M_1, M_2, H, D, C, U, A, GO, R, W) = \\ \quad eq?(U, A) \rightarrow \\ \qquad IDLE(M_1, M_2, H, D, C, U, A, GO, \text{true}, not(W)), \\ \qquad W \rightarrow \\ \qquad\quad NEXT(M_1, M_2, read(M_2, U), D, C, U, A, GO, R, W), \\ \qquad\quad NEXT(M_1, M_2, read(M_1, U), D, C, U, A, GO, R, W) \end{array}$$

and similarly for the whole of Figure 5–1. Since all memory operations are predicated by a test on W, the size of the specification is nearly doubled. The manner of introducing W reflects the planned implementation. W might be seen as an input to a TWOMEMORY component abstraction and would then be added as an argument to *read* and *write*. However, the PALs provide enough gates to switch the data paths internally. Making W a selection predicate leads to this implementation.

The design's selector, *SEL*, in Figure 5–2, and the initial system description, in Figure 5–3, are both mechanically derived. The grouping of predicates, called *STATUS* in Figure 5–2, was done manually to condense Figure 5–3. This combination is automatically encapsulated later in the derivation process.

Five system factorizations are applied to Figure 5–3, resulting in Figures 5–4 and 5–5. These are signal factorizations of M_1, M_2, and C; and two general factorizations of arithmetic. Subject terms for one of the general factorizations are indicated by boxes in Figure 5–3. The transformation system must correctly draw these terms into a single combination and synthesize an abstract component specification.

Factorization of signal M_1 (and M_2 similarly) leaves residual signals $M_1.I$, $M_1.A$, and $M_1.D$ (instruction, address, and data).

$$
\begin{aligned}
M_1 &= MEMORY(\phi, M_1.I, M_1.A, M_1.D)\\
M_1.I &= SEL(\,P,\ \mathsf{@},\ \mathsf{@},\ \mathsf{@},\ \mathsf{@},\ \mathsf{R},\ \mathsf{R},\ \mathsf{@},\ \mathsf{@},\ \mathsf{@},\ \mathsf{@},\ \mathsf{W},\ \mathsf{W},\ \mathsf{W},\ \mathsf{@},\ \mathsf{W},\ \mathsf{@},\ \mathsf{W},\\
&\qquad \mathsf{@},\ \mathsf{R},\ \mathsf{W},\ \mathsf{@},\ \mathsf{W},\ \mathsf{R},\ \mathsf{W},\ \mathsf{W},\ \mathsf{@},\ \mathsf{R},\ \mathsf{W},\ \mathsf{R},\ \mathsf{W},\ \mathsf{W},\ \mathsf{@},\ \mathsf{R},\ \mathsf{W}\,)\\
M_1.A &= SEL(\,P,\ \phi,\ \phi,\ \phi,\ \phi,\ U,\ H,\ \phi,\ \phi,\ \phi,\ \phi,\ U,\ H,\ A,\ \phi,\ U,\ \phi,\ A,\ \phi,\\
&\qquad Z_2,\ U,\ \phi,\ A\ ,\ Z_1,\ A,\ H,\ \phi,\ Z_2,\ Z_1,\ Z_1,\ U,\ H,\ \phi,\ Z_2,\ Z_1\,)\\
M_1.D &= SEL(\,P,\ \phi,\ \phi,\ \phi,\ \phi,\ \phi,\ \phi,\ \phi,\ \phi,\ \phi,\ \phi,\ cell(H,D),\ cell(\mathsf{fwd},A),\\
&\qquad D,\ \phi,\ cell(H,A),\ \phi,\ D,\ \phi,\ \phi,\ cell(H,A),\ \phi,\ D,\ \phi,\ D,\\
&\qquad cell(\mathsf{fwd},A),\ \phi,\ \phi,\ D,\ \phi,\ cell(H,A),\ cell(\mathsf{fwd},A),\ \phi,\ \phi,\\
&\qquad D\,)
\end{aligned}
$$

These connect to the abstract component *MEMORY*, whose description in Figure 5–5 is a byproduct of the factorization. Instructions to a *MEMORY* are @ (do nothing), R (read), and W (write).

$$
\begin{aligned}
&MEMORY(\grave{m}, Inst, Addr, Data) \stackrel{\mathrm{d}}{=} read(M)\\
&\quad \text{where}\\
&\qquad M = \grave{m}\ !\ Interpret(M, Inst, Addr, Data)\\
&\quad \text{and}\\
&Interpret(m, inst, addr, data) \stackrel{\mathrm{d}}{=} per\ inst\ viz\ \ \mathsf{@} : m\\
&\qquad\qquad\qquad\qquad\qquad\qquad\qquad\qquad\quad \mathsf{R} : m\\
&\qquad\qquad\qquad\qquad\qquad\qquad\qquad\qquad\quad \mathsf{W} : write(m, addr, data)
\end{aligned}
$$

$$
\begin{array}{l}
SEL([p_0, p_1, p_2, W, p_4, p_5, p_6, p_7, p_8, p_9, p_{10}, p_{11}], \\
\quad v_0, v_1, v_2, v_3, v_4, v_5, v_6, v_7, v_8, v_9, v_{10}, v_{11}, v_{12}, v_{13}, v_{14}, v_{15}, v_{16}, v_{17}, \\
\quad v_{18}, v_{19}, v_{20}, v_{21}, v_{22}, v_{23}, v_{24}, v_{25}, v_{26}, v_{27}, v_{28}, v_{29}, v_{30}, v_{31}, v_{32}, v_{33}) \\
\quad \stackrel{d}{=} \ per\ p_0 \ \mathbf{viz} \quad
\begin{array}{rl}
\text{idle}: & p_1 \rightarrow v_0,\ v_1 \\
\text{driver}: & p_2 \rightarrow v_2, \\
 & \qquad W \rightarrow v_3,\ v_4 \\
\text{next}: & p_4 \rightarrow W \rightarrow v_5,\ v_6, \\
 & p_5 \rightarrow v_7,\ v_8 \\
\text{obj}: & p_6 \rightarrow W \rightarrow v_9,\ v_{10}, \\
 & p_7 \rightarrow W \rightarrow v_{11},\ v_{12}, \\
 & p_8 \rightarrow W \rightarrow v_{13},\ v_{14}, \\
 & p_9 \rightarrow W \rightarrow v_{15},\ v_{16}, \\
 & p_{10} \rightarrow v_{17},\ \phi \\
\text{pair1}: & W \rightarrow v_{18},\ v_{19} \\
\text{pair2}: & W \rightarrow v_{20},\ v_{21} \\
\text{vec}: & W \rightarrow v_{22},\ v_{23} \\
\text{vloop}: & p_{11} \rightarrow W \rightarrow v_{24},\ v_{25}, \\
 & \qquad\quad W \rightarrow v_{26},\ v_{27} \\
\text{bvec}: & W \rightarrow v_{28},\ v_{29} \\
\text{bloop}: & p_{11} \rightarrow W \rightarrow v_{30},\ v_{31}, \\
 & \qquad\quad W \rightarrow v_{32},\ v_{33}
\end{array}
\end{array}
$$

$$
\begin{array}{ll}
STATUS(S, H, D, C, U, A, GO, W) \stackrel{d}{=} & [\ S \\
 & \ \ GO \\
 & \ \ eq?(U, A) \\
 & \ \ W \\
 & \ \ pointer?(H) \\
 & \ \ bvec?(H) \\
 & \ \ eq?(\text{fwd}, tag(D)) \\
 & \ \ eq?(\text{pair}, tag(H)) \\
 & \ \ eq?(\text{vec}, tag(H)) \\
 & \ \ eq?(\text{bvec}, tag(H)) \\
 & \ \ eq?(\text{fbvec}, tag(H)) \\
 & \ \ eq?(C, \text{-1}) \\
 &]
\end{array}
$$

FIGURE 5–2. SELECTOR COMBINATION AND STATUS GROUPING

$$P = STATUS(S, H, D, C, C, U, A, GO, W)$$

$$S \overset{\phi!}{=} SEL(\text{P, next, idle, idle, next, next, obj, obj, driver, driver, driver, driver, pair1, pair1, vec1, vec1, bvec1, bvec1, driver, pair2, pair2, driver, driver, vloop, vloop, driver, driver, vloop, vloop, bloop, bloop, driver, driver, bloop, bloop})$$

$$M1 \overset{\phi!}{=} SEL(\text{P}, M1, M1, M1, M1, M1, M1, M1, M1, M1, M1, write(M1, U, cell(H, D)), write(M1, H, cell(\text{fwd}, A)), write(M1, A, D), M1, write(M1, U, cell(H, A)), M1, write(M1, A, D), M1, M1, write(M1, U, cell(H, A)), M1, write(M1, A, D), M1, write(M1, A, D), write(M1, H, cell(\text{fwd}, A)), M1, M1, write(M1, \boxed{incadd}(C, A), D), M1, write(M1, U, cell(H, A)), write(M1, H, cell(\text{fwd}, A)), M1, M1, write(M1, \boxed{incadd}(C, A), D))$$

$$M2 \overset{\phi!}{=} SEL(\text{P}, M2, M2, M2, M2, M2, M2, M2, M2, M2, write(M2, U, cell(H, D)), M2, write(M2, A, D), write(M2, H, cell(\text{fwd}, A)), write(M2, U, cell(H, A)), M2, write(M2, A, D), M2, M2, write(M2, U, cell(H, A)), M2, write(M2, A, D), M2, write(M2, A, D), M2, M2, write(M2, H, cell(\text{fwd}, A)), write(M2, \boxed{incadd}(C, A), D), M2, write(M2, U, cell(H, A)), M2, M2, write(M2, H, cell(\text{fwd}, A)), write(M2, \boxed{incadd}(C, A), D), M2)$$

$$H \overset{\phi!}{=} SEL(\text{P, *H*, H, H}, read(M2, U), read(M1, U), \text{H, H, H, H, H, H, H, H, H, H, H, H, H, H, H, H, H, H, H, H, H, H, H, H, H, H, H, H, H})$$

$$D \overset{\phi!}{=} SEL(\text{P, D, D, D, D, D}, read(M1, H), read(M2, H), \text{D, D, D, D, D, D, D, D}, cell(D, btow(ptr(D))), cell(D, btow(ptr(D))), \text{D}, read(M1, \boxed{inc}(ptr(H))), read(M2, \boxed{inc}(ptr(H))), \text{D, D}, read(M1, \boxed{add}(ptr(H), C)), read(M2, \boxed{add}(ptr(H), C)), \text{D, D}, read(M1, \boxed{add}(ptr(H), C)), read(M2, \boxed{add}(ptr(H), C)), read(M1, \boxed{add}(ptr(H), ptr(D))), read(M2, \boxed{add}(ptr(H), ptr(D))), \text{D, D}, read(M1, \boxed{add}(ptr(H), C)), read(M2, \boxed{add}(ptr(H), C)))$$

$$C \overset{\phi!}{=} SEL(\text{P, C, C, C, C, C, C, C, C, C, C, C, C, C}, ptr(D), ptr(D), btow(ptr(D)), btow(ptr(D)), \text{C, C, C, C, C}, dcr(C), dcr(C), \text{C, C}, dcr(C), dcr(C), dcr(C), dcr(C), \text{C, C}, dcr(C), dcr(C))$$

$$U \overset{\phi!}{=} SEL(\text{P, 0, U, U, U, U, U, U}, \boxed{addinc}(U, btow(ptr(H))), \boxed{inc}(U), \boxed{inc}(U), \boxed{inc}(U), \text{U, U}, \boxed{inc}(U), \boxed{inc}(U), \text{U, U}, \boxed{inc}(U), \text{U, U}, \boxed{inc}(U), \boxed{inc}(U), \text{U, U, U, U, U, U}, \boxed{inc}(U), \boxed{inc}(U), \text{U, U, U, U})$$

$$A \overset{\phi!}{=} SEL(\text{P, 0, A, A, A, A, A, A, A, A, A, A, A, A, A, A, A, A, A}, \boxed{inc}(A), \boxed{inc}(A), \boxed{inc}(A), \boxed{inc}(A), \text{A, A}, \boxed{incadd}(A, ptr(D)), \boxed{incadd}(A, ptr(D)), \text{A, A, A, A}, \boxed{incadd}(A, btow(ptr(D))), \boxed{incadd}(A, btow(ptr(D))), \text{A, A})$$

$$R \overset{\phi!}{=} SEL(\text{P, false, true, true, R, R, R, R, R, R, R, R, R, R, R, R, R, R, R, R, R, R, R, R, R, R, R, R, R, R, R, R, R, R, R})$$

$$W \overset{\phi!}{=} SEL(\text{P, W, W}, not(W), \text{W, W, W, W, W, W, W, W, W, W, W, W, W, W, W, W, W, W, W, W, W, W, W, W, W, W, W, W, W, W, W})$$

FIGURE 5–3. INITIAL SYSTEM DESCRIPTION

$$S \stackrel{\phi!}{=} SEL(P, \text{next}, \text{idle}, \text{idle}, \text{next}, \text{next}, \text{obj}, \text{obj}, \text{driver}, \text{driver}, \text{driver}, \text{driver}, \text{pair1}, \text{pair1}, \text{vec1}, \text{vec1}, \text{bvec1}, \text{bvec1}, \text{driver}, \text{pair2}, \text{pair2}, \text{driver}, \text{driver}, \text{vloop}, \text{vloop}, \text{driver}, \text{driver}, \text{vloop}, \text{vloop}, \text{bloop}, \text{bloop}, \text{driver}, \text{driver}, \text{bloop}, \text{bloop})$$

$$H \stackrel{\phi!}{=} SEL(P, \text{*H*}, H, H, M_2, M_1, H, H, H, H, H, H, H, H, H, H, H, H, H, H, H, H, H, H, H, H, H, H, H, H, H, H, H, H, H)$$

$$D \stackrel{\phi!}{=} SEL(P, D, D, D, D, D, M_1, M_2, D, D, D, D, D, D, D, D, \text{cell}(D, Z_3), \text{cell}(D, Z_3), D, M_1, M_2, D, D, M_1, M_2, D, D, M_1, M_2, M_1, M_2, D, D, M_1, M_2)$$

$$U \stackrel{\phi!}{=} SEL(P, 0, U, U, U, U, U, U, Z_1, Z_1, Z_1, Z_1, U, U, Z_1, Z_1, U, U, Z_1, U, U, Z_2, Z_2, U, U, U, U, U, U, Z_2, Z_2, U, U, U, U)$$

$$A \stackrel{\phi!}{=} SEL(P, 0, A, A, A, A, A, A, A, A, A, A, A, A, A, A, A, A, A, Z_1, Z_1, Z_1, Z_1, A, A, Z_1, Z_1, A, A, A, A, Z_1, Z_1, A, A)$$

$$R \stackrel{\phi!}{=} SEL(P, \text{false}, \text{true}, \text{true}, R, R, R, R, R, R, R, R, R, R, R, R, R, R, R, R, R, R, R, R, R, R, R, R, R, R, R, R, R, R, R)$$

$$W \stackrel{\phi!}{=} SEL(P, W, W, \text{not}(W), W, W, W, W, W, W, W, W, W, W, W, W, W, W, W, W, W, W, W, W, W, W, W, W, W, W, W, W, W, W, W)$$

$$C = COUNT(\phi, C.I, C.D)$$

$$C.I = SEL(P, @, @, @, @, @, @, @, @, @, @, @, @, @, \text{L}, \text{L}, \text{L}, \text{L}, @, @, @, @, @, \text{D}, \text{D}, @, @, \text{D}, \text{D}, \text{D}, \text{D}, @, @, \text{D}, \text{D})$$

$$C.D = SEL(P, \phi, \phi, \phi, \phi, \phi, \phi, \phi, \phi, \phi, \phi, \phi, \phi, \phi, \text{ptr}(D), \text{ptr}(D), Z_3, Z_3, \phi, \phi, \phi, \phi, \phi, \phi, \phi, \phi, \phi, \phi, \phi, \phi, \phi, \phi, \phi, \phi, \phi)$$

$$M_1 = MEMORY(\phi, M_1.I, M_1.A, M_1.D)$$

$$M_1.I = SEL(P, @, @, @, @, \text{R}, \text{R}, @, @, @, @, \text{W}, \text{W}, \text{W}, @, \text{W}, @, \text{W}, @, \text{R}, \text{W}, @, \text{W}, \text{R}, \text{W}, \text{W}, @, \text{R}, \text{W}, \text{R}, \text{W}, \text{W}, @, \text{R}, \text{W})$$

$$M_1.A = SEL(P, \phi, \phi, \phi, \phi, U, H, \phi, \phi, \phi, \phi, U, H, A, \phi, U, \phi, A, \phi, Z_2, U, \phi, A, Z_1, A, H, \phi, Z_2, Z_1, Z_1, U, H, \phi, Z_2, Z_1)$$

$$M_1.D = SEL(P, \phi, \phi, \phi, \phi, \phi, \phi, \phi, \phi, \phi, \phi, \text{cell}(H, D), \text{cell}(\text{fwd}, A), D, \phi, \text{cell}(H, A), \phi, D, \phi, \phi, \text{cell}(II, A), \phi, D, \phi, D, \text{cell}(\text{fwd}, A), \phi, \phi, D, \phi, \text{cell}(H, A), \text{cell}(\text{fwd}, A), \phi, \phi, D)$$

FIGURE 5–4. FACTORED SYSTEM DESCRIPTION

(continued on next page)

$M_2 = MEMORY(\phi, M_2.I, M_2.A, M_2.D)$

$M_2.I = SEL(P$, Q, Q, Q, R, Q, Q, R, Q, Q, W, Q, W, W, W, Q, W, Q, Q, W, R, W, Q, W, R, Q, W, W, R, W, R, Q, W, W, R)

$M_2.A = SEL(P, \phi, \phi, \phi, U, \phi, \phi, H, \phi, \phi, U, \phi, A, H, U, \phi, A, \phi, \phi, U, Z_2, A, \phi, A, Z_1, \phi, H, Z_1, Z_2, U, Z_1, \phi, H, Z_1, Z_2)$

$M_2.D = SEL(P, \phi, \phi, \phi, \phi, \phi, \phi, \phi, \phi, \phi, \mathrm{cell}(H, D), \phi, D, \mathrm{cell}(\mathrm{fwd}, A), \mathrm{cell}(H, A), \phi, D, \phi, \phi, \mathrm{cell}(H, A), \phi, D, \phi, D, \phi, \phi, \mathrm{cell}(\mathrm{fwd}, A), D, \phi, \mathrm{cell}(H, A), \phi, \phi, \mathrm{cell}(\mathrm{fwd}, A), D, \phi)$

$Z_1 = ALU1(Z_1.I, Z_1.A, Z_1.B)$

$Z_1.I = SEL(P$, C, C, C, C, C, C, C, AI, I, I, I, C, C, I, I, C, C, I, I, I, I, I, A, A, IA, IA, IA, IA, A, A, IA, IA, IA, IA)

$Z_1.A = SEL(P, \phi, \phi, \phi, \phi, \phi, \phi, \phi, U, U, U, U, \phi, \phi, U, U, \phi, \phi, U, A, A, A, A, \mathrm{ptr}(H), \mathrm{ptr}(H), A, A, C, C, \mathrm{ptr}(H), \mathrm{ptr}(H), A, A, C, C)$

$Z_1.B = SEL(P, \phi, \phi, \phi, \phi, \phi, \phi, \phi, Z_3, \phi, \phi, \phi, \phi, \phi, \phi, \phi, \phi, \phi, \phi, \phi, \phi, \phi, \phi, C, C, \mathrm{ptr}(D), \mathrm{ptr}(D), A, A, \mathrm{ptr}(D), \mathrm{ptr}(D), Z_3, Z_3, A, A)$

$Z_2 = ALU2(Z_2.I, Z_2.A, Z_2.B)$

$Z_2.I = SEL(P$, C, C, C, C, C, C, C, C, C, C, C, C, C, C, C, C, C, C, I, I, I, I, C, C, C, C, A, A, I, I, C, C, A, A)

$Z_2.A = SEL(P, \phi, \phi, \phi, \phi, \phi, \phi, \phi, \phi, \phi, \phi, \phi, \phi, \phi, \phi, \phi, \phi, \phi, \phi, \mathrm{ptr}(H), \mathrm{ptr}(H), U, U, \phi, \phi, \phi, \phi, \mathrm{ptr}(H), \mathrm{ptr}(H), U, U, \phi, \phi, \mathrm{ptr}(H), \mathrm{ptr}(H))$

$Z_2.B = SEL(P, \phi, C, C, \phi, \phi, \phi, \phi, C, C)$

$Z_3 = BYTETOWORD(Z_3.I, Z_3.B)$

$Z_3.I = SEL(P$, C, C, C, C, C, C, C, B, C, C, C, C, C, C, C, B, B, C, C, C, C, C, C, C, C, C, C, C, C, C, B, B, C, C)

$Z_3.B = SEL(P, \phi, \phi, \phi, \phi, \phi, \phi, \phi, \mathrm{ptr}(H), \phi, \phi, \phi, \phi, \phi, \phi, \phi, \mathrm{ptr}(D), \mathrm{ptr}(D), \phi, \phi, \phi, \phi, \phi, \phi, \phi, \phi, \phi, \phi, \phi, \phi, \phi, \mathrm{ptr}(D), \mathrm{ptr}(D), \phi, \phi)$

FIGURE 5–4 (CONTINUED). FACTORED SYSTEM DESCRIPTION

$$MEMORY(\check{m}, Inst, Addr, Data) \stackrel{d}{=} \mathrm{read}(M)$$

where

$$M = \check{m} \;!\; Interpret(M, Inst, Addr, Data)$$

and

$$Interpret(m, inst, addr, data) \stackrel{d}{=}$$

$$\textit{per } inst \textit{ viz } \begin{array}{l} @ : m \\ \mathsf{R} : m \\ \mathsf{W} : \mathrm{write}(m, addr, data) \end{array}$$

$$ALU1(Inst, A, B) \stackrel{d}{=} X$$

where

$$X = \textit{per } Inst \textit{ viz } \begin{array}{r} \mathsf{C} : 0 \\ \mathsf{AI} : addinc(A, B) \\ \mathsf{I} : inc(A) \\ \mathsf{A} : add(A, B) \\ \mathsf{IA} : incadd(A, B) \end{array}$$

$$ALU2(Inst, A, B) \stackrel{d}{=} X$$

where

$$X = \textit{per } Inst \textit{ viz } \begin{array}{r} \mathsf{C} : 0 \\ \mathsf{I} : inc(A) \\ \mathsf{A} : add(A, B) \end{array}$$

$$COUNT(\check{x}, Inst, Data) \stackrel{d}{=} X$$

where

$$X = \check{x} \;!\; Interpret(X, Inst, Data)$$

and

$$Interpret(x, inst, data) \stackrel{d}{=} \textit{per } inst \textit{ viz } \begin{array}{l} @ : x \\ \mathsf{L} : data \\ \mathsf{D} : dcr(x) \end{array}$$

$$BYTETOWORD(Inst, Byte) \stackrel{d}{=} X$$

where

$$X = \textit{per } Inst \textit{ viz } \begin{array}{l} \mathsf{C} : 0 \\ \mathsf{B} : btow(Byte) \end{array}$$

FIGURE 5–5. FACTORED COMPONENT SPECIFICATIONS

Factorization of signal C encapsulates the combinational operations on register C; these constitute a counter.

Next, two general factorizations of arithmetic yield three abstract components. The operation *btow* is hidden in a component *BYTETOWORD*. Finally, subject terms are developed from the set

$$\{add, inc, dcr, addinc, incadd\}.$$

As is illustrated in Section 4, generalizations of the subject terms must be collated into signal equations. In this case, the attempt to collate leads to clashes, which means that more than one adder is needed. When this happens, the transformation system must partition the subject terms to generate additional component abstractions. Partitioning is interactive because it involves tactical design considerations. For example, the residual signal $Z_2.B$ in Figure 5–4 simplifies to C as a result of the partitioning choice.

Two issues of concern in this work are the degree to which specifications must anticipate factorizations, and the effort needed to revise descriptions when factorizations don't satisfy engineering intentions. More serialization of addition might be coded in the specification, but two adders are tolerated because the intent is to keep the memories busy. On a design of this size, it is reasonable to experiment rather than to analyze. In fact, the first derivation attempt led to *three* adders, and the specification was subsequently changed. Re-execution of the derivation went through quickly. A more timely serialization, through transformations on Figure 5.3 for example, is worth study but is beyond the current capabilities of our system.

Subsequent algebra is directed toward a physical organization and consequently toward a boolean representation of the symbolic basis. One reason for directing this derivation toward a bit-slice implementation is the lack of mechanisms for imposing representations. The derivation strategy minimizes the impact of this gap in automation. This is a central issue in the continuation of this work, which we discuss briefly in Section 6.

Figure 5–4 is a description in which all substantial operations are factored out. The remaining equations are in terms of selection and record manipulation, and the latter are realized implicitly in the architecture. For example, the operation *ptr* is simply the extraction of a 24-bit ADDRESS field from a 32-bit CONTENT. Each bit slice represents the projection of an n-bit quantity to a single bit. That is, the signals in Figure 5–4, excluding Q, can be interpreted as single-bit projections; field manipulations project to identity functions and can be ignored. What remains is a system of boolean selections that assemble to PAL fuse maps (Recall the generic PAL description in Section 3.2).

In this interpretation, a bit-sliced collector is described by n instances of Figure 5–4, connected in parallel to the component abstractions. However, some equations are irrelevant in certain slices. The projection of bit $N+m$ of an N-bit vector is vacuous; equations for N-bit signals need not appear in the description of slice $N+m$.

Figure 5–6 shows the physical composition of the collector. Six distinct PAL descriptions are developed, each implementing a subset of equations in Figure 5–4, with some differing only in their representation of constants. Internal descriptions of components *PALa* through *PALf* can be inferred from the equations that use them. For instance, the equation

$$[H_{24}, D_{24}, M_1.D_{24}, M_2.D_{24}] = PALb(CMD, M_1, M_2, R)$$

indicates that *PALb* is described by equations for H, D, $M_1.D$, and $M_2.D$ in Figure 5–4. A broadcast command signal is specified as:

$$CMD = SEL(Q, 0,\ 1,\ \cdots,\ 33)$$

The encoding is automatically synthesized to obtain description *PALsel*. The decoding of *CMD* is manifested in the remaining components, as illustrated in Section 3.2. Signals S, W, R, $M_1.I$, $M_2.I$, $C.I$, $Z_1.I$, and $Z_2.I$ are packaged in a single component *PALinst*. The equations for multiple-bit signals are manually replicated and constants assigned. The boolean description of *PALinst* is then automatically generated.

All PAL descriptions are generated, reduced, and translated to Altera Design Files. Simplification is done by Espresso [19]. Thirty-four instances of eight descriptions, *PALsel*, *PALinst*, *PALa*, ..., *PALf* are assembled to Altera PALs to implement the majority of the collector's architecture. Figure 5–7 shows a custom PLA realization of the *PALa* bit slice, the largest of eight boolean descriptions. It is generated from the same design file by the `MPLA` function of the Berkeley VLSI tools [19]. Four register feedback paths are added manually in `Magic`.

The *STATUS* combination (Figure 5–2) and two dynamic-RAM memories are built by hand. Standard LSI is used to realize specifications of the arithmetic abstract components. The prototype is wire-wrapped on a Logic Engine, where a microprogrammed serial interface cross-loads heap images from an Apollo workstation. The workstation hosts the benchmark Scheme implementation.

$$\textbf{COLLECTOR}(GO) \stackrel{\mathrm{d}}{=} R \quad \text{where}$$

$$\begin{aligned}
P &= STATUS(S,\ H,\ D,\ C,\ U,\ A,\ GO,\ W)\\
CMD &= PALsel(P)
\end{aligned}$$

$$[\,S, R, W, M_1.I,\ M_2.I,\ Z_1.I,\ Z_2.I,\ C.I,\ Z_3.I\,] = PALinst(CMD)$$

$$\begin{aligned}
M_1 &= MEMORY(\phi, M_1.I,\ [\,M_1.A_{23} \cdots M_1.A_0\,],\ [\,M_1.D_{31} \cdots M_1.D_0\,])\\
M_2 &= MEMORY(\phi, M_2.I,\ [\,M_2.A_{23} \cdots M_2.A_0\,],\ [\,M_2.D_{31} \cdots M_2.D_0\,])\\
Z_1 &= ALU1(Z_1.I,\ [\,Z_1.A_{23} \cdots Z_1.A_0\,],\ [\,Z_1.B_{23} \cdots Z_1.B_0\,])\\
Z_2 &= ALU2(Z_2.I,\ [\,Z_2.A_{23} \cdots Z_2.A_0\,],\ [\,Z_2.B_{23} \cdots Z_2.B_0\,])\\
C &= COUNT(\phi, C.I,\ [\,C.D_{23} \cdots C.D_0\,])\\
Z_3 &= BYTETOWORD(Z_3.I,\ [\,Z_3.B_{23} \cdots Z_3.B_0\,])
\end{aligned}$$

$$\begin{aligned}
H &\stackrel{\phi!}{=} [\,H_{31} \cdots H_0\,]\\
D &\stackrel{\phi!}{=} [\,D_{31} \cdots D_0\,]\\
U &\stackrel{\phi!}{=} [\,U_{23} \cdots U_0\,]\\
A &\stackrel{\phi!}{=} [\,U_{23} \cdots U_0\,]
\end{aligned}$$

$$\begin{aligned}
&[\,H_0, D_0, U_0, A_0, M_1.A_0, M_1.D_0, M_2.A_0, M_2.D_0\\
&\qquad\qquad Z_1.A_0, Z_1.B_0, Z_2.A_0, Z_2.B_0, C.D_0, Z_3.B_0\,]\\
&\quad = PALa(CMD, M_1, M_2, Z_1, Z_2, C, Z_3, R)\\
&\qquad \vdots\\
&[\,H_{23}, D_{23}, U_{23}, A_{23}, M_1.A_{23}, M_1.D_{23}, M_2.A_{23}, M_2.D_{23},\\
&\qquad\qquad Z_1.A_{23}, Z_1.B_{23}, Z_2.A_{23}, Z_2.B_{23}, C.D_{23}, Z_3.B_{23}\,]\\
&\quad = PALa(CMD, M_1, M_2, Z_1, Z_2, C, Z_3, R)
\end{aligned}$$

$$\begin{aligned}
&[\,H_{24}, D_{24}, M_1.D_{24}, M_2.D_{24}\,] = PALb(CMD, M_1, M_2, R)\\
&\qquad \vdots\\
&[\,H_{27}, D_{27}, M_1.D_{27}, M_2.D_{27}\,] = PALb(CMD, M_1, M_2, R)\\
&[\,H_{28}, D_{28}, M_1.D_{28}, M_2.D_{28}\,] = PALc(CMD, M_1, M_2, R)\\
&[\,H_{29}, D_{29}, M_1.D_{29}, M_2.D_{29}\,] = PALd(CMD, M_1, M_2, R)\\
&[\,H_{30}, D_{30}, M_1.D_{30}, M_2.D_{30}\,] = PALe(CMD, M_1, M_2, R)\\
&[\,H_{31}, D_{31}, M_1.D_{31}, M_2.D_{31}\,] = PALf(CMD, M_1, M_2, R)
\end{aligned}$$

FIGURE 5–6. GARBAGE COLLECTOR CIRCUIT DESCRIPTION

6. Summary and Directions

A digital engineer balances many considerations when developing an implementation. A need to decompose descriptions in orthogonal hierarchies is fundamental, and system factorizations are a basic mechanism for syntactic decomposition. We are just beginning to explore what parameters are involved in factorization. This paper looks at impositions of logical and physical structure, including the control/architecture decomposition, the use of complex types in specification, and physical partitionings.

The transformation system discussed here establishes a path to physical implementation. It is envisaged as a vehicle for exploring how functional modeling techniques can be brought to bear on a wider range of target technologies. We see a working relationship between this approach and hardware verification research. At the same time, a transformational discipline seems practical without being tied to formal correctness. The frontier of our work is to confront the problem of representing abstract bases in low level digital descriptions.

Derived implementations are correct provided that (1) their specifications are correct, (2) the transformations preserve correctness, and (3) the physical elements used accurately model the theory of description. These are three ways that formal verification methods integrate with a transformational design. In the other direction, hardware verification needs automation to compose "theorems" because hardware proofs (e.g., [8], [11], [15], [6]), almost necessarily, follow the logical structure of design. We use Hunt's FM8501 proof [11] as an example of the relationship. The proof has two levels, one addressing the implementation of an arithmetic basis, the other a realization of instruction semantics. To get a physical description of FM8501, it remains to derive a sequential-system description from its register-transfer specification, to incorporate the representation of the ground type, and to develop an appropriate physical organization of the whole. It is toward these kinds of manipulations that our transformations are directed. For the garbage collector derived in Section 5, integrating Hunt's proof of arithmetic primitives would secure a "more correct" implementation.

There was no digital debugging of the garbage collector derived in Section 5. Though mistakes were made in assembling the prototype, none were found in the programmed hardware. The collector worked on its first full test. The control algorithm was well understood. The specification and intermediate forms (e.g., Figures 5–1 through 5–5) are executable; implementation proceeded directly from a tested specification. The later stages of derivation could also have been used to explore the design, but were not in this exercise.

In addition, Figure 5-1 served almost immediately as a simulation of the algorithm. On representative heaps, the derived collector is 62 to 98 times faster than a compatible 68000 implementation. The hardware collects at a rate of about 5M bytes per second. The main factors in speed improvement are hardware support of tagged data, broader data paths, parallel activity on two memories, and the usual gains over software implementation of an algorithm [4]. The derivation process exposed optimizations in physical organization and these led to revisions (i.e., manual transformations) on the initial description. One instance discussed was the decision to switch memory paths internally in PALs, which lead to a more conventional memory abstraction. The derived collector is comparable to a design developed manually using structured digital design techniques for the same class of components. The experience convinces us that engineering can be done in the linear notation exhibited in this paper. That is, the transformation system presented an adequate global view of the design while factorizations isolated facets of immediate interest.

Targeting the garbage collector for a bit-slice PAL implementation was an expedient derivation strategy. A number of practical problems were circumvented. At the same time, a fundamental formal issue was also bypassed; it is the problem of incorporating representations. In the figurative sense of Section 1, the "concrete" type Γ is replaced by a suitable representation, G

$$\begin{array}{ccccccc} \mathcal{S}_{\Theta/\Gamma} & \longrightarrow & \mathcal{I}_{\Theta/\Gamma} & \longrightarrow & \mathcal{C}_{\Theta/\Gamma} & \longrightarrow & \mathcal{C}'_{\Gamma}\|\Theta_{\Gamma} & & \\ & & & & & & \Downarrow & & \\ & & & & & & \mathcal{C}'_{G}\|\Theta_{G} & \longrightarrow & \cdots \end{array}$$

The derivation in Section 5 is tailored to make this step superficial. Bit slices are an extreme example of orthogonal logical and physical decompositions. They project the logical structure into each physical component. For a glimpse at this issue, let us consider incorporating addition in the garbage collector's data path. Using the *Adder* defined in Section 2, it is not hard to see that *BitSum* combinations would be incorporated in each projection. A different physical organization, such as partitioning for 4-bit slices, would also have to be imposed on *Adder*. One PAL description, *PALinst* in Figure 5–6, was interactively developed for a similar reason.

Incorporating representations can be highly exacting and therefore needs automation. However, a general treatment should not be restricted to the simple types involved in this paper's examples. Discussions in Section 3 suggest that abstraction and hierarchy are useful in this dimension of derivation. Considered as a component for a list processing system, the garbage collector implements the complex type HEAP in a processor specification [21]. Theories

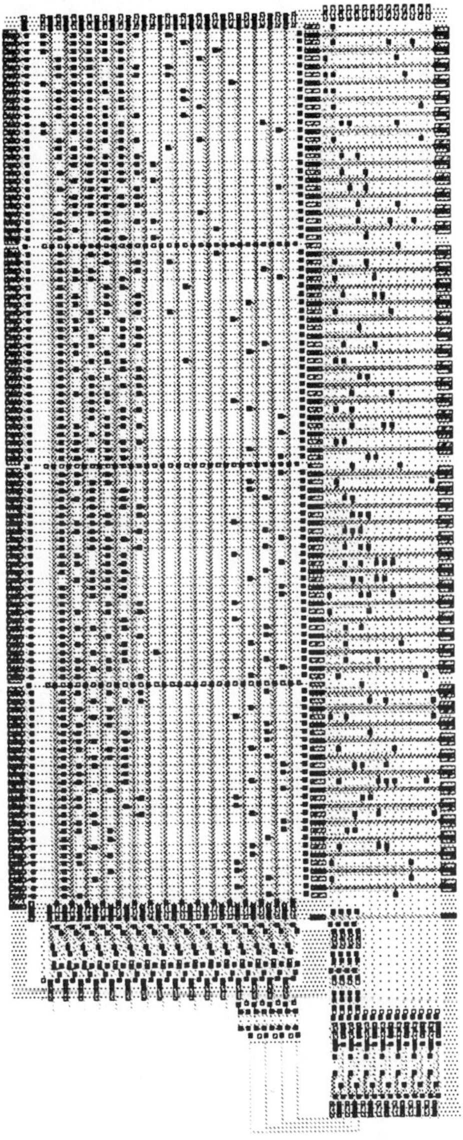

FIGURE 5–7. PLA IMPLEMENTATION OF COMPONENT *PALa*

for representation are developing that will have an impact for both hardware and software engineering methodology.

This work reflects a view of digital engineering as a creative endeavor. Automated support should present intuitive facets of design while correctly maintaining the relationships among them. We seek characterizations of basic abstractions, to be codified in an algebraic framework. Our early experience with this approach suggests that reasonable and correct implementations can be derived through a manageable set of general transformations. The immediate challenge is to broaden the range of implementations while preserving generality in the algebra.

Acknowledgments

Will Clinger made numerous contributions to the design exercise reported in Section 5, including his implementation of the Scheme test bed. Eric Ost helped establish the test environment. Bob Wehrmeister provided essential support in the integration of design and fabrication tools. We are grateful to David Winkel and Franklin Prosser for their insights in methodology and their help in constructing the prototype.

References

[1] Altera Corporation, Altera Programmable Logic User System User Guide (Version 4.0), Altera Corporation, Santa Clara, 1985,

[2] Bose, Bhaskar, *Hardware Derivation from a High Level Specification: an Automated Transformation System Implementation,* in progress.

[3] Boute, R. T., Current Work on the Semantics of Digital Systems, in G. Milne and P. Subrahmanyam (eds.) *Formal Aspects of VLSI Design,* North-Holland, Amsterdam, 1986, 99–112.

[4] Boyer, C. David, *A Transformationally Correct Hardware Garbage Collector Implementation,* in progress.

[5] Clinger, William C., The Scheme 312 Version 4 Reference Manual, 1985 (unpublished).

[6] Cohn, Avra, A Proof of Correctness of the Viper Microprocessor: The First Level, *this volume.*

[7] Fenichel, R., and J. Yochelson, A LISP garbage-collector for virtual-memory computer systems, *Comm. ACM* **12**(11):611–612 (1969).

[8] Gordon, Mike, Proving a Computer Correct, University of Cambridge Computer Laboratory Technical Report No. 42, Cambridge, 1983.

[9] Harel, David, On folk theorems *Comm. ACM* **23** (7):379–389, (1980)

[10] Hill, F. J. and G. R. Peterson, *Introduction to Switching Theory and Logical Design* (Third Ed.), John Wiley&Sons, New York, 1981.

[11] Hunt, Warren A., Jr., *FM8501: A verified Microprocessor,* Ph.D. dissertation, Technical Report 47, Institute for Computing Science, The University of Texas at Austin, 1985.

[12] Johnson, Steven D., Digital Design in a Functional Calculus, in G. Milne and P. Subrahmanyam (eds.) *Formal Aspects of VLSI Design,* North-Holland, Amsterdam, 1986, 153–178.

[13] Johnson, Steven D., Applicative Programming and Digital Design, *Proc. Eleventh Annual ACM SIGACT-SIGPLAN Symposium on Principles of Programming Languages* (1984), 218–227.

[14] Johnson, Steven D., *Synthesis of Digital Designs from Recursion Equations, The ACM Distinguished Dissertation Series,* The MIT Press, 1984.

[15] Joyce, Jeffrey, Formal Verification and Implementation of a Microprocessor, *this volume.*

[16] O'Donnell, John T., Hardware Description with Recursion Equations, *Proc. 8th International Symposium on Computer Hardware Description Languages and their Applications,* Amsterdam, April, 1987.

[17] Prosser, Franklin P., and David E. Winkel, The Logic Engine Development System—Support for Microprogrammed Bit-Slice Development, *Proc. Micro 16,* 84–91.

[18] Rees, Jonathan and William C. Clinger (eds.), Revised^3 Report on the Algorithmic Language Scheme, Indiana University Computer Science Department Technical Report No. 174, Bloomington, December, 1986.

[19] Scott, Walter S., Robert N. Mayo, Gordon Hamachi, and John K. Ousterhout (eds.), 1986 VLSI Tools, Report No. UCB/CSD 86/272, Computer Science Division (EECS), University of California at Berkeley, 1985.

[20] Winkel, David E., What Next for PALs. Technical Report No. 188, Indiana Univ. Computer Science Dept., Bloomington, Indiana, February, 1986.

[21] Winkel, David E. and Christopher T. Haynes, Hardware Design Using Functionally Connected Units, Indiana University Computer Science Dept. Technical Report No. 219, *submitted for publication.*

[22] Winkel, David E., and Franklin P. Prosser. *The Art of Digital Design, 2nd Edition* Prentice-Hall, Englewood Cliffs, New Jersey, 1986.

13

VERIFICATION OF ASYNCHRONOUS CIRCUITS : BEHAVIORS, CONSTRAINTS AND SPECIFICATIONS

C. Berthet and E. Cerny
Département d'Informatique et Recherche Opérationnelle
Université de Montréal, C.P. 6128, Succ. A
Montréal, Québec, H3C 3J7, Canada

ABSTRACT

An algebraic methodology based on Characteristic Functions (CFs) and allowing to compare switch-level circuits with higher-level specifications is presented.

Switch-level networks, 'user' behaviors, input constraints and specifications are modelled as asynchronous automata which form a Boolean algebra.

Machine composition corresponds to the product of the Boolean algebra. Internal variables can be abstracted, and more generally, a component automaton can be reduced by projection under the presence of a local domain constraint. The constraints are validated by comparison with the actual user of each component.

Formal verification implies comparison with a high-level specification which may either be complete or consists of a collection of properties. Verification of safety and liveness with respect to a property reduces to the validation of Boolean inequalities.

This methodology is applied to a new structure of mutual exclusion ensuring fairness.

INTRODUCTION

Formal verification techniques for hardware design are being developed as an alternative to simulation. Whereas simulation compares results, proofs of correctness involve comparison of functional descriptions. The advantage of formal verification over simulation is that it is a complete method, however, its cost is high computational complexity. Numerous theoretical models have been proposed: algebraic [12] and axiomatic [16] models, denotational models using recursive expressions [8,15], predicate logic [1], higher-order logic [9] and various forms of temporal logic [3,7].

The main problems of formal verification are the following : a) How to obtain a functional description from the circuit structure ? b) How to express the high level specification ? c) How to compare the extracted description with the specification ?

Verification requires the definition of a formal model, including the specification of primitive elements (axioms), and composition and abstraction laws. Specifications can be expressed in the form of high-level automata or as a collection of properties. The formalism used to express specifications must be as close as possible to the formal model used to extract circuit's behavior.

Our model is algebraic. The main advantage of the approach is that abstractions can be made before composition, permitting to limit the combinational explosion of states. Circuit and user behaviors as well as constraints are modeled as asynchronous non-deterministic automata, uniformly represented using a pair of Characteristic Functions (CFs). The model of time is such that delay is considered as arbitrary but finite, which allows to verify whether circuits satisfy the property of speed-independence.

In Section I, the algebraic model of CFs and a Boolean algebra of asynchronous machines is defined. In Section II, this algebra is shown to cover the behaviors of switch-level circuits and external users, and also input constraints. The methodology of behavioral extraction, is described in Sections III (composition under constraint) and IV (reduction under constraint). Finally, in Section V, comparison with a specification is approached by means of an example : a new switch-level structure of a fair mutual exclusion element. The specification is given as a collection of properties, each property being represented by an asynchronous automaton. Verification is performed by showing that the given circuit is safe and live with respect to each property, which, in our model, is expressed through Boolean relations ($\leq,\geq$).

I) A BOOLEAN ALGEBRA OF ASYNCHRONOUS MACHINES

We consider vectors of binary variables. The values of the vector **x**, which are elements of $\{0,1\}^{|\mathbf{x}|}$ are noted x, or using subscripts: x_0, x_1,.. A subvector is defined in the usual way. In the sequel, **x** denotes the input vector, **y** the state vector and **z** the output vector. In our model , **z** is a sub-vector of **y**, $\mathbf{z} \subseteq \mathbf{y}$.

A Characteristic Function (CF) **f**(**x**) is a Boolean function defined over a vector **x**: $\{0,1\}^{|\mathbf{x}|} \rightarrow \{0,1\}$. The set of all such CFs forms a Boolean algebra. Sum (OR) and product (AND) are noted + and $\cdot$, the extended Boolean summation (OR) and product (AND) are represented by $\sum$ and $\prod$. The complement of a CF **f** is noted $\overline{\mathbf{f}}$. The partial order is: $\mathbf{f} \leq \mathbf{g} \Leftrightarrow \mathbf{f} \cdot \mathbf{g} = \mathbf{f} \Leftrightarrow \mathbf{f} \cdot \overline{\mathbf{g}} = \mathbf{0}$.

A CF **f**(**x**) defined over a vector **x**: $\{0,1\}^{|\mathbf{x}|} \rightarrow \{0,1\}$ can be trivially extended over any vector of which **x** is a subvector: The value of the extended CF is determined by **x**, regardless of the values of the additional variables.

We denote $\sum_{\mathbf{t}} \mathbf{f}$ the Boolean summation of **f** over all values of the vector **t**. It corresponds to the deletion of all occurences of the variables of **t** in an algebraic expression of **f** [5].

Definition 1: An asynchronous automaton (or machine) is a tuple $(\mathbf{x},\mathbf{y},\mathbf{z},\Phi,\Psi)$ where **x** is the input vector, **y** the state vector and $\mathbf{z} \subseteq \mathbf{y}$ the output vector. The CF $\Phi(\mathbf{x},\mathbf{y})$: $\{0,1\}^{|\mathbf{x}|}\times\{0,1\}^{|\mathbf{y}|}\rightarrow\{0,1\}$ represents the stable states: $\Phi(x,y)=1$ iff y is stable with input x. The CF $\Psi(\mathbf{x},\mathbf{y},\mathbf{y}')$: $\{0,1\}^{|\mathbf{x}|}\times\{0,1\}^{|\mathbf{y}|}\times\{0,1\}^{|\mathbf{y}|}\rightarrow\{0,1\}$ represents the transitions: $\Psi(x,y_1,y_2)=1$ iff y_2 may follow y_1 with input x. The pair (Φ,Ψ) of CFs is called a Dynamic CF (DCF).

In any transition $<x,y_1,y_2>$, x is the input value at time t, y_1 the state value at time t, and y_2 the state value at time $t+\delta$, where δ represents an arbitrary finite quantum of time. A transition $<x,y,y>$ is a self-loop. A self-loop $<x,y,y>$ is stable if $\Phi(x,y) = 1$. Otherwise, the self-loop is unstable. In a stable self-loop, the number of consecutive passages through the loop may be infinite, while in an unstable self-loop it is always finite.

The algebraic expression of a DCF is equivalent to the state transition diagram of the automaton. An arrow in the diagram from state (node) y_1 to state (node) y_2 corresponds to all x such that $<x,y_1,y_2>$ is a solution of the equation $\Psi(\mathbf{x},\mathbf{y},\mathbf{y}')=1$. Graphically, stable and unstable self-loops are distinguished using solid and dotted lines, respectively.

DCFs permit modelling two different forms of non-determinism:

a) $\exists(x,y_1,y_2\neq y_1)$ such that $\Psi(x,y,y_1)=1$ and $\Psi(x,y,y_2)=1$: The next state of y with input x is either y_1 or y_2, without an a priori knowledge of which one will be selected.

b) $\exists(x,y,y_1\neq y)$ such that $\Phi(x,y)=1$, $\Psi(x,y,y)=1$ and $\Psi(x,y,y_1)=1$: With input x, the automaton can remain forever in the same stable state y or change to state y_1 at any time, without an a priori knowledge of when this may happen.

Definition 2: A closed (inputless) asynchronous automaton is defined by the tuple $(\mathbf{y},\mathbf{z},\Phi,\Psi)$, where **y** is the state vector, $\mathbf{z} \subseteq \mathbf{y}$ the output vector, $\Phi(\mathbf{y})$: $\{0,1\}^{|\mathbf{y}|}\rightarrow\{0,1\}$, $\Phi(y)=1$ iff y can remain stable forever and $\Psi(\mathbf{y},\mathbf{y}')$: $\{0,1\}^{|\mathbf{y}|}\times\{0,1\}^{|\mathbf{y}|}\rightarrow\{0,1\}$, such that $<y_1,y_2>$ is a transition from y_1 to y_2 iff $\Psi(y_1,y_2)=1$.

A closed automaton can be interpreted as having a one-symbol input alphabet. In the sequel, this symbol will be ignored. In the corresponding diagram, there is an arrow, without label, from state y_1 to state y_2 if $\Psi(y_1,y_2)=1$. Furthermore, a self-loop around a state y is marked with a solid line if $\Phi(y)=1$ (stable state), otherwise the line is dotted (unstable state).

The following inputless automaton with $y=z=R_1 \in \{0,1\}$, can remain forever in the stable state $R_1=0$ or it can change. In contrast, $R_1=1$ can be sustained only during a finite period of time.

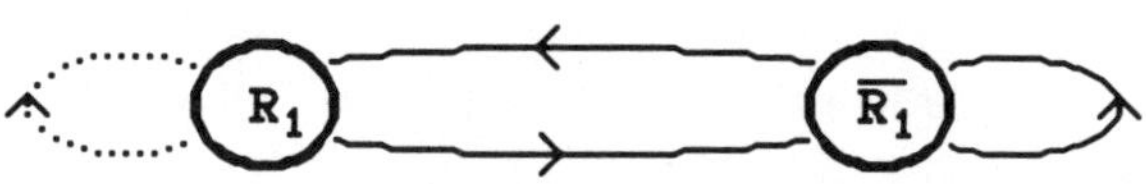

The automaton's DCF is : $(\overline{R}_1, 1)$. It will be seen in Section V that this DCF may be used to specify fairness.

Let (Φ_1,Ψ_1) be a DCF over input $\mathbf{x}_1$ and state $\mathbf{y}_1$, and (Φ_2,Ψ_2) a DCF over input $\mathbf{x}_2$ and state $\mathbf{y}_2$. Each DCF can be trivially extended over the union of both the input and the state vectors.

The <u>product</u> of DCFs is : $(\Phi_1,\Psi_1) \cdot (\Phi_2,\Psi_2) = (\Phi_1 \cdot \Phi_2, \Psi_1 \cdot \Psi_2)$.

The resulting automaton may be more deterministic (and restricted) than the components. The machine is stable when both machines are stable, and similarly for transitions and self-loops.

The <u>sum</u> of DCFs is : $(\Phi_1,\Psi_1) + (\Phi_2,\Psi_2) = (\Phi_1 + \Phi_2, \Psi_1 + \Psi_2)$.

The sum may increase non-determinism. The machine is stable when either machine is stable (or both). Similarly for transitions and self-loops.

A particular form of the sum of DCFs is the abstraction of a state subvector **t**. The obtained DCF is : $(\sum_{\mathbf{t}} \Phi, \sum_{\mathbf{t},\mathbf{t}'} \Psi)$. This summation consists of deleting any occurence of a variable of **t** (resp. **t**, **t'**) in Φ (resp. Ψ).

A <u>partial order</u> is defined as : $(\Phi_1,\Psi_1) \leq (\Phi_2,\Psi_2) \Leftrightarrow \Phi_1 \leq \Phi_2$ and $\Psi_1 \leq \Psi_2$.

The <u>complement</u> of a DCF (Φ,Ψ) is: $(\overline{\Phi}, \overline{\Psi})$.

The set of DCFs over input **x** and state **y** forms a Boolean algebra.

II) CIRCUITS, USERS AND CONSTRAINTS

The above model allows representing switch-level circuits as well as deterministic and non-deterministic behaviors of the environment.

II.1) Switch-level circuits

A network can be divided into transistor groups (TG) [4,6], within which signal propagation is bidirectional. Between groups, propagation is unidirectionnal.

The most widely used model to find the steady state signals in a TG represents each signal as a logic value (0, 1 or U for undefined) associated with a strength [4,10]. The signal strength, an integer, takes into account the resistive effects of transistors and the effects of charge storage. The set of gates controlled by an output node is represented by a charge well. The state of a TG is defined by the values of the charge wells.

In such a model, each node is either driven by one or several static sources, or it is isolated from them (permitting dynamic memory through charge storage). The resulting signal values on the internal nodes can be obtained by path algebra methods [4], or decision diagrams [6].

In formal verification, we are only interested in representing the behavior of a TG with defined (i.e., binary) input and internal state values. The node values may become undefined due to shorting of sources or charge-sharing. When this happens on output nodes, we propose to use a static constraint C (subset of $\{0,1\}^{|\mathbf{x}|}$) which eliminates the input values which lead to undefined signals.

The behavior of a TG with input vector **x**, state (charge wells) vector **y**, and node vector **Y**, under the (static) constraint C can be represented by a CF **F**: $\{0,1\}^{|\mathbf{x}|} \times \{0,1\}^{|\mathbf{y}|} \times \{0,1\}^{|\mathbf{Y}|} \rightarrow \{0,1\}$, such that **F**(x,y,Y)=1 if the resulting node value is Y given the state (charge wells) y and input $x \in C$; otherwise **F**(x,y,Y)=0.

For a single node TG (node Y and well y), the behavior is represented by a quadruple of FCs (**f**,**g**,**h**,**i**), representing a partition of $\{0,1\}^{|\mathbf{x}|}$, where **f** corresponds to the input values for which the node is VDD-driven, **g** to the input values for which the node is GND-driven, **h** the values for which the node is isolated (dynamic memory), and **i** corresponds to the values for which the node is undefined (U). Thus: $\mathbf{F} = Y\,\mathbf{f} + \overline{Y}\,\mathbf{g} + \mathbf{h}\,(Y\,y + \overline{Y}\,\overline{y})$ and $\overline{\mathbf{i}} = \mathbf{f}+\mathbf{g}+\mathbf{h}$ is the CF equivalent to C.

The above model is quite restrictive, because any output is either driven by a static source or keeps its (0,1) charge. Its advantage is that a TG behavior is described by a Boolean function.

The preceding CF **F** is a static expression relating steady states. In order to model dynamic behavior of a TG, an inertial delay element is attached to any charge well. The gates controlled by the charge well belong to an equipotential, isochronic region. The delay corresponds to charging or discharging the region as a whole.

A delay can be of any finite duration, and it is expressed as an unknown but finite multiple of a time quantum δ. All potential state transitions are captured. This model has been used for studying speed-independent [14] and self-timed circuits [13].

Inertial delay of a charge well with input Y (node) and state y (charge well) can be described by an asynchronous automaton whose DCF is:

$$(\Phi,\Psi)_\Delta = (yY + \overline{y}\ \overline{Y},\ y(y' + \overline{Y}) + \overline{y}(\overline{y'} + Y)),$$

where y and Y are the values of state and input at time t and y' the next state (at time t+δ). The state y can become and remain unstable ('excited') for an unknown but finite time. Thereafter, it changes value if the excitation persists. The corresponding diagram is :

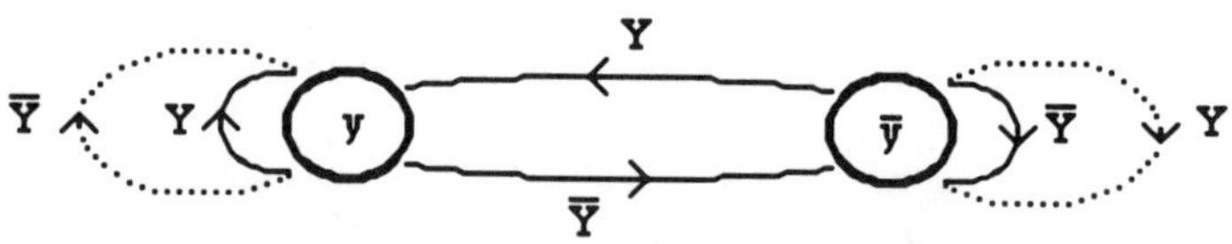

A TG is thus represented by an asynchronous automaton, whose DCF is obtained by composition of the delays' DCFs with the static CF **F**. When the TG is represented by **(f,g,h,i)**, its DCF is:

$$(\Phi,\Psi) = (y\,\mathbf{f} + \overline{y}\,\mathbf{g} + \mathbf{h},\ y\,(y'\,\overline{\mathbf{i}} + \mathbf{g}) + \overline{y}\,(\overline{y'}\,\overline{\mathbf{i}} + \mathbf{f})).$$

When TGs are interconnected in a network, the DCF of the network as a whole is obtained as a product of the DCFs of components TGs. The resulting DCF represents an asynchronous automaton in which there is a state variable for each isochronic region. The interconnection of TGs is valid only if the individual constraints C_i are verified by the network. A global constraint C is derived. The automaton is incompletely defined if $C \neq \{0,1\}^{|\mathbf{x}|}$.

II.2) Users

The behavior of the environment of a circuit, i.e., the user, can also be described formally using the DCF notation. The user's behavior can be represented using an asynchronous automaton with DCF $(\Phi_\varepsilon,\Psi_\varepsilon)$.

The user's input (resp. output) vector is the circuit's output (resp. input) vector. A user may contain internal state variables : The circuit's input vector is a subvector of the user's state vector.

A user is generally non-deterministic. Certain states of the user may remain stable (possibly for an infinite duration) or change, thus possibly changing the circuit's input value, with no knowledge of when this might occur.

For instance, consider a user involved in a four-cycle signaling (FCS) [13] with a circuit. The user can stay idle (possibly for an infinite time) or emit a request by rising its output y. When the circuit activates the user input x indicating acknowledgement, the user must, after some finite time, lower its output. A new request can occur only after the acknowledgement signal x is lowered.

The user's input vector is x, the user's state (and output) vector is y. The corresponding DCF is:

$$(\Phi_\varepsilon,\Psi_\varepsilon) = (\overline{y} + \overline{x}\ ,\ y\,(y'+x) + \overline{y}\,(\,\overline{y'} + \overline{x}\,)\,).$$

Its diagram is :

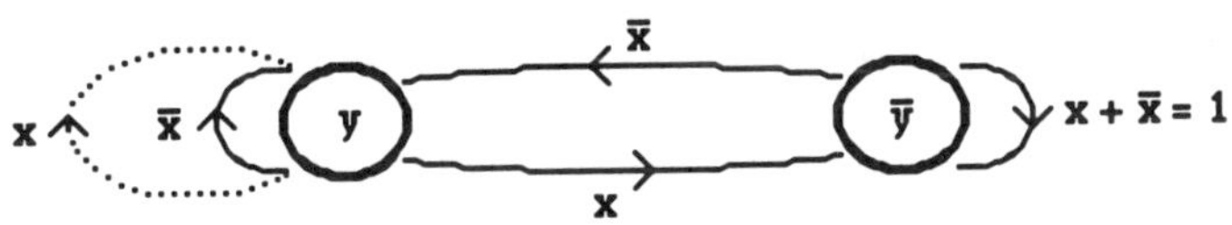

II.3) Domain constraints

When analysing the individual components of a circuit, we usually do not know what is exactly the behavior of the user of each component. Every component is itself a part of the user of the other components. Nevertheless, when analysing a component, assumptions are usually made about the external behavior.

Constraints on possible input transitions are generally present, even if implicitly. In synchronous systems, inputs can change only in certain relation to clock transitions ('set-up time', 'hold time' for a clocked flip-flip). For asynchronous circuits, the fundamental mode hypothesis is usually assumed: Only one input variable can change and only after circuit stabilization. In the context of speed-independent circuits [14], a circuit can have independent asynchronous inputs, each of which can change when the circuit is still unstable. However, it is usually assumed that certain rules are respected.

Such constraints on a component of a circuit usually represent the maximal admissible external behavior ensuring that the component will work correctly, i.e. without hazards, races or oscillations. Moreover, assuming these constraints, useless or redundant internal variables can be removed. For example, if not externally used (i.e. non-observable), the complemented output of an asynchronous latch can be removed when fundamental mode hypothesis is assumed. The main problem facing a verifier (or a designer) is the following : Does the actual environment of a component satisfy the constraints used when analysing the component (and possibly reducing its behavioral description) ?

This verification is usually done in an informal way. We propose to use DCFs to this end.

The constraint of a circuit with input **x**, output **z**, subvector of state **y**, is represented by a DCF (**C**,**Θ**), where **C**(**x**): $\{0,1\}^{|\mathbf{x}|} \rightarrow \{0,1\}$, and **Θ**(**z**,**x**,**x'**): $\{0,1\}^{|\mathbf{z}|} \times \{0,1\}^{|\mathbf{x}|} \times \{0,1\}^{|\mathbf{x}|} \rightarrow \{0,1\}$.

The stable CF, **C**, represents the allowed combinations of the circuit input **x**, which may remain stable indefinitely, independently of the circuit's output. Usually, **C** is the CF corresponding to the static constraint C: If $C=\{0,1\}^{|\mathbf{x}|}$ then **C**=**1**.

The transition CF, **Θ**, expresses restrictions on input transitions. It is a temporal extension of the stable part **C**, indicating that certain input transitions can occur only when some output conditions hold.

The constraint must be well-formed, i.e., its diagram must be connected.

The following diagram represents a possible constraint on the inputs R ('Reset') and S ('Set') of an asynchronous latch (with output y). Both inputs may not be high simultaneously. Only single input transitions are allowed. An input can rise at any time. However, the input R (resp. input S) may fall only if the output y is low (resp. high).

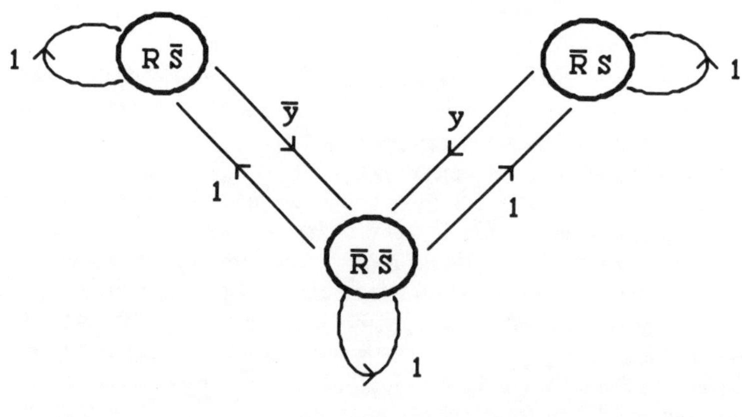

This constraint is more permissive than allowed by the fundamental mode hypothesis, since no assumption is made about the stability of the output variable y or any possible internal variable.

A domain constraint must be satisfied by all external components feeding the circuit, i.e., it is the maximal admissible behavior of a user. Since the state vector of the constraint (i.e., the input vector of the circuit) is a sub-vector of the user state vector, it is possible to compare directly the constraint and the user DCFs. A user $(\Phi_\varepsilon,\Psi_\varepsilon)$ satisfies the constraint (C,Θ) iff $(\Phi_\varepsilon,\Psi_\varepsilon) \leq (C,\Theta)$. If (C,Θ) is well-formed, this relation reduces to $\Psi_\varepsilon \leq \Theta$.

In contrast to users, constraints do not possess internal state variables. Their inclusion would require to have a more precise knowledge of the environment of the component under consideration. Furthermore, the comparison between the user and the constraint would not be a simple Boolean relation anymore.

III) COMPOSITION UNDER CONSTRAINT

The switch-level circuit DCF (Φ,Ψ) as obtained in Section II.1 describes the dynamic behavior of a circuit submitted to a static constraint C which insures that all internal signals are defined. However, this does not imply any constraint on the input transitions, i.e., all input transitions from C to C are possible at any time. When this DCF is associated with an input constraint DCF (C,Θ), such as defined in Section II.3, the behavior may be simplified. The stronger the constraint is, the more partial the description may become, and thus more internal state variables may be abstracted from.

A circuit combined with its input constraint forms a <u>closed system</u> whose internal state vector is the union of the circuit's state and input vectors. Similarly, a circuit combined with its user forms a closed system whose internal state vector is the union of both state vectors.

This is an extension of the so-called "complete circuits" [14]: it encompasses TGs (which are more general than logic gates), non-deterministic users and constraints.

A closed system is algebraically represented by an inputless automaton. The closed system DCF resulting from the combination of the circuit's DCF $(\Phi(\mathbf{x},\mathbf{y}), \Psi(\mathbf{x},\mathbf{y},\mathbf{y}'))$ and its constraint $(C(\mathbf{x}), \Theta(\mathbf{z},\mathbf{x},\mathbf{x}'))$ is obtained as : $(\Phi,\Psi) \cdot (C,\Theta) = (\Phi \cdot C, \Psi \cdot \Theta)$. In the following we suppose that the constraint is well-formed. The closed system's state vector is $\mathbf{x}\cup\mathbf{y}$, of which only $\mathbf{x}\cup\mathbf{z}$ is observed.

<u>Definition 3</u>: A path $<y_0,\ldots,y_n>$ of a closed automaton is a sequence of states such that $<y_i,y_{i+1}>$ is a transition, $\forall i\in[0,n-1]$.

A closed system DCF may include transitions from unstable states, which are inaccessible by paths from a stable state. We only consider such systems in which all transitions are reachable from a stable state. Hence we define a reachability operator $\Re$ (implemented by a path analysis algorithm) that eliminates all transitions not reachable from stable states : $\Re[\Psi](y,y') = 1$ if $\exists$ a path $\langle y_0,..,y,y'\rangle$ in Ψ and $\Phi(y_0)=1$, $\Re[\Psi](y,y') = 0$ otherwise. The automaton $(\Phi,\Re[\Psi])$ does not include unstable source states. It can leave a stable state only if it contains a non-deterministic element (user or constraint), otherwise all states are stable and represent deadlock situations.

Definition 4: The circuit's partial DCF under the constraint (C,Θ) is:

$$(\tilde{\Phi},\tilde{\Psi}) = (\Phi\cdot C\ ,\ \sum_{\mathbf{x}'}\Re[\Psi\cdot\Theta]).$$

In this DCF, certain transitions and states may have been eliminated, since they are inaccessible from a stable state under the constraint (C,Θ). The Boolean summation over $\mathbf{x}'$ corresponds to the fact that information about the next permissible input is disregarded.

The following lemma enables to establish that circuit's behavior is equally decribed by the original DCF (Φ,Ψ) or the partial DCF $(\tilde{\Phi},\tilde{\Psi})$ under the constraint.

Lemma 1: $\tilde{\Psi}\cdot\Theta = \Re[\tilde{\Psi}\cdot\Theta] = \Re[\Psi\cdot\Theta]$.

Theorem 1 extends this result to any user behavior contained in the constraint.

Theorem 1: Let $(\Phi_\varepsilon,\Psi_\varepsilon)$ be a user DCF such that $\Theta \geq \Psi_\varepsilon$. Then :

$$\tilde{\Phi}\cdot\Phi_\varepsilon = \Phi\cdot\Phi_\varepsilon \quad \text{and} \quad \Re[\tilde{\Psi}\cdot\Psi_\varepsilon] = \Re[\Psi\cdot\Psi_\varepsilon].$$

Although any circuit must be described using two DCFs (behavior and constraint), the resulting reduction in the complexity of the description can be important, especially in the case of the composition operation.

Let m interconnected components form a closed system. Each circuit is described by its DCF: (Φ_j,Ψ_j), $j\in[1,m]$, input vector $\mathbf{x}^j$ and state vector $\mathbf{y}^j = \mathbf{z}^j \cup \mathbf{t}^j$. The sub-vector $\mathbf{t}^j$ represents the externally non-observable state variables of the component and $\mathbf{z}^j$ represents interconnection variables. The state vector of the closed system is $\mathbf{y} = \mathbf{z} \cup \mathbf{t}$:

$$\mathbf{z} = \bigcup_{j\in[1,m]} \mathbf{z}^j \quad \text{and} \quad \mathbf{t} = \bigcup_{j\in[1,m]} \mathbf{t}^j.$$

We suppose that the observed vector of the closed system is **z**.

Let $(\tilde{\Phi}_j,\tilde{\Psi}_j)$, $j\in[1,m]$, be the partial DCFs under some domain constraints (C_j,Θ_j), such that $C_j \geq \Phi_j$ (to cover all stable states). We also suppose that each constraint is well-formed.

The following theorem holds:

<u>Theorem 2</u>:

$$\forall j\in[1,m], \Theta_j \geq \prod_{\substack{k\in[1,m]\\ k\neq j}} \tilde{\Psi}_k$$

$$\Rightarrow \Re\Big[\prod_{j\in[1,m]} \tilde{\Psi}_j\Big] = \Re\Big[\prod_{j\in[1,m]} \Psi_j\Big].$$

In other words, if the local constraint of every component is satisfied by the composition of partial DCFs of the other components acting as its user, then the overall system behavior can be obtained by the product of the partial DCFs rather than using their more complex original forms.

In the case m=2, the verification is straightforward, since the condition becomes: $\Theta_1 \geq \tilde{\Psi}_2$ and $\Theta_2 \geq \tilde{\Psi}_1$.

For the general case, verification of the condition can be tedious since it implies m products of (m-1) DCFs and m Boolean comparisons.

Some form of inductive process is very advantageous. If the m components belong to several different classes of components, only one product and one Boolean comparison per class is necessary.

IV) REDUCTION UNDER CONSTRAINT

An asynchronous automaton can be reduced by projection using a partition of the state set. Reduction depends on the external behavior : the more externally constrained the automaton is, the more its description may be reduced.

The abstraction of internal variables is a particular case of reduction. It is possible to remove any state variable that does not appear in the specification of the circuit or the user behavior (or constraint), provided that its elimination does not change the observed behavior of the closed system.

The reduction problem consists of a search of an asynchronous automaton which has the same observable behavior and a lower number of states. It is close to the problem of reducing non-deterministic abstract automata [17]. The main difference lies in the treatment of time. In our model, an arbitrary delay is associated with any transition. Thus, any arbitrary but finite sequence of unstable states each having the same observed value is externally equivalent to any other arbitrary but finite sequence, and in particular to a single unstable state.

In the following, we consider the reduction of asynchronous automata of closed systems (inputless automata, Cf. Definition 2). Let (Φ,Ψ) be the DCF of a closed system with state vector $\mathbf{y}$, consisting of an observed sub-vector $\mathbf{z}$ and an internal non-observed sub-vector $\mathbf{t}$. We consider partitions of the state set, such that all states of a partition block have the same observed value. A block becomes a meta-state, since it results from merging of states. The asynchronous automaton built on the meta-states is called a meta-automaton.

Definition 5: A partition is output-consistent (or simply consistent) iff the meta-automaton corresponding to the partition has the same observable behavior as the original automaton.

The reduction problem is to verify whether a state set partition preserves the original observable behavior. A possible solution is to find a condition on the partition which insures that no observable sequence is added without performing an exhaustive comparison.

Such a condition may be defined as follows: If any state of a block can reach (by a path within the block) all the blocks reachable from the other states of the block then the partition of the state set is consistent. An algorithm can be derived which generates this type of a partition. A similar principle of consistency was also used in [11] for protocol projections.

The following condition ensures that an internal variable can be abstracted from the description without modifying the observable behavior. Such a variable is called non-essential. Let (Φ,Ψ) be the DCF of a closed system with state vector $\mathbf{y}$ as before, and let $\mathbf{w}$ be the subvector such that $\mathbf{y} = \mathbf{w} \cup t$, where t is an internal variable. Furthermore, let $\mathbf{m_y}$ be a minterm function : $\mathbf{m_y}(y)=1$ if $\mathbf{y}=y$, and 0 otherwise.

Definition 6: A variable t of (Φ,Ψ) is non-essential iff :
$\forall$ w, w $\in \{0,1\}^{|w|}$ such that: $\Psi(wt,wt)=1$ and $\Psi(w\,\bar{t},w\,\bar{t})=1$,
$\forall$ u, u $\in \{t, \bar{t}\}$, and v= $\bar{u}$, if $\Psi(wu,wv)=0$ then :

$$\sum_{t,t'} (\Psi(y,y') \cdot \mathbf{m}_{wv}(y)) \le \sum_{t,t'} (\Psi(y,y') \cdot \mathbf{m}_{wu}(y)).$$

In others words, the next possible values of vector **w**, for each value w, do not depend on the value of the variable t. The corresponding state partition is consistent, because the observable sequences are the same as those obtained before abstraction of the variable t.

Suppose that a consistent state partition of a closed system (state **y**) is known. Let **z** be the observable subvector of state **y** and **t** be the non-observable subvector ($y=t\cup z$). The partition can be coded using some vector **s**, and the corresponding projection is defined by the CF **P**(z,t,s): $\{0,1\}^{|z|} \times \{0,1\}^{|t|} \times \{0,1\}^{|s|} \to \{0,1\}$, such that **P**(z,t,s)=1 iff state zt belongs to block zs. We denote **P'**= **P**(z',t',s').

Definition 7: The DCF reduced with partition **P** under the constraint is:

$$(\sum_{t} (\Phi \cdot \mathbf{P}) + cycle_t(\Psi)\ ,\ \sum_{t,t'} (\Psi \cdot \mathbf{P} \cdot \mathbf{P'})).$$

The CF $cycle_t(\Psi)$: $\{0,1\}^{|z|}\times\{0,1\}^{|s|}\to\{0,1\}$ represents the fact that if a block of the partition contains a cycle, then the corresponding meta-state is stable.

The projection of DCF (Φ,Ψ) over a subvector **u** of the state vector **y** corresponds to **P**=1 and s=∅.

Definition 8: The DCF obtained by projection over the subvector **u** is:

$$(\sum_{y-u} \Phi + cycle_{y-u}(\Psi)\ ,\ \sum_{y-u,\ y-u'} \Psi).$$

In the following the projection of (Φ,Ψ) over subvector **u** is denoted : $\sum_{y-u}(\Phi,\Psi)$.

We now consider a circuit described by a partial DCF $(\tilde{\Phi},\tilde{\Psi})$ simplified under a well-formed constraint (C,Θ). **t** is the circuit's internal sub-vector.

Theorem 3: If the partition **P** is consistent in $\tilde{\Psi} \cdot \Theta$, then for any user $(\Phi_\varepsilon,\Psi_\varepsilon)$ such that $\Theta \ge \Psi_\varepsilon$, the following statements are valid:

a) **P** is consistent in $\Re[\Psi \cdot \Psi_\varepsilon]$.

b) Initialisation and reduction operators commute.

c) Reduction can be done before product with external behavior, i.e.:

$$\sum_{t,t'} (\Re[\Psi \cdot \Psi_\varepsilon] \cdot \mathbf{P} \cdot \mathbf{P'}) \;=\; \Re[\; (\sum_{t,t'} \tilde{\Psi} \cdot \mathbf{P} \cdot \mathbf{P'}) \cdot \Psi_\varepsilon \;].$$

Abstraction of a subvector **r** of the internal vector **t** is a particular case of projection. If the corresponding partition is consistent, **r** is called non-essential.

Definition 9: The DCF abstracted from **r** is:

$$\sum_{\mathbf{r}} (\tilde{\Phi}, \tilde{\Psi}) = (\sum_{\mathbf{r}} \tilde{\Phi} + \mathrm{cycle}_{\mathbf{r}}(\tilde{\Psi}) \;,\; \sum_{\mathbf{r},\mathbf{r}'} \tilde{\Psi}).$$

Theorem 4: If **r** is non-essential in $\tilde{\Psi} \cdot \Theta$, then for any user $(\Phi_\varepsilon, \Psi_\varepsilon)$ such that $\Theta \geq \Psi_\varepsilon$, the following applies :

a) **r** is non-essential in $\Re[\Psi \cdot \Psi_\varepsilon]$.

b) Initialisation and abstraction operators commute.

c) Abstraction can be done before product with external behavior :

$$\sum_{\mathbf{r},\mathbf{r}'} \Re[\Psi \cdot \Psi_\varepsilon] = \Re[(\sum_{\mathbf{r},\mathbf{r}'} \tilde{\Psi}) \cdot \Psi_\varepsilon].$$

When interconnections of multiple modules are considered, the abstracted DCFs may be used. Let the m components be described by (Φ_j,Ψ_j), $j\in[1,m]$, input $\mathbf{x}^j$, state $\mathbf{y}^j = \mathbf{z}^j \cup \mathbf{t}^j$ ($\mathbf{t}^j$: non-observed). Under some constraint (C_j,Θ_j), such that $C_j \geq \Phi_j$, a simplified DCF $(\tilde{\Phi}_j, \tilde{\Psi}_j)$ is obtained. Let $\mathbf{r}^j$, $j\in[1,m]$, be a non-essential sub-vector of $\mathbf{t}^j$ in each component. Let the abstracted DCF be : $\sum_{\mathbf{r}^j} (\tilde{\Phi}_j,\tilde{\Psi}_j)$.

Theorem 5:

$$\forall\, j\in[1,m],\; \Theta_j \;\geq \prod_{\substack{k\in[1,m]\\ k\neq j}} \Big(\sum_{\mathbf{r}^k,\mathbf{r}^{k'}} \tilde{\Psi}_k \Big)$$

$$\Rightarrow\; \Re\Big[\prod_j \Big(\sum_{\mathbf{r}^j,\mathbf{r}^{j'}} \tilde{\Psi}_j \Big) \Big] = \sum_{\mathbf{r},\mathbf{r}'} \Re\Big[\prod_j \Psi_j\Big].$$

If for every component, the constraint is satisfied by the product of abstracted DCFs of the other components acting as its user, the following holds :

a) Every local abstraction remains valid,

b) The product of abstracted DCFs is equal to the global behavior after abstraction of vector $\mathbf{r} = \bigcup_{j\in[1,m]} \mathbf{r}^j$, i.e., abstraction can be done before composition, thus reducing its complexity.

V) COMPARISON WITH SPECIFICATION

A DCF (Φ,Ψ) extracted from the structure of a circuit under an external constraint (C,Θ) can be compared with the circuit's specification. This specification must include both the circuit's behavior specification and the 'user' specification. The comparison between extracted and specified behaviors is possible if the 'user' specification $(\Phi,\Psi)_\varepsilon$ satisfies: $(\Phi,\Psi)_\varepsilon \leq (C,\Theta)$.

If the specification can be expressed as an asynchronous automaton with DCF $(\Phi,\Psi)_s$, verification consists of showing that the two closed systems $(\Phi,\Psi) \cdot (\Phi,\Psi)_\varepsilon$ and $(\Phi,\Psi)_s \cdot (\Phi,\Psi)_\varepsilon$ have the same observable behavior, i.e., it corresponds to determining the equivalence of two automata.

A specification can also be expressed as a collection of properties, each property describing a particular view of the behavior of the closed system 'circuit-user'. Let $(\Phi,\Psi)_p$ describe the property P. We suppose that the properties do not possess internal state variables, i.e. $(\Phi,\Psi)_p$ is defined over a subvector $\mathbf{u}$ of the observable vector $\mathbf{z}$. A more general form of comparison is necessary if $(\Phi,\Psi)_p$ contains internal state variables.

Verification is done by showing that the closed system is safe and live with respect to each property. For example, an interconnection verifies the FCS (Four Cycle Signaling) scheme if it can be shown that it is safe with respect to FCS, i.e., that it does not exhibit transitions or stable states that do not belong to FCS, and that it is live with respect to FCS, i.e., that it has the capability of going through all the transitions.

Let (Φ,Ψ) be a closed system (circuit-user) DCF defined over state vector $\mathbf{y}$ (observed vector $\mathbf{z} \subseteq \mathbf{y}$). Let $(\Phi,\Psi)_p$ be a DCF representing a property P, over a subvector $\mathbf{u}$ of $\mathbf{z}$. $\sum_{\mathbf{y}-\mathbf{u}} (\Phi,\Psi)$ is the DCF obtained by projection over subvector $\mathbf{u}$.

Definition 8: The closed system is safe with respect to property P if:

$$\sum_{\mathbf{y}-\mathbf{u}} (\Phi,\Psi) \leq (\Phi,\Psi)_p.$$

The extracted behavior must be included in the specified behavior.

Definition 9: The closed system is live with respect to property P if:

a) The projection over $\mathbf{u}$ is consistent, and b) $\sum_{\mathbf{y}-\mathbf{u}} (\Phi,\Psi) \geq (\Phi,\Psi)_p$.

In this definition, part a) insures that the projection respects the observable behavior. Part b) then verifies that the extracted behavior has at least all the specified transitions.

As an example, consider a mutual exclusion element (ME) connected to two FCS users. Such an element is usually contained in the arbiter circuit [3,7]. ME has two inputs R_1 and R_2, and two outputs A_1 and A_2 such that both outputs are never high simultaneously.

An implementation of ME was given in [13]. The internal structure and symbol are the following :

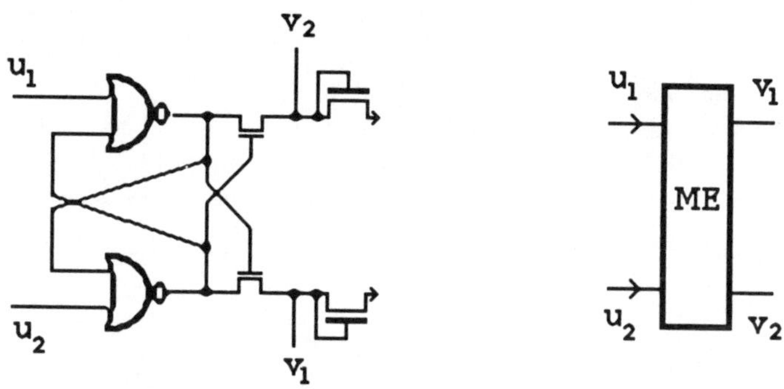

The closed system under consideration is:

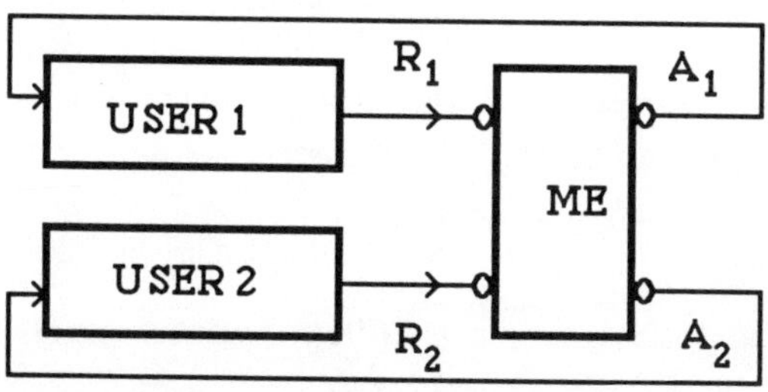

The observable vector is (R_1,A_1,R_2,A_2).

Let the specification of mutual exclusion be described using the following properties a) and b). (Since it will not involve any consideration of delay, this specification is speed-independent.)

a) Both interconnections R_i,A_i $i=1,2$ respect the FCS scheme, i.e., the DCF: $(\Phi,\Psi)_{FCS}$.

For the interconnection R_1-A_1 the DCF's diagram is:

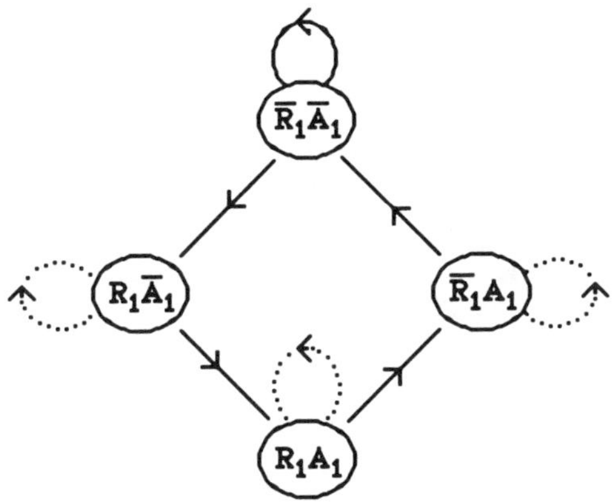

b) Both outputs are never high simultaneously.
This is expressed by the DCF: ($\overline{A}_1\ \overline{A}_2$, $\overline{A}_1\ \overline{A}_1' + \overline{A}_2\ \overline{A}_2'$), whose diagram is:

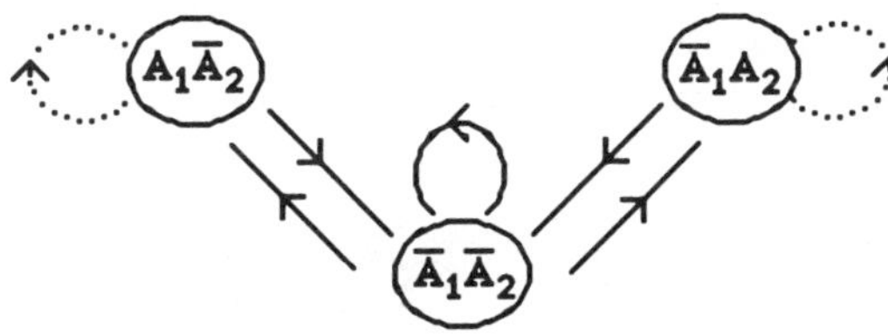

The DCF $(\Phi,\Psi)_{ME}$ as extracted from the ME's structure is live and safe with respect to the two properties. Its behavior is speed-independent. However, as it was noticed by several authors [3,7], this structure is not fair in the sense that a user may appropriate the ME element, implying that a request of the other user may never be satisfied.

Fairness of a mutual exclusion element can be expressed by the two following DCFs: ($\overline{R}_i$,1), i=1,2, meaning that R_i will not remain stable high indefinitely (cf. Diagram in Section I). Hence, any request will inevitably be satisfied.

The extracted DCF $(\Phi,\Psi)_{ME}$ can be compared with the properties expressing fairness: The consistent projection of $(\Phi,\Psi)_{ME}$ over R_1 (R_2) introduces a stable high state due to a possible cycle in the service of user 2 (1).

The DCF obtained by projection is **(1,1)**. It is clear that ME is live but not safe with respect to fairness, i.e., it may be fair but is not necessarily.

The following figure is a state diagram of the complemented ME element:

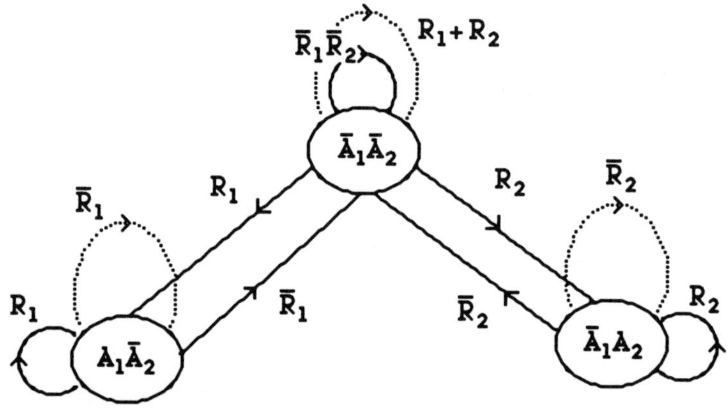

The unstable self-loop around state $A_1= A_2=0$ allows a new request from the same user to occur and be selected even if the other user has a pending request.

In order to be fair, a mutual exclusion element should give the priority to the other user in such a case. It should incorporate internal (non-observable) variables permitting to memorize which was the last user served. We thus propose a new structure of mutual exclusion (FME) which is both fair and speed-independent:

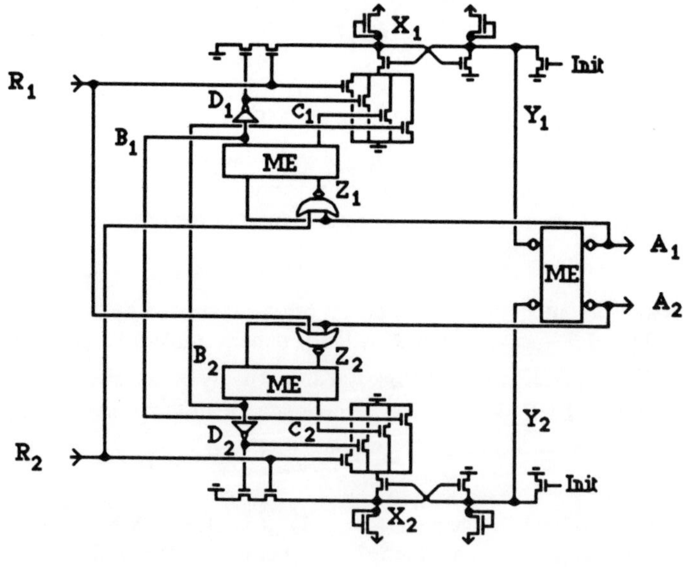

FME is constructed using 'unfair' ME elements. For each user, it incorporates a blocking mechanism which prevents a possible new access to an ME element (vertical in the figure) if the other user has emitted a request before the end of its last transaction. Note that two ME elements (horizontal in the figure) are necessary to prevent erroneous behavior due to a potential pulse caused by the occurence of a request before the blocking mechanism is disabled.

The methodology developped in Sections III and IV was applied to this circuit connected with 2 FCS-users. Abstractions are possible in components. Finally, a reduced DCF $(\Phi,\Psi)_{FME}$ of the closed system can be obtained. The extracted DCF $(\Phi,\Psi)_{FME}$ can be shown safe and live not only with respect to the two previous properties representing mutual exclusion but also with respect to fairness, i.e., a request is always satisfied.

It can be argued that the specification used to represent fairness is a weak form of fairness : it only implies that any request will eventually be satisfied after some finite time. It does not say anything about the exact time when this may happen. In fact, it would be possible to express a stronger form of fairness: Any request occuring during the transaction of the other user will be the next request satisfied. In order to express such a form of fairness, it would be necessary to use internal variables, and our model does not allow, at the moment, to compare two asynchronous automata defined over different state vectors. Note, however, that the proposed FME is fair in this stronger sense too !

CONCLUSION

The DCF model and algebra allow representing circuits with unknown inertial delays such as switch-level networks, and non-deterministic user behaviors, domain constraints and specifications. Composition is the product operation of a Boolean algebra. Under domain constraints, circuit behaviors can be reduced.

The partial order of the algebra permits comparing constraints with actual users and reduced behaviors with specifications.

Our methodology enables to compare a switch-level structure with a collection of properties. With this methodology, we have verified a new structure of FIFO queue element [2] and a fair mutual exclusion element.

Presently, constraints and specifications cannot be described using internal state variables. A more general method to compare asynchronous automata is a part of our current research.

ACKNOWLEDGMENTS

This research was supported by scholarships from EUMC (Entraide Universitaire Mondiale du Canada) and CRIM (Centre de Recherches Informatiques de Montréal), and in part by an NSERC Canada Grant No A3879.

REFERENCES

[1] Barrow H.G., "Verify: A Program for Proving Correctness of Digital Hardware Designs", Artificial Intelligence, Vol 24,pp 437-491, 1984.

[2] Berthet C., Cerny E., "An algebraic model for asynchronous circuits verification", to appear in IEEE Transactions on Computers.

[3] Bochmann G., "Hardware Specifications with Temporal Logic : an Example", IEEE Transactions on Computers, Vol C-31, No 3, 1982.

[4] Bryant R.E., "A Switch-Level Model and Simulator for MOS Digital Systems", IEEE Transactions on Computers, Vol C-33, No 2, 1984.

[5] Cerny E., "An Approach to Unified Methodology of Combinational Switching Circuits", IEEE Transactions on Computers, Vol C-26, No 8, 1977.

[6] Cerny E., Gecsei J., "Simulation of MOS Circuits by Decision Diagrams", IEEE Transactions on Computer-Aided Design, Vol CAD-4, No 4, pp. 685-693, October 1985.

[7] Dill D.L., Clarke E. M., "Automatic Verification of Asynchronous Circuits using Temporal Logic", Chapel Hill Conference on VLSI, 1985.

[8] Gordon M., "A Model of Register Transfert Systems with Applications to Microcode and VLSI Correctness", Computer Science Report, No CSR-82-81, University of Edinburgh, 1981.

[9] Gordon M., "Why higher-order logic is a good formalism for specifying and verifying hardware", Computer laboratory, Cambridge, September 1985.

[10] Hayes J.P., "A Unified Switching Theory with Applications to VLSI Design", Proceedings of the IEEE, Vol 70, No 10,pp 1140-1151, 1982.

[11] Lam S.S., Sankhar A.U., "Protocol Verification Via Projections", IEEE Transactions on Software Engineering, Vol. SE-10, No. 4, July 1984.

[12] Leinwand S.M., Automatic Design Verification and Test Generation for Digital Systems, Ph.D. Thesis, Weizmann Institute of Science, Rehovot, 1980.

[13] Mead C.A., Conway L.A., Introduction to VLSI Systems, Addison-Wesley, 1980.

[14] Miller R.E., Switching theory, John Wiley and sons, 1965.

[15] Milne G.J., "A Model for Hardware Description and Verification", 21st Design Automation Conference, 1984.

[16] Shostack R., "Formal Verification of Circuit Design", Proc. VI International Conference on Hardware Description Langages, IFIP 1983.

[17] Starke P.H., Abstract Automata, North-Holland, 1972.